2 Modern Mathematics for Schools

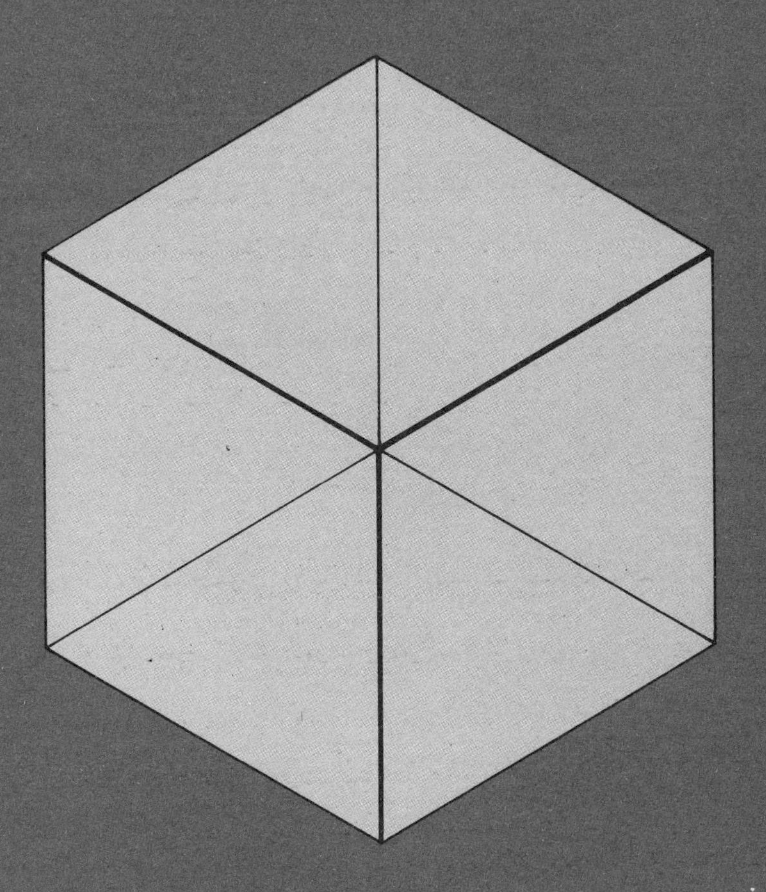

Modern Mathematics for Schools

Second Edition

Scottish Mathematics Group

Blackie

Chambers

Blackie & Son Limited

Bishopbriggs · Glasgow G64 2NZ
Furnival House · 14–18 High Holborn
London WC1V 6BX

W & R Chambers Limited

43–5 Annandale Street · Edinburgh EH7 4AZ

First Published 1971
Reprinted 1985

Designed by James W. Murray

International Standard Book Numbers
Pupils' Book
Blackie 0 216 89404 2
Chambers 0 550 75912 3
Teachers' Book
Blackie 0 216 89405 0
Chambers 0 550 75922 0

Printed in Great Britain by
Thomson Litho Ltd, East Kilbride, Scotland
Set in 10pt Monotype Times Roman

Members associated with this book

W. T. Blackburn
Dundee College of Education

Brenda I. Briggs
Formerly of Mary Erskine School for Girls

W. Brodie
Trinity Academy

C. Clark
Formerly of Lenzie Academy

D. Donald
Formerly of Robert Gordon's College

R. A. Finlayson
Allan Glen's School

Elizabeth K. Henderson
Westbourne School for Girls

J. L. Hodge
Madras College

J. Hunter
University of Glasgow

T. K. McIntyre
High School of Stirling

R. McKendrick
Langside College

W. More
Formerly of High School of Dundee

Helen C. Murdoch
Hutchesons' Girls' Grammar School

A. G. Robertson
John Neilson High School

A. G. Sillitto
Formerly of Jordanhill College of Education

A. A. Sturrock
Grove Academy

Rev. J. Taylor
St. Aloysius' College

E. B. C. Thornton
Bishop Otter College

J. A. Walker
Dollar Academy

P. Whyte
Hutchesons' Boys' Grammar School

H. S. Wylie
Govan High School

Only with the arithmetic section

R. D. Walton
Dumfries Academy

Book 1 of the original series *Modern Mathematics for Schools* was first published in July 1965. This revised series has been produced in order to take advantage of the experience gained in the classroom with the original textbooks and to reflect the changing mathematical needs in recent years, particularly as a result of the general move towards some form of comprehensive education.

Throughout the whole series, the text and exercises have been cut or augmented wherever this was considered to be necessary, and nearly every chapter has been completely rewritten. In order to cater more adequately for the wider range of pupils now taking certificate-oriented courses, the pace has been slowed down in the earlier books in particular, and parallel sets of A and B exercises have been widely introduced. The A sets are easier than the B sets, and provide straightforward but comprehensive practice; the B sets have been designed for the more able pupils, and may be taken in addition to, or instead of, the A sets. Occasionally a basic exercise, which should be taken by all pupils, is followed by a harder one on the same work; in such a case the numbering is, for example, Exercise 2 followed by Exercise 2B. It is hoped that this arrangement, along with the 'Graph Workbook for Modern Mathematics', will allow considerable flexibility of use, so that while all the pupils in a class may be studying the same topic, each pupil may be working examples which are appropriate to his or her aptitude and ability.

Each chapter is backed up by a summary, and by A and B revision exercises; in addition, cumulative summaries and exercises have been introduced at the end of alternate books. A new feature is the series of Computer Topics from Book 4 onwards. These form an elementary introduction to computer studies, and are primarily intended to give pupils some appreciation of the applications and influence of computers in modern society.

Books 1 to 7 provide a suitable course for many modern Ordinary Level and Ordinary Grade syllabuses in mathematics, including the University of London GCE Syllabus C, the Associated Examining Board Syllabus C, the Cambridge Local Syndicate Syllabus C, and the Scottish Certificate of Education. Books 8 and 9 complete the work for the Scottish Higher Grade Syllabus, and provide a good preparation for all Advanced Level and Sixth Year Syllabuses, both new and traditional.

Related to this revised series of textbooks are the *Modern Mathematics Newsletters*, the *Teacher's Editions* of the textbooks, the *Graph Workbook for Modern Mathematics*, the *Three-Figure Tables for Modern Mathematics*, and the booklets of *Progress Papers for Modern Mathematics*. These new Progress Papers consist of short, quickly marked objective tests closely connected with the textbooks. There is one booklet for each textbook, containing A and B tests on each chapter, so that teachers can readily assess their pupils' attainments, and pupils can be encouraged in their progress through the course.

The separate headings of Algebra, Geometry, Arithmetic, and later Trigonometry and Calculus, have been retained in order to allow teachers to develop the course in the way they consider best. Throughout, however, ideas, material and method are integrated *within* each branch of mathematics and *across* the branches; the opportunity to do this is indeed one of the more obvious reasons for teaching this kind of mathematics in the schools—for it is *mathematics* as a whole that is presented.

Pupils are encouraged to find out facts and discover results for themselves, to observe and study the themes and patterns that pervade mathematics today. As a course based on this series of books progresses, a certain amount of equipment will be helpful, particularly in the development of geometry. The use of calculating machines, slide rules, and computers is advocated where appropriate, but these instruments are not an essential feature of the work.

While fundamental principles are emphasized, and reasonable attention is paid to the matter of structure, the width of the course should be sufficient to provide a useful experience of mathematics for those pupils who do not pursue the study of the subject beyond school level. An effort has been made throughout to arouse the interest of all pupils and at the same time to keep in mind the needs of the future mathematician.

The introduction of mathematics in the Primary School and recent changes in courses at Colleges and Universities have been taken into account. In addition, the aims, methods, and writing of these books have been influenced by national and international discussions about the purpose and content of courses in mathematics, held under the auspices of the Organization for Economic Co-operation and Development and other organizations.

The authors wish to express their gratitude to the many teachers who have offered suggestions and criticisms concerning the original series of textbooks; they are confident that as a result of these contacts the new series will be more useful than it would otherwise have been.

Algebra

1 *Replacements and Formulae* 3

Replacements; using formulae; constructing formulae; summary.

2 *Inequalities and Inequations* 14

Inequalities; the number line; inequations—1;
inequations—2; writing mathematical sentences; summary.

3 *Negative Numbers* 27

An extension of the number line; the set of integers;
order on the number line;
addition of positive and negative numbers;
the negative (or additive inverse) of a number;
subtraction;
sets of numbers—natural numbers, whole numbers, integers, rational numbers;
standard form for small numbers;
using positive and negative integers; summary.

4 *Distributive Law* 50

Expressing a product of factors as a sum or difference of terms;
expressing a sum or difference of terms as a product of factors;
collecting like terms; application of the distributive law to the solution of equations and inequations;
illustrations of the law; summary.

Revision Exercises 64

Cumulative Revision Section (Books 1 and 2) 79

Geometry

1 Rectangle and Square 91

Looking back, and looking ahead; fitting shapes; the rectangle; coordinates; the square; axes of symmetry; drawing rectangles and squares; the cube and cuboid; summary.

2 Triangles 114

Introduction; right-angled triangles; isosceles triangles; drawing isosceles triangles; coordinates; the sum of the angles of a triangle; equilateral triangles; the area of a triangle; the construction of triangles; summary.

Topic to Explore 136

Revision Exercises 137

Cumulative Revision Section (Books 1 and 2) 146

Arithmetic

1 Decimals and the Metric System 157

Notation; the decimal point; measurement; approximation—significant figures, rounding off, decimal places; addition and subtraction; multiplication; division; rough estimates of answers; relations between decimal and common fractions; standard form (scientific notation); indices; very large numbers; the metric system; calculation of percentages of money; summary.

2 Computers and Binary Arithmetic 187

The binary system—counting in twos; models that show numbers in binary form; addition; subtraction; multiplication; division; electronic computers; punched cards; punched tape; miscellaneous questions; summary.

3 Introduction to Statistics 207

Pictographs; bar charts; pie charts; line graphs; summary.

Revision Exercises 227

Cumulative Revision Section (Books 1 and 2) 238

Answers 249

Notation

Sets of numbers

Different countries and different authors
give different notations and definitions
for the various sets of numbers.
In this series the following are used:

E The universal set

ϕ The empty set

N The set of natural numbers $\{1, 2, 3, \ldots\}$

W The set of whole numbers $\{0, 1, 2, 3, \ldots\}$

Z The set of integers $\{\ldots, -2, -1, 0, 1, 2, \ldots\}$

Q The set of rational numbers

R The set of real numbers

The set of prime numbers $\{2, 3, 5, 7, 11, \ldots\}$

Algebra

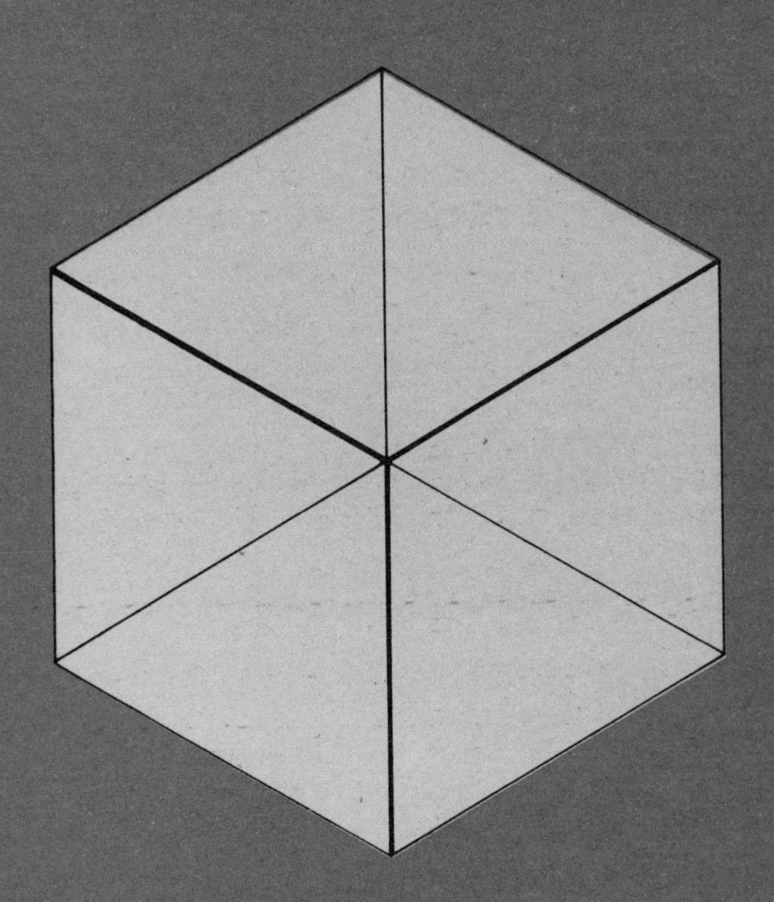

Algebra

Replacements and Formulae

1 Replacements

Suppose that x is a variable on the set $\{1, 2, 3, 4\}$; that is, x can be replaced by 1, 2, 3 or 4.

If $n = 2x-1$, what set of values of n can be obtained?

Replacing x by 1, $n = (2\times 1)-1 = 1$
Replacing x by 2, $n = (2\times 2)-1 = 3$
Replacing x by 3, $n = (2\times 3)-1 = 5$
Replacing x by 4, $n = (2\times 4)-1 = 7$

The set of values of n is $\{1, 3, 5, 7\}$, the first four odd numbers. Replacing x by members of the set of natural numbers $\{1, 2, 3, ...\}$ will give the set of odd numbers $\{1, 3, 5, ...\}$.

We say that $n = 2x-1$ is a *formula* which gives the set of odd numbers.

Can you write a formula which gives the set of even numbers? Test your formula by replacing x by 1, 2, 3 and 4.

Reminders. $2a = a+a$; $a^2 = a\times a$; $3a^2 = 3\times a\times a$
$3a = a+a+a$; $a^3 = a\times a\times a$; $2a+3a = 5a$

Example.—Find the value of $2a^2+5a-3$ when $a = 4$.

$$\begin{aligned} & 2a^2+5a-3 \\ = & (2\times 4\times 4)+(5\times 4)-3 \\ = & 32+20-3 \\ = & 49 \end{aligned}$$

Exercise 1A

1 Find the value of each of the following when $x = 3$:

a $x+2$ *b* $x-2$ *c* $5x$ *d* $4x-7$
e $2+3x$ *f* $10-2x$ *g* x^2 *h* x^3

2 Find the value of each of the following when $t = 5$:

a $2t$ *b* $4t$ *c* t^2 *d* $10t^2$ *e* t^3 *f* $2t^3$

3 Find the value of each of the following:

a $n+9$, when $n = 8$ *b* $2c+1$, when $c = 5$

c $n-7$, when $n = 15$ *d* $11t$, when $t = 12$

e $15-m$, when $m = 9$ *f* y^2-25, when $y = 5$

g $p+q$, when $p = 16$ and $q = 15$ *h* $r-s$, when $r = 35$ and $s = 17$

4 Given that $a = 1$, $b = 3$ and $c = 5$, calculate the values of:

a $a+b+c$ *b* $2a+3b+4c$ *c* $b+c-8a$

d abc *e* $10a+10b+10c$ *f* $ab+bc+ca$

5 If $p = 3$ and $q = 2$, find the values of:

a p^2 *b* p^2+q^2 *c* $(p-q)^2$ *d* p^2-q^2

e $(p+q)^2$ *f* $5p^2$ *g* $4q^2$ *h* $2p^2+3q^2$

6 Say which of the following are true, and which are false:

a $a+5 = 10$, when $a = 5$ *b* $a^2 = 10$, when $a = 5$

c $2b-1 = 3$, when $b = 2$ *d* $3c^2 = 12$, when $c = 2$

e $4d+0 = 4$, when $d = 1$ *f* $4d\times 0 = 4$, when $d = 1$

7 x is a variable on the set {1, 2, 3, 4, 5}. Find the set of values of n given by the formula $n = 3x$. Describe this set in words.

8 x is a variable on the set {1, 2, 3, 4, 5}. Find the set of values of n given by $n = 6x-1$. Describe the kind of numbers you obtain.

If x is a variable on the set {1, 2, 3, ..., 10}, would your description still be true? Give a reason for your answer.

9 Repeat question ***8*** for the formula $n = 6x+1$.

10 $v = 10+2t^2$, and t is a variable on the set {0, 1, 2, 3, 4, 5}. Find the set of all possible values of v.

Exercise 1B

In questions ***1*** to ***10***, state whether each is true or false:

1 $a+4 = 12$, when $a = 8$ **2** $b+4 = 4$, when $b = 0$

3 $2c+8 = 12$, when $c = 3$ **4** $3d-1 = 10$, when $d = 3$

5 $2p = p+3$, when $p = 3$ **6** $5q+4 = 3q+12$, when $q = 4$

7 $\frac{1}{2}a = 15$, when $a = 30$ **8** $\frac{1}{3}b = 30$, when $b = 10$

9 $\frac{1}{2}(a+5) = 6\frac{1}{2}$, when $a = 8$ **10** $\frac{1}{3}(b-4) = 3$, when $b = 10$

11 If x is a variable on the set {1, 2, 3, 4, 5, 6}, find the set of values of each of the following:

a $2x$ *b* $2x+3$ *c* x^2 *d* x^2+1

12 If x and y are variables on the set $\{0, 1, 2, 3\}$, find the set of values of:

a $3x+4$ *b* x^2 *c* $5x^2$ *d* y^3 *e* $x+y$

What is (*1*) the greatest value of $x+y$ (*2*) the least value of $x+y$?

13 If x can be replaced only by 2 or 4, and y can be replaced only by 1 or 3, find all the possible values of $x+2y$.

14 Using the formula $v = 50-9{\cdot}8t$, calculate v when:

a $t = 5$ *b* $t = 1{\cdot}5$

15 The distance s km travelled by an air liner flying at a steady speed of v km/h for t hours is given by $s = vt$.

If $v = 1600$, find s when $t = 4$, and also when $t = 1\frac{1}{2}$.

16 A distance M miles can be expressed approximately as K kilometres by the formula $K = 1{\cdot}6\ M$.

a Find K when $M = 250$.

b Use the formula to express a speed of 30 miles per hour in kilometres per hour.

17 The number of measurements M which are required to fix the size and shape of a plane figure of n sides is given by the formula $M = 2n-3$.

a Find M when $n = 3$, and also when $n = 7$.

b If $M = 21$, find n.

18 The distance h metres fallen from rest by an object during t seconds is given by the formula $h = 4{\cdot}9\ t^2$.

A stone is dropped from a cliff top and takes 2·5 seconds to reach the base of the cliff. Using the above information, calculate the height of the cliff to the nearest metre.

2 *Using formulae*

Exercise 2

Study these formulae, and calculate the items asked:

	FORMULA	MEANING	CALCULATION
1	$P = 4s$	P cm = Perimeter of square s cm = length of side	The perimeter of a square with ***a*** $s = 12$ ***b*** $s = 15$
2	$A = s^2$	A cm² = Area of square s cm = length of side	The area of a square with: ***a*** $s = 10$ ***b*** $s = 22$
3	$A = lb$	A cm² = Area of rectangle l cm = length of rectangle b cm = breadth of rectangle	The area of a rectangle with: ***a*** $l = 15,\ b = 8$ ***b*** $l = 2{\cdot}5,\ b = 1{\cdot}5$
4	$V = lbh$	V m³ = Volume of cuboid l m = length of cuboid b m = breadth of cuboid h m = height of cuboid	The volume of a cuboid with: ***a*** $l = 15, b = 8, h = 2$ ***b*** $l = b = h = 10$
5	$c = 90 - x$	$c°$ is the complement of the angle $x°$.	The complement of angles of: 10°, 55° and 79°
6	$s = 180 - y$	$s°$ is the supplement of the angle $y°$.	The supplement of angles of: 100°, 55° and 123°
7	$D = ST$	D km is the distance travelled at S km/h in T hours.	***a*** D when $S = 65, T = 4$ ***b*** S when $D = 250, T = 5$
8	$v = \frac{1}{2}(m+n)$	v is the average of two numbers represented by m and n.	The average of: ***a*** 12 and 16 ***b*** 23 and 37 ***c*** 8·7 and 5·5
9	$C = \frac{83n}{100}$	£C is the total cost of n books costing 83 pence each.	The total cost of: ***a*** 10 books ***b*** 80 books
10	$T = 42 - 2N - 2P$	T hours is the time after midday for month N (Jan = 1, Dec = 12) given by the position P of the 'pointer' stars of the Great Bear.	The time after midday in: ***a*** December, if $P = 5$ ***b*** April, if $P = 7$

3 Constructing formulae

a Draw a rectangle with length 5 cm and breadth 3 cm.
Underneath your diagram, copy and complete the following:

(*1*) Total length of the two long sides = ... cm
(*2*) Total length of the two short sides = ... cm
(*3*) Perimeter of rectangle = ... cm

b Figure 1 shows a rectangle with length x cm and breadth y cm.

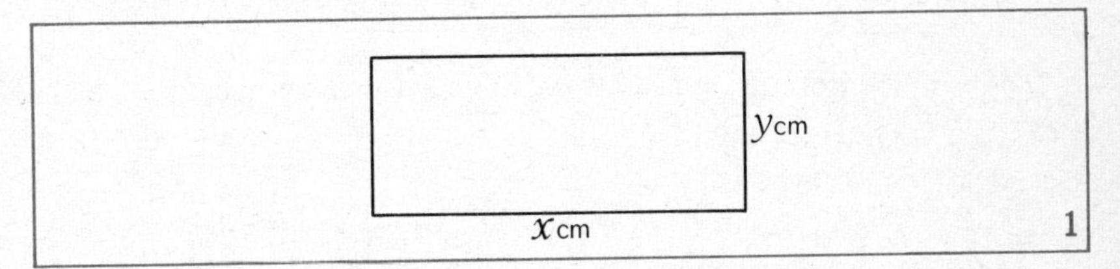

Copy and complete the following:

(*1*) Total length of the two long sides = ... cm
(*2*) Total length of the two short sides = ... cm
(*3*) Perimeter of rectangle = ... cm

Using P for the number of centimetres in the perimeter, we can write $P = 2x+2y$, which is a *formula* for finding the perimeter of a rectangle.

c Can you discover another way of finding the perimeter of the rectangle?

Exercise 3

1 Copy and complete the following table with the help of a formula for the perimeter of a rectangle:

Length (cm)	3	8	14	6·5	4	l
Breadth (cm)	2	5	6	2·5	b	b
Perimeter (cm)	.	.	.	.	.	.

2 Write down a formula for the perimeter P of each of the shapes in Figure 2. All the units are centimetres.

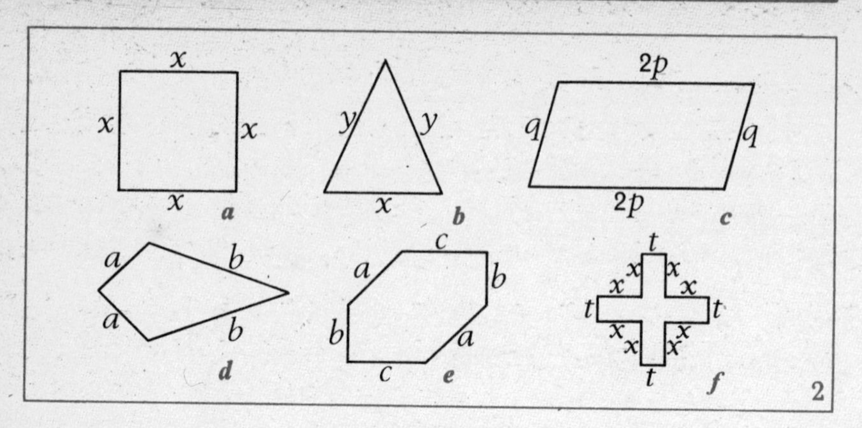

3 The perimeter of a triangle is P cm, and each side has length m cm.

a Find a formula for the perimeter.

b Hence calculate P when:

(*1*) $m = 8$ (*2*) $m = 5{\cdot}5$ (*3*) $m = 10{\cdot}4$

4 Figure 3 shows the route of an aircraft which flew x km east, then y km north. The total distance it flew from O to E to N was d km.

a Write down a formula for d.

b Hence calculate d if $x = 840$ and $y = 460$.

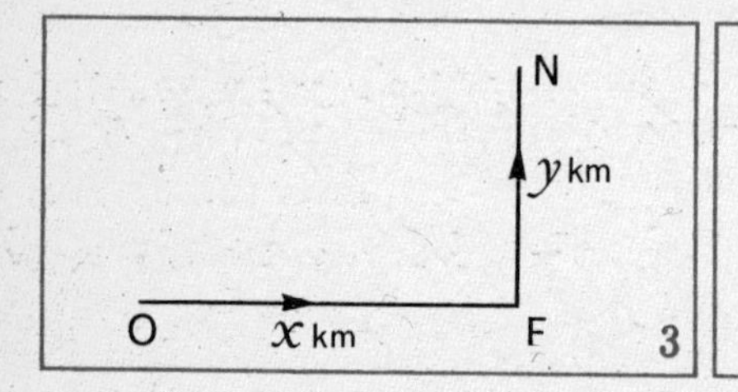

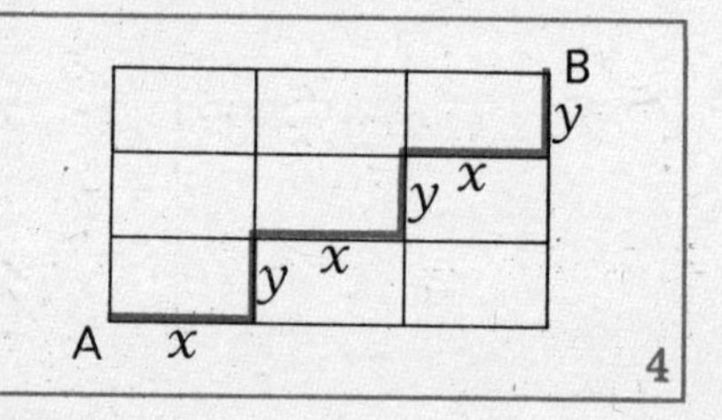

5 Figure 4 shows a network of streets in a town.

a Find a formula for the distance d, along the streets shown, from A to B. All the units are metres.

b Calculate d if $x = 325$ and $y = 180$.

6 A lorry weighs x tonnes when empty. If y tonnes of coal are added, its loaded weight is W tonnes. Write down a formula for W.

a Find W when $x = 2{\cdot}75$ and $y = 2{\cdot}50$.

b Find y when $x = 3\frac{1}{2}$ and $W = 6$.

7 A plank of wood is p metres long. When q metres have been sawn off, the remaining length is L metres. Write down a formula for L, and hence find L when $p = 5$ and $q = 1{\cdot}5$.

8 For Figure 5 write down a formula for P in terms of x, y, and z.

a Find P when $x = 35$, $y = 15$, and $z = 12$.
b Find z when $P = 64$, $x = 12$, and $y = 10$.
c Find x when $P = 36$, $y = 5$, and $z = 12$.

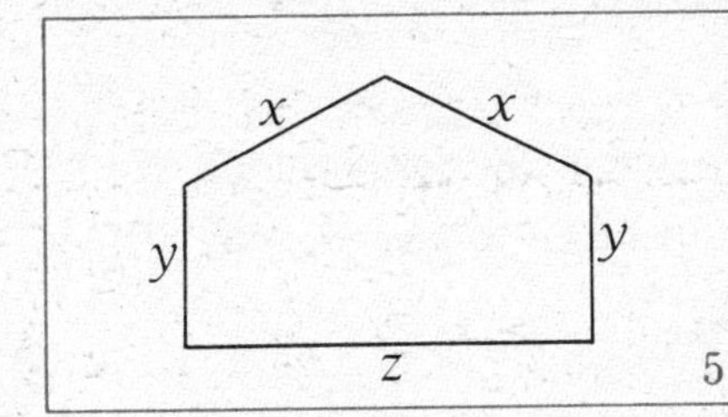

5

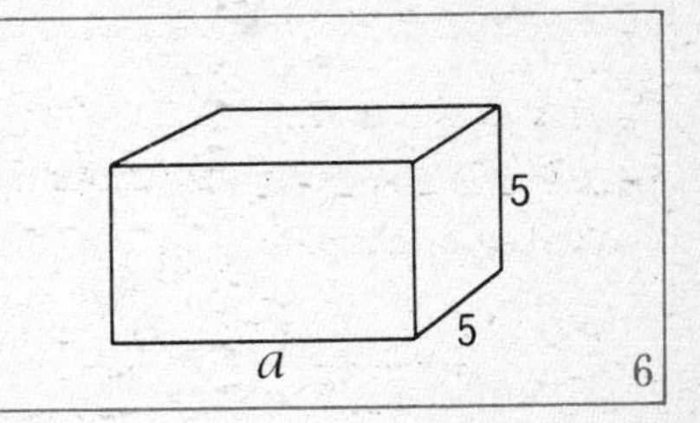

6

9 Figure 6 shows a cuboid with length a mm, breadth 5 mm and height 5 mm.

a Find a formula for the sum S mm of the lengths of all the edges of the cuboid.
b Calculate S if $a = 15$.

10 A rectangle is $4k$ cm long and k cm broad. With the aid of a sketch, find a formula for:

a its perpendicular P cm *b* its area A cm^2
Calculate the perimeter and area of the rectangle for which $k = 7{\cdot}5$.

Exercise 3B

1 Figure 7 on page 10 shows three squares of side x cm. All the units are centimetres.

a Copy and complete the following, working along each row from left to right:

	Area of whole square (cm^2)	*Area of unshaded part* (cm^2)	*Area of shaded part* (cm^2)	*Formula for area of shaded part*
(*1*)	x^2	$4x$	x^2-4x	$A = x^2-4x$
(2)	x^2	...	...	...
(*3*)	...	...	...	...

b Using your formulae, find in each case the area of the shaded part if $x = 10$, $y = 4$, and $z = 7$.

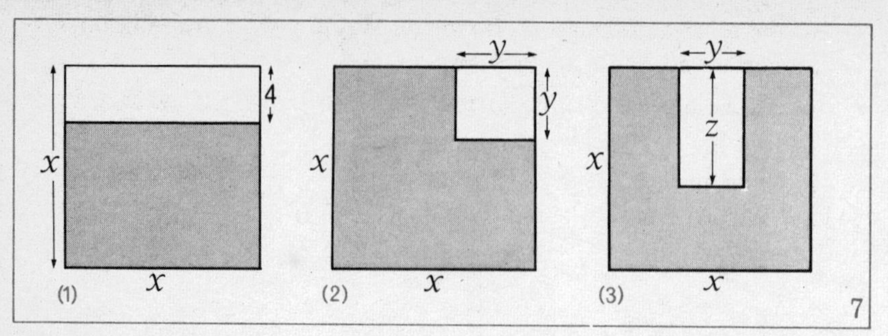

2 Figure 8 shows three rectangles of length a mm and breadth b mm.

a Make up a table as in question *1* to find a formula for the area (A mm^2) of the shaded part in each case.

b Use your formulae to find each shaded area if $a = 15$, $b = 12$, $c = 10$, $d = 6$, $e = 6$, $f = 9$, $g = 4$.

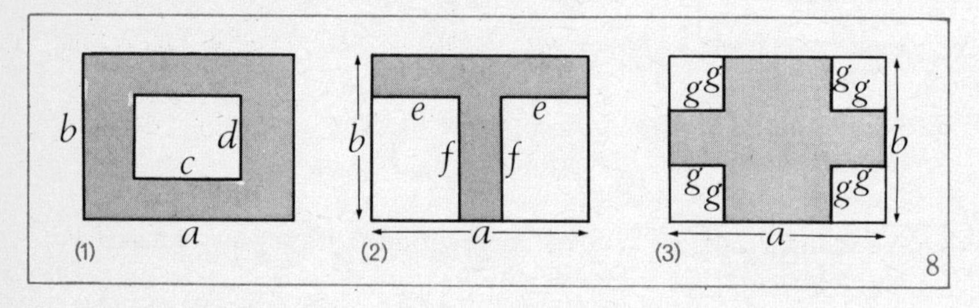

3 An aircraft weighs t tonnes when fully loaded. It uses f tonnes of fuel per hour.

a How many tonnes of fuel does it use up in *(1)* 2 hours *(2)* 3 hours *(3)* h hours?

b If W tonnes is the weight of the aircraft after h hours, write down a formula for W, assuming that the aircraft was fully loaded at the start.

c Use your formula to calculate W when $t = 12$, $f = \frac{1}{4}$, and $h = 5$.

4 The practical rule for expressing a Fahrenheit temperature as a Celsius (Centigrade) temperature is 'Subtract 32, and multiply the result by $\frac{5}{9}$'.

a Use this rule to find the reading (°C) on a Celsius thermometer when the reading (°F) on a Fahrenheit thermometer is:
(1) 86° (a hot day) *(2)* 50° (a cold day) *(3)* 32° (freezing-point of water).

b Give a formula for C in terms of F. Use it to calculate C when $F = 212$, and when $F = 1004$.

5 *a* Find the 5th, 6th, and 7th numbers of the sequence 3, 7, 11, 15, ...
b Can you write down the 100th number?
c The nth number N is given by the formula $N = 4n - 1$. Check that this formula is true for the 5th, 6th, and 7th numbers, and use it to find the 100th and the 1000th numbers.

6 Figure 9 shows a trellis of squares, each of side 1 cm.
a How many squares of side 1 cm are there?
b How many points are there where two lines cross or meet?
c If there were x squares across and y squares up,

(*1*) find a formula for the number S of squares of side 1 cm,
(*2*) find a formula for the number P of points where two lines cross or meet, and
(*3*) find S and P for the case when $x = 29$ and $y = 30$.

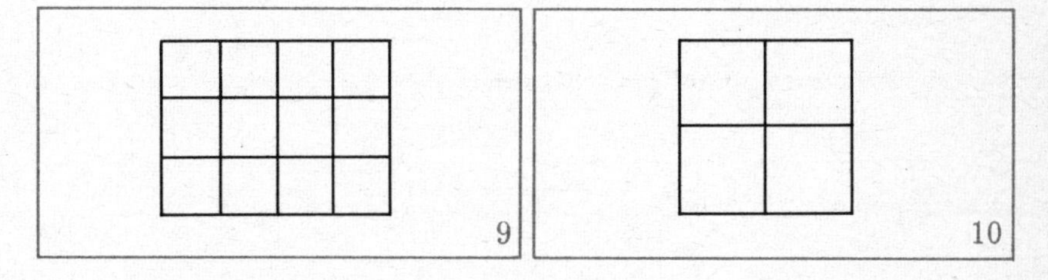

9 10

7 In Figure 10 how many squares are there altogether in the 2×2 square tiling shown? (Not four!)
a Draw a 3×3 tiling. How many squares are in it? (Not nine or ten!)
b For an $a \times a$ tiling, the number of squares N is given by the formula $N = \frac{1}{6}a(a+1)(2a+1)$. Use this formula to check your answers for the 2×2 and 3×3 tilings.
c Calculate the number of squares in a 100×100 tiling.

8 Figure 11 shows a quadrilateral (four sides) and a pentagon (five sides), with all the diagonals drawn.

a Write down the total number of diagonals in each figure.

b Draw a hexagon (six sides) with all its diagonals. How many diagonals are there?

c For a figure with n sides (a polygon), the number of diagonals N is given by the formula $N = \frac{1}{2}n(n-3)$.
Check your answers to ***a*** and ***b*** by means of this formula.

d How many diagonals would a figure of twenty sides have?

e How many sides would a polygon with 54 diagonals have?

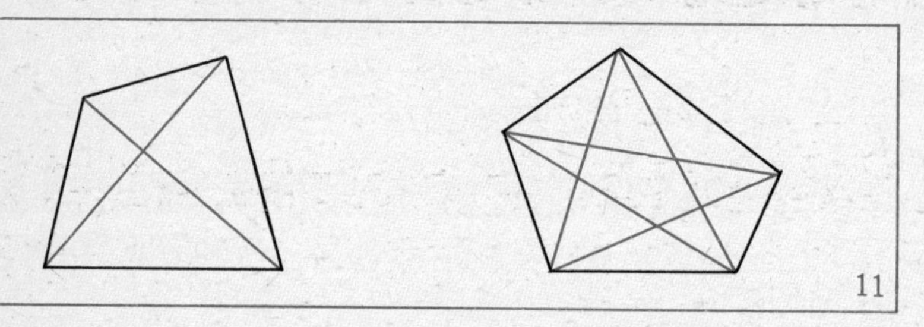

11

Summary

1 *Replacements* can be made for the variable in an open sentence.

2 *Formulae* are useful when the same kind of calculation has to be repeated with different sets of numbers.

Examples are:

Perimeter of rectangle. $P = 2l+2b$, or $P = 2(l+b)$
Area of rectangle. $A = lb$
Volume of cuboid. $V = lbh$

Inequalities and Inequations

1 Inequalities

The numbers used for counting and measuring have *order*.

For example, $5 = 4+1$ and $6 = 5+1$, so

5 *is greater than* 4 and 5 *is less than* 6

Using > for 'is greater than', and < for 'is less than', we can write these *inequalities* like this:

$$5>4 \quad \text{and} \quad 5<6$$

(Notice that the smaller end of the arrowhead is nearer the smaller number.)

If a and b represent any two numbers, then:

$$a>b, \quad \text{or} \quad a=b, \quad \text{or} \quad a<b$$

If a is not equal to b, we can write $a \neq b$.

Exercise 1A

1 Write in shorter form, using the symbols >, < :

a 7 is greater than 4   *b* 2 is less than 3

c 5 is greater than 1   *d* 10 is less than 100

2 Write in words:

a $2<5$ *b* $10>6$ *c* $0<4$ *d* $8 \neq 3$

3 Which of the following sentences are true?

a $5>1$ *b* $4<4$ *c* $2<4$ *d* $15 \neq 5$

4 Copy the following as given, and insert one of the symbols $>$, $=$, $<$ between each pair of numbers to make a true sentence:

a 3 ... 4 *b* 10 ... 10 *c* 8 ... 7 *d* 1 ... 0
e 0 ... 10 *f* 4+1 ... 5−3 *g* 99 ... 99 *h* 99 ... 101

5 Which of the following sentences is false?

a 1 m > 10 cm *b* £12 ≠ £15 *c* 1 kg > 1000 g

6 List the whole numbers which are:

a greater than 2 and less than 5 *b* greater than 0 and less than 8

7 The inequalities $1<2$ and $2<3$ can be combined as $1<2<3$ (read '1 is less than 2 is less than 3'). Combine in the same way:

a $5<6$ and $6<7$ *b* $2<4$ and $4<6$
c $8>7$ and $7>6$ *d* $9>5$ and $5>1$

Exercise 1B

1 Copy the following as given, and insert one of the symbols $>$, $=$, $<$ between each pair of numbers to make a true sentence:

a 199 ... 200 *b* 2×7 ... $13+1$ *c* 2^3 ... 3^2
d 0 ... 1 *e* 91 ... 19 *f* 35 ... 34
g $0{\cdot}1$ ... $(0{\cdot}1)^2$ *h* $\frac{1}{4}$ ... $\frac{1}{2}$ *i* $3+0$ ... 3×0

2 State whether each of the following is true or false:

a $9+7>8+8$ *b* $5\times 1>5\times 0$ *c* £12 = 1200p
d $6\times 5\times 4 = 4\times 5\times 6$ *e* $\frac{3}{2} \neq \frac{2}{3}$ *f* 2·5 cm > 25 mm

3 The inequalities $3<5$ and $5<8$ can be combined into one sentence $3<5<8$ which is read '3 is less than 5 is less than 8'. This also means that 5 is between 3 and 8. Combine the following inequalities:

a $1<2$ and $2<5$ *b* $0<5$ and $5<10$
c $2<3$ and $5>3$ *d* $10>5$ and $10<20$

4 Write as inequalities:

a 6 is between 3 and 8 *b* 10 is between 1 and 12
c x is between 1 and 5 *d* x is between a and b

5 In question ***3***, instead of using $<$, we could use $>$ and write $8>5>3$. Express each of the four parts of question ***3*** in this way.

6 Arrange each of the following 'triples' in order, first using the symbol $<$ only, and then using $>$ only.
For example, 4, 3, 7 give $3<4<7$ *or* $7>4>3$.

a 3, 1, 5 *b* 0, 2, 1 *c* 1 cm, 1 mm, 1 m
d 2 g, 2kg, 2 mg *e* 0·2, 0·02, 2·0 *f* 50p, 5p, 10p

2 *The number line*

The *order* of the whole numbers 0, 1, 2, ... may be shown by points marked at equal intervals on a straight line as in Figure 1.

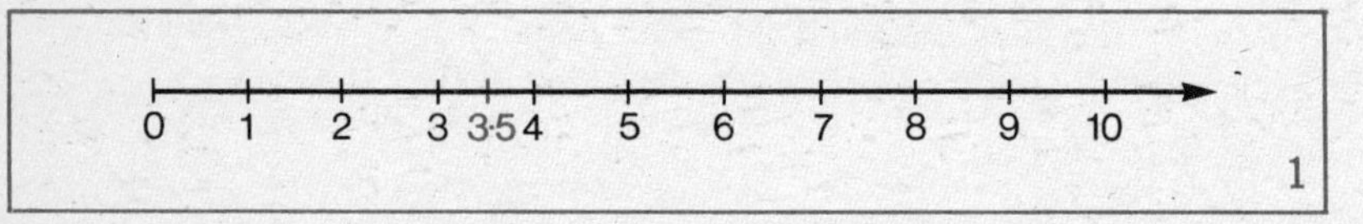

1

This picture is called a *number line*. Fractions may be placed on the number line; also the number 3·5 is shown by marking a point midway between 3 and 4.

From the number line, we can tell at a glance whether or not a number is greater than another one. For example:

Number line	*Inequality*
5 is to the right of 4	$5>4$
5 is to the left of 6	$5<6$

Note ***a*** $n \geqslant 4$ means 'n is greater than or equal to 4'
b $n \leqslant 2$ means 'n is less than or equal to 2'.

If n is a whole number less than 10, then for ***a*** possible numbers are 4, 5, 6, 7, 8, 9; and for ***b***, 0, 1, 2.

Figure 2 illustrates the above.

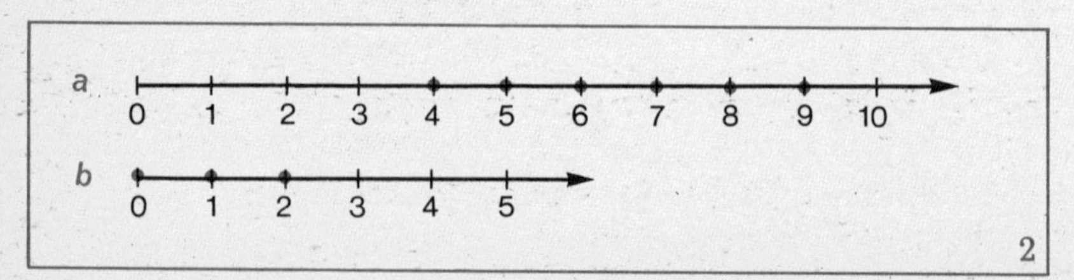

2

Example.—Show on a number line the subset {2, 3, 4, 6} of the set of whole numbers.

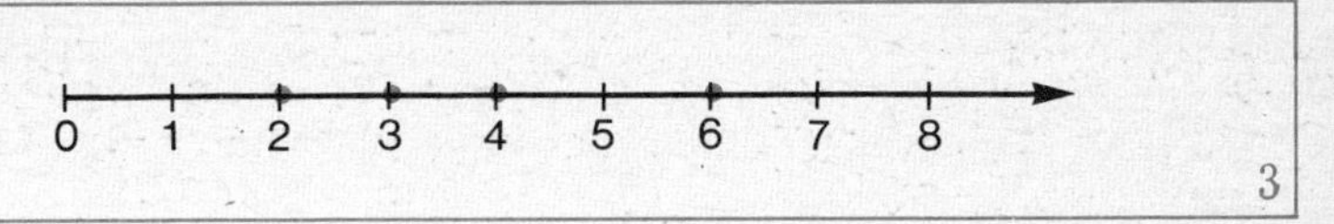

3

The resulting picture, shown in Figure 3, is called the *graph* of the subset.

Exercise 2

1 a Draw a number line showing the positions of whole numbers from 0 to 10.

b Show by dots the even numbers between 1 and 7, and the number 9.

c Show also by crosses the positions of 7·5, 0·5, 8·2.

2 On separate number lines show the graphs of the following subsets of the set of whole numbers:

a {1, 3, 5, 7, 9} *b* {0, 2, 4, 6, 8, 10}

c {1, 4, 7, 10} *d* {1, 2, 3, 5, 8}

3 With the aid of a number line complete the following:

a If number $a > 4$, a lies to the ... of 4.

b If number $b < 7$, b lies to the ... of 7.

c If number $c = 5$, c lies ... 4 and 6.

4 If n represents a whole number, mark all possible positions of n on a number line for each of the following:

a $n < 4$ *b* $n = 6$ *c* $7 < n < 10$

5 a Show the set $E = \{1, 2, 3, ..., 10\}$ on a number line.

b (*1*) Mark on the diagram for E the members of the subset $\{p \in E: 1 < p < 5\}$ (read 'p is a member of E, such that p is greater than 1 and less than 5').

(*2*) Similarly mark the subset $\{p \in E: 8 \leqslant p \leqslant 10\}$.

6 $E = \{0, 1, 2, 3, 4, 5, 6\}$, and two subsets of E are shown by the red dots in Figure 4.

Express these subsets in the form used in question ***5b*** (*2*).

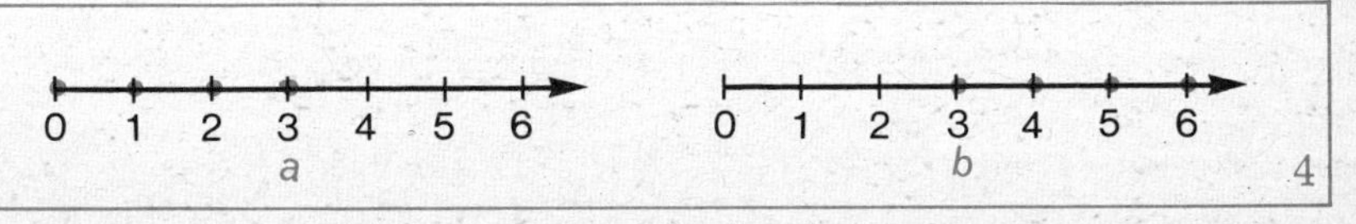

4

3 Inequations—1

Certain mathematical sentences contain the verb 'is equal to'. We call them *equations*.

$x = 3$ and $x+2 = 6$ are equations.

With the verb 'is greater than', or 'is less than', the sentence is called an *inequation*.

$x > 3$ and $x+2 < 6$ are inequations.

Example 1. If x is a variable on the set {0, 1, 2, 3, 4, 5, 6}, find the solution set of the inequation $x > 3$.

Replacing x by 0, 1, 2 or 3 we obtain *false* sentences.
Replacing x by 4, 5 or 6 we obtain *true* sentences.
So 4, 5, 6 are solutions of the inequation $x > 3$.
Hence the solution set is {4, 5, 6}.
The solution set of the inequation $x > 3$ is shown on the number line in Figure 5.

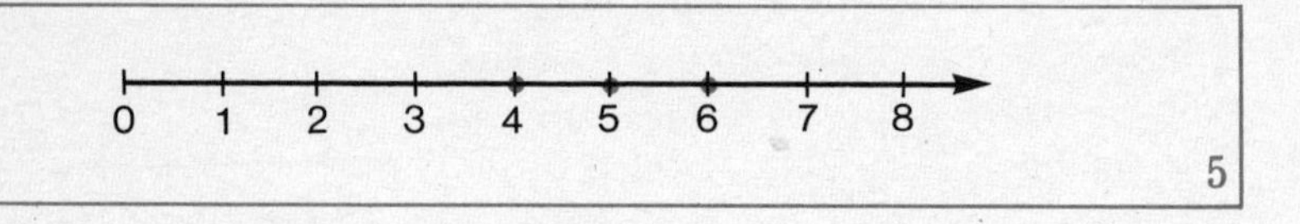

5

This picture of the solution set on the number line is called the *graph* of the solution set of the inequation $x > 3$, which we shorten to 'the graph of $x > 3$'.

Example 2. Given that x is a variable on the set {0, 1, 2, ..., 9, 10}, solve the inequation $x+2 \leqslant 6$.

Replacing x by 0, 1, 2, 3, 4 in $x+2 \leqslant 6$, we obtain *true* sentences.
The remaining replacements 5, ..., 10 give *false* sentences.
Hence the solution set is {0, 1, 2, 3, 4}.
The solution set is shown on the number line in Figure 6.

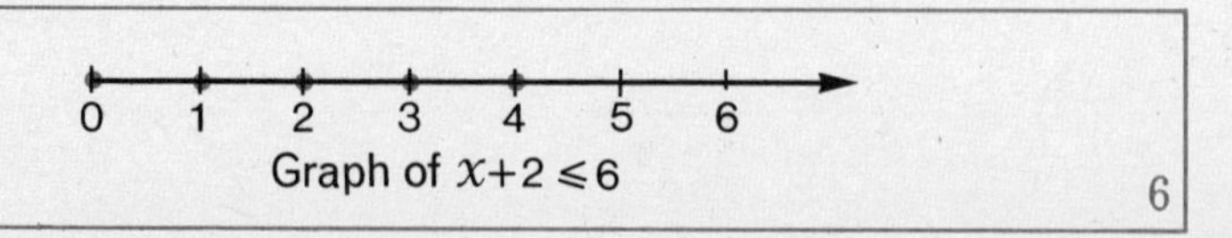

Graph of $x+2 \leqslant 6$

6

Note.—If no replacement from the given set makes the sentence true, then the solution set is empty. Thus, the solution set of $x+2 < 1$ is ø.

Exercise 3A

In questions ***1*** to ***12***, x is a variable on the set {0, 1, 2, 3, 4, 5, 6}. Find the solution set of each of the following inequations:

1 $x > 2$ *2* $x < 3$ *3* $x \leqslant 3$

4 $x < 1$ *5* $x \geqslant 7$ *6* $8 > x$

7 $x > 1$ and $x < 5$ *8* $2 < x < 6$ *9* $2 \leqslant x \leqslant 6$

10 $4 > x > 2$ *11* $5 > x > 1$ *12* $x > 0$

In questions ***13*** to ***21***, x is a variable on the set {0, 1, 2, ..., 10}. Write down the solution set of each inequation, and show the solution set on a number line.

13 $x+1 < 5$ *14* $x+4 < 6$ *15* $x+5 < 10$

16 $4+x \geqslant 8$ *17* $x+5 > 5$ *18* $x-1 > 7$

19 $x-3 < 6$ *20* $x-1 \leqslant 2$ *21* $3 < x-4$

In questions ***22*** to ***27***, x is a variable on the set {1, 2, 3, ..., 10}. Solve:

22 $2x > 8$ *23* $2x < 10$ *24* $3x < 10$

25 $4x > 19$ *26* $5x < 18$ *27* $x^2 < 20$

28 Using the symbol > or <, write in shorter form:

a p is greater than 5.
b n is less than 10.
c When 10 is added to x, the sum is greater than 20.
d When 1 is subtracted from c, the difference is less than 13.
e Twice y is less than 30.
f p plus q is greater than or equal to r.

* * * * *

Example 3. x is a variable on the set {1, 2, 3, ..., 9, 10}. Find the solution set of the inequation $2x+1 > 7$.

Replacing x by 1 we obtain $(2\times 1)+1 > 7$, i.e. $3 > 7$
Replacing x by 2 we obtain $(2\times 2)+1 > 7$, i.e. $5 > 7$
Replacing x by 3 we obtain $(2\times 3)+1 > 7$, i.e. $7 > 7$

Replacing x by 4 we obtain $(2\times4)+1 > 7$, i.e. $9 > 7$

Replacing x by 5 we obtain $(2\times5)+1 > 7$, i.e. $11 > 7$,
and so on.

It is clear that the only replacements for x which give true sentences are 4, 5, 6, 7, 8, 9, 10.

Hence, the solution set is $\{4, 5, 6, ..., 10\}$.

Exercise 3B

1 If x is a variable on the set $\{0, 1, 2, 3, 4, 5\}$, find the solution set of each of the following, and show the graph of each on the number line:

a $2x+1 > 5$ *b* $2x-1 < 5$ *c* $2x+5 < 15$
d $2x+8 < 9$ *e* $2x < x+4$ *f* $4x+5 > 9$

2 Given that m is a variable on the set $\{1, 2, 3, ..., 10\}$, find the solution set of each of the following:

a $11+3m \leqslant 26$ *b* $11+3m < 19$ *c* $11-3m < 9$
d $11-3m > 9$ *e* $3m^2 > 20$ *f* $m^3 < 50$

3 If x is a variable on the set $\{0, 1, 2, 3\}$, find the solution set of each of the following:

a $2x+3 < 7$ *b* $2x < x+1$ *c* $19x < 42-x$
d $2(3-x) > 1$ *e* $18 > 5x+4$ *f* $21 \leqslant 7x+2$

4 x is a variable on the set $\{0, 1, 2, ..., 10\}$. Write down the solution set of each inequation below. Hence find the intersection of the solution sets in each pair of inequations.

a $x > 3, x < 8$ *b* $x < 7, x > 4$
c $x > 5, x < 4$ *d* $x+2 < 6, x-1 > 1$

5 Using the symbol $>$ or $<$, write each of the following as an inequation in shorter form:

a p exceeds 50.
b g is less than or equal to 10.
c Three times t is less than 25.
d If 2 is added to x, the sum is less than y.
e y is greater than 5 and less than 15.
f h is less than 10 and greater than 0.

4 *Inequations—2*

Example.—Find the solution set of the inequation $3x-4 > 7$, where x is a variable on the set of whole numbers.

Replacing x by 0 we obtain $(3\times 0)-4 > 7$, i.e. $-4 > 7$, which is false.

Replacing x by 1 we obtain $(3\times 1)-4 > 7$, i.e. $3-4 > 7$, which is false.

Replacing x by 2 we obtain $(3\times 2)-4 > 7$, i.e. $2 > 7$, which is false.

Replacing x by 3 we obtain $(3\times 3)-4 > 7$, i.e. $5 > 7$, which is false.

Replacing x by 4 we obtain $(3\times 4)-4 > 7$, i.e. $8 > 7$, which is true.

Replacing x by 5 we obtain $(3\times 5)-4 > 7$, i.e. $11 > 7$, which is true.

From the above, x must be greater than 3, and so the solution set is $\{4, 5, 6, \ldots\}$.

Note that ... means 'and so on'.

Exercise 4

Find the solution set of each of the following inequations (from the set of whole numbers):

1 $x+7 > 11$ *2* $x+11 > 7$ *3* $x-11 > 7$

4 $x-7 < 11$ *5* $11-x < 7$ *6* $7-x < 11$

7 $x+11 < 7$ *8* $p+2 \geqslant 5$ *9* $y+4 > 15$

10 $w+25 < 29$ *11* $n-2 \leqslant 5$ *12* $m-6 \geqslant 6$

13 $5 > x+4$ *14* $x > x+1$ *15* $x+8 > 4$

16 $x+8 < 4$ *17* $2x+1 \leqslant 5$ *18* $2x+1 \geqslant 5$

19 $2x+1 < 5$ *20* $2x > 15$ *21* $13 > 2x$

22 $20 \geqslant 2x-10$ *23* $4x-1 < 21$ *24* $7x+2 \geqslant 30$

Exercise 4B

1 Fill in the blank spaces:

a $8 > 3$ because $8 = 3 + \dots$
b $34 > 19$ because $34 = \dots + \dots$
c $7 < 11$ because $11 = \dots + \dots$
d $5 < 9$ because $\dots = \dots + \dots$
e $1 > 0$ because $\dots = \dots + \dots$

2 p and q are variables on the set of natural numbers. Put the correct symbol $>$, $<$, or $=$ in each blank space.

a $p = q+5$, so $p \dots q$ *b* $p-q = 0$, so $p \dots q$
c $p+1 = q$, so $p \dots q$ *d* $p+p = q$, so $p \dots q$
e $p = 11, q = 15$, so $p \dots q$ *f* $q = p+3$, so $p \dots q$

3 a and b are variables on the set of whole numbers. Put the correct symbol $>$, $=$, or $<$ in each blank space.

a $2a < 2b$, so $a \dots b$ *b* $a-1 > b$, so $a \dots b$
c $a+1 < b$, so $a \dots b$ *d* $a-b = 7$, so $a \dots b$
e $a+5 > b+5$, so $a \dots b$ *f* $a+7 = b+10$, so $a \dots b$

4 Find the whole number c such that the solution set of the inequation $x+c < 20$ is $\{0, 1, 2, \dots, 14\}$.

5 Given that x is a variable on the set $\{0, 1, 2, 3, 4, 5\}$, find the solution set of:

a $x^2 < 15$ *b* $x^2 \geqslant 3x$ *c* $x^2+4 > 4x$

6 x is a variable on the set of whole numbers. Copy and complete the following:

a $2x+1 = 7$, $2x = \dots$, $x = \dots$
b $2x+1 > 7$, $2x > \dots$, $x > \dots$
c $4x+5 = 9$, $4x = \dots$, $x = \dots$
d $4x+5 < 9$, $4x < \dots$, $x < \dots$
e $3x+2 = 20$, $3x = \dots$, $x = \dots$
f $3x+2 \geqslant 20$, $3x \geqslant \dots$, $x \geqslant \dots$
g $2x+7 = 15$, $2x = \dots$, $x = \dots$
h $2x+7 \leqslant 15$, $2x \leqslant \dots$, $x \leqslant \dots$

7 A rectangle has an area of 24 cm². If its length is p cm and its breadth is q cm, copy and complete this table:

p	1	2	3	4	6	8	12	24
q	24	12						

Using the table, write down and complete $p+q \geqslant \dots$ and $p+q \leqslant \dots$.

8 Repeat question *7* for a rectangle of area 36 cm^2, p and q being variables on the set {1, 2, 3, 4, 6, 9, 12, 18, 36}.

9 The perimeter of a rectangle is 16 cm, i.e. the sum of its four sides is 16 cm. If its length and breadth are p cm and q cm respectively, copy and complete the following:

a $2p+2q = ...$ *b* $p+q = ...$ *c* $p < ...$

d

p	1	2	3	4	5	6	7
q	7	6	.	.	.	.	.

e $pq \geqslant ...$ and $pq \leqslant ...$

10 Repeat question *9* for a rectangle whose perimeter is 12 cm; for the table assume that p and q are variables on the set {1, 2, 3, 4, 5}.

11 A man walked from a place A to a place B, covering 24 km. His average speed was greater than 4 km per hour but less than 6 km per hour. If he took t hours for the journey, copy and complete the inequations: *a* $t > ...$ *b* $t < ...$

12 From a square of side 8 units, equal squares of side x units are cut from the four corners, where $x \in \{1, 2, 3, 4\}$. In each case this gives the net of an open box as shown in Figure 7. When the box is made up, it can be exactly filled with n unit cubes.

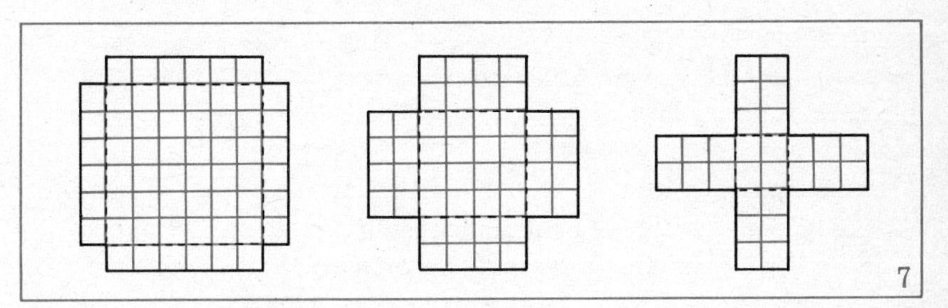
7

Copy and complete the following:

a

x	1	2	3	4
n	36			

b $... \leqslant n \leqslant ...$

5 *Writing mathematical sentences*

In everyday life, many statements and symbols include ideas that can be expressed by means of equalities or inequalities. For example, the road sign shown here means that vehicles passing it must not exceed the speed of 50 km/h.

8

For a speed of S km/h, we can express the instruction in the form of an inequation, $S \leqslant 50$.

S could be a variable on the set of whole numbers and fractions.

Exercise 5

In each of the following examples give a mathematical sentence containing the variable indicated to express the idea in the English sentence.

1 In throwing a die, a score (S) of 5 or over wins.

2 Police recruits must be at least 172 centimetres in height (H metres).

3 The minimum weight of a certain packet of oatcakes is 326 grammes. Take W grammes for the weight of a packet.

4 The crowd limit (L) at a football match was 130 000.

5 In a golf competition, a score (S) of 275 or less was required to qualify for the final rounds.

6 The maximum number of passengers allowed on a bus is the sum of 38 downstairs, 28 upstairs and 8 standing. Take N for the number of passengers.

7 The minimum flying speed of an aeroplane is 180 km/h and the maximum speed is 1250 km/h. The speed of the aeroplane is m km/h.

8 A slot in a coin-operated machine will not accept a coin with diameter (d cm) equal to or greater than 2·4 cm, and the machine will not work if the diameter of the coin is 2·2 cm or less.

9 Part of the Beaufort Scale gives wind velocity (v km/h) at a height of 11 metres as follows:

Strong gale: 75–87 km/h; whole gale: 88–101 km/h;
storm: 102–121 km/h; hurricane: over 121 km/h.

Give the limits of v for: ***a*** a whole gale ***b*** a hurricane.

10 A factory makes pistons for motor-car engines. The diameter of each piston has to be 8 cm approximately; if the diameter is 0·002 cm greater or less than this the piston is not accepted. Take d cm for the diameter of the piston.

11a Two six-sided dice are thrown. Write down the set S of possible scores on throwing the dice.

b In a game, the winning score is 7 or over. If x is a variable on S, write down a suitable inequation to express this condition, and state its solution set.

12 The sum of the lengths of any two sides of a triangle is always greater than the length of the third side.
If the sides are x cm, x cm and 10 cm, form an inequation in x, and express it in the form '$x >$ (a number)'.

Summary

1 The numbers used for counting and measuring have *order*, which can be shown on the number line.

2 If a and b are any two numbers, then:

$$a > b, \quad \text{or} \quad a = b, \quad \text{or} \quad a < b$$

3

Inequality symbols	*Meaning*
$>$	is greater than
$\geqslant$	is greater than or equal to
$<$	is less than
$\leqslant$	is less than or equal to
$\neq$	is not equal to

4 An *inequation* is an open sentence containing one of the verbs in ***3*** above.

	Inequation
$5 > 3$ (True)	$x > 3$ (neither true nor false)
$5 < 3$ (False)	

5 A *solution* of an inequation is a replacement for the variable which gives a true sentence, e.g. 2 is a solution of $x < 3$.

The *solution set* is the set of all such solutions in a given set E.

6 The *graph* of an inequation in one variable shows its solution set on the number line, e.g. for $x < 3$, where $x \in W$, we have:

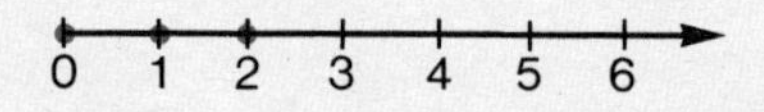

Negative Numbers

1 An extension of the number line

In the Celsius (or Centigrade) scale of temperature the boiling-point of water is 100°, and the freezing-point is 0°. For temperatures below the freezing-point we need to extend the scale back beyond zero. A temperature of two degrees below zero is written -2°C, and spoken 'negative 2°Celsius'.

* * * * *

Do you know how to give the coordinates of the points B, C and D in Figure 1 to indicate the positions of the points?

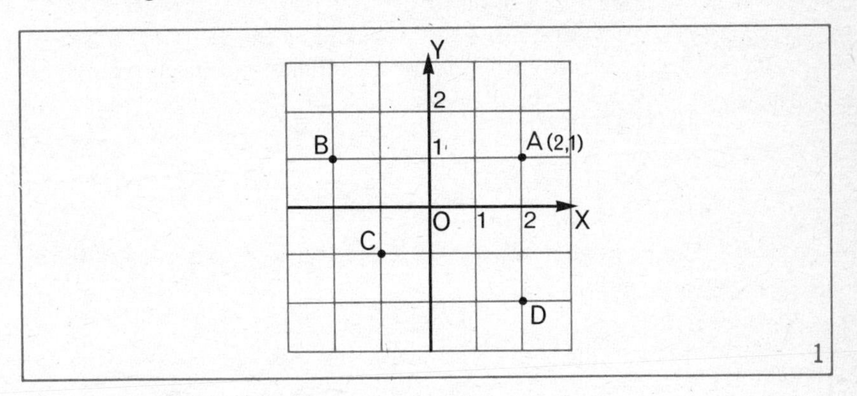

1

Figure 2 shows part of a number line, with some of the whole numbers marked in. The number corresponding to each point is the number of units of length from zero to the point, measured to the *right*. What numbers do you suppose correspond to those points to the *left* of zero?

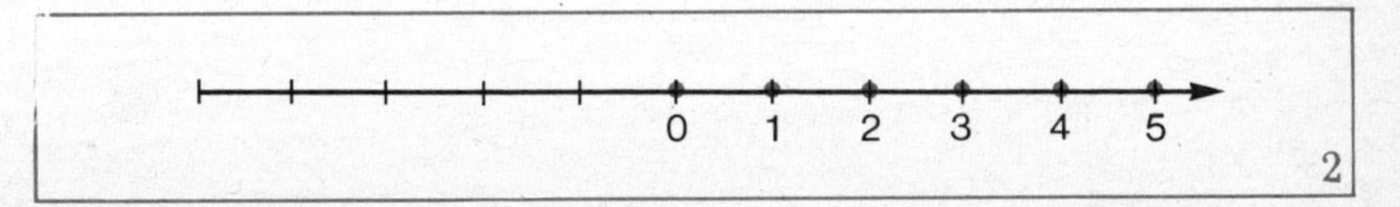

2

Figure 3 shows some of the numbers corresponding to those points. The point one unit to the *left* of zero corresponds to the number (-1), read 'negative one'. The point two units to the left corresponds to (-2), and so on. The numbers $(-1), (-2), (-3), \ldots$ are called *negative integers.*

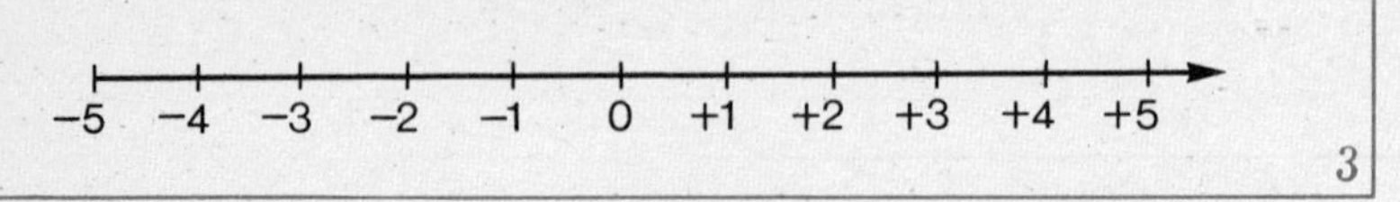

3

Notice that in Figure 3, the numbers to the right of zero are written $+1, +2, +3, \ldots$ These numbers are called *positive integers.*

The set of positive integers, negative integers and zero is called the *set of integers*, and is named Z.

The positive integers $+1, +2, +3, \ldots$ have all the properties of the natural numbers which you have already studied and are usually written like natural numbers, without the + sign in front of them. The negative integers are often written $-1, -2, -3, \ldots$ without the brackets. You may also hear them spoken as 'minus one', 'minus two', etc., but there are good reasons for calling them 'negative one', 'negative two', and so on. Thus we may write:

$$Z = \{\ldots, -3, -2, -1, 0, 1, 2, 3, \ldots\}$$

Exercise 1A

1 Write down the three numbers that should come next in each of the following sequences:

a 12, 9, 6, 3, ..., ..., ... *b* 5, 3, 1, ..., ..., ...
c $-12, -10, -8$, ..., ..., ... *d* $-13, -9, -5$, ..., ..., ...

2 Figure 4 shows part of a Celsius thermometer scale. Copy the diagram and complete the marking of the scale. What temperature is the thermometer showing?

a How would you write a temperature of:
(*1*) 5° below zero (*2*) 5° above zero?

b What would be the meaning of a temperature of:
(*1*) $-10°$ (*2*) 0° (*3*) 20° (*4*) 8° of frost?

c Which reading indicates a warmer temperature:
(*1*) 0° or 3° (*2*) $-15°$ or $-10°$?

Fig. 4

3 The countdown to the time of launch of a rocket is given in seconds by the sequence 10, 9, 8, ...
Write the rest of the sequence to the time of firing, and continue the sequence for a further four numbers.

4 Figure 5 shows the key to shading on a map of Holland.

a Copy the figure and complete the missing numbers.

b What can you say about the land marked as in:
(*1*) the highest rectangle
(*2*) the lowest rectangle?

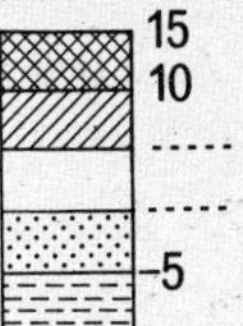

Fig. 5

5 *a* If +10 km means 10 km *north* of a place, what will −10 km mean?

b If +15 newtons means a force of 15 newtons to the *right*, what will a force of −15 newtons indicate?

c If +10 metres per second means a speed of 10 metres per second *upwards*, what will a speed of −25 metres per second mean?

d Suggest a way of showing that your watch is:
(*1*) 1 minute fast
(*2*) 2 minutes slow.

6 At 4 pm the temperature in Glasgow was 5°C. At 6 pm a cold front passed over, and the temperature fell by 7°C. What was the temperature then? (Your thermometer drawing for question **2** would help you here.)

Later the temperature rose by 2°C. What was the temperature then?

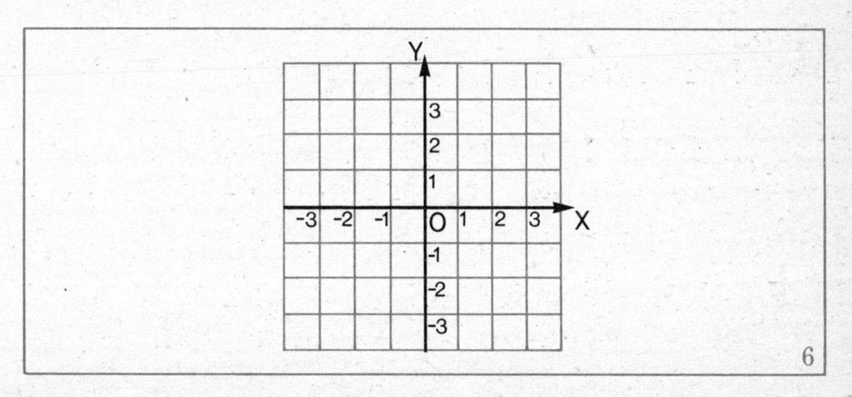

7 *a* Copy the axes, marked as shown in Figure 6 on page 29.
b Plot the points A(3, 3), B(2, 2), C(1, 1), O(0, 0), P(−1, −1), Q(−2, −2), R(−3, −3).
c Write down the coordinates of the next three points in this sequence, and plot the points.

8 Suppose that your diagram for question 7 is folded along OY, so that A, B and C touch the paper at A′, B′ and C′.
Write down the coordinates of A′, B′ and C′. (A′, B′ and C′ are the *images* of A, B and C in the *y*-axis.)

9 Suppose that your diagram for question 7 is folded along OX, so that P, Q and R touch the paper at P′, Q′ and R′.
Write down the coordinates of P′, Q′ and R′ (the images of P, Q and R in the *x*-axis).

10 A is the point (4, 2), B(−1, 1), C(−3, −5), D(6, 0). Give the image of each point: *a* in the *y*-axis *b* in the *x*-axis.

Exercise 1B

1 Write down the three numbers that should come next in each of the following sequences:
a 12, 8, 4, ..., ..., ... *b* 4, 1, −2, ..., ..., ...
c −9, −7, −5, ..., ..., ...

2 The temperature on a thermometer was 2° Celsius. What would the reading be if the temperature:
a rose 2° *b* fell 2° from the first reading
c fell 9° from the first reading?

3 *a* Copy the axes, marked as shown in Figure 6.
b Draw the triangle ABC with vertices A(1, 1), B(3, 1), C(2, 4).
c If your diagram is folded about OY until A, B and C touch the paper at A′, B′ and C′, then A′, B′ and C′ are the images of A, B and C in the *y*-axis. Write down the coordinates of A′, B′ and C′.

4 Write down the images of A, B and C of question **3** in the *x*-axis.

5 *a* Plot the points P(−3, 1), Q(1, 1), R(1, 2), S(−3, 2) on squared paper.
b What shape is PQRS?
c Show the image of PQRS in the *x*-axis, and write down the coordinates of its vertices.

d Show the image of PQRS in the y-axis, and write down the co-ordinates of its vertices.

6 Mr Smith has a credit balance of £10 in the bank. Write down his new balance

a if he drew out £8

b if he drew out £12 (the bank allows overdrafts).

7 When the regulator on a clock was set to $+1$, the clock kept time. When it was set to $+4$, the clock gained one minute a day. How much would it gain or lose if the regulator was set to -2?

8 List some more situations involving negative integers.

2 *Order on the number line*

In Chapter 2 we studied *order* in the set of whole numbers, the order relations being shown on a number line. Thus if one number is greater than another, then the first number appears to the right of the second number on the number line.

For example, $5 > 4$ (read 'five is greater than four') is shown as follows:

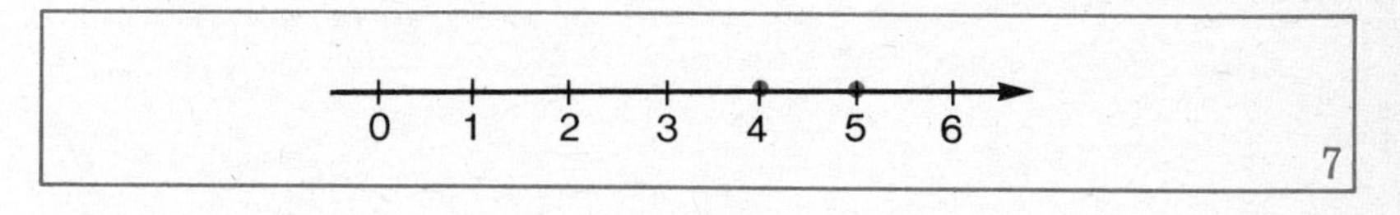

7

This idea may be extended to the set of integers. We see that -1 is to the right of -3 so that -1 is *greater* than -3, and we write $-1 > -3$.

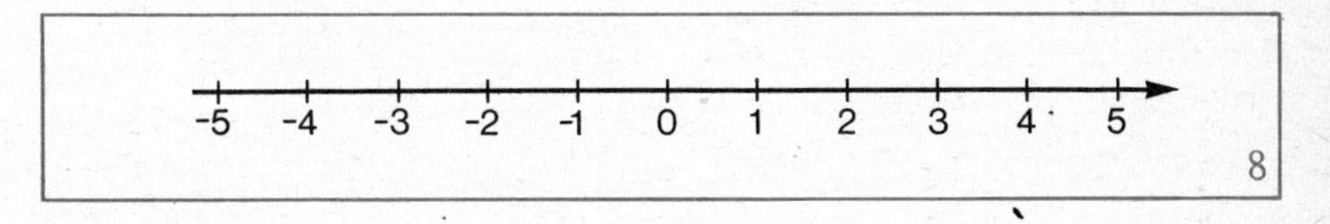

8

Again, -4 is to the left of -2 so that -4 is *less* than -2, and we write $-4 < -2$.

Note also that zero is *less* than every positive integer and *greater* than every negative integer. The point assigned to zero is called the *origin*. Positive integers are shown to the right of the origin and negative integers to the left of the origin.

Exercise 2A

1 Draw a number line as in Figure 8 on page 31. With its aid, write down the greater number in each pair:

a 4, 3 *b* 0, 1 *c* $-1, 0$ *d* $-4, -5$

2 In each of the following pairs say which is the greater temperature:

a 15°C, 12°C *b* 3°C, 0°C *c* -3°C, -2°C *d* -1°C, 1°C

3 In each of the following pairs say which is the greater height above sea level:

a 200 m, 100 m *b* 50 m, -50 m
c -40 m, -30 m *d* 0 m, -10 m

4 Insert the symbol $>$ in each of the following pairs of numbers, *arranged so as to make true sentences:*

a 5, 3 *b* $0, -1$ *c* 1, 3 *d* 0, 1
e $\frac{1}{2}, -2$ *f* $-2, 2$ *g* $-3, -4$ *h* $-20, -30$
i $-2, 0$ *j* $-5, -1$ *k* $\frac{3}{4}, \frac{1}{4}$ *l* $-1, 10$

5 Insert the symbol $<$ in each of the following pairs of numbers, *arranged so as to make true sentences:*

a 5, 12 *b* $-9, 9$ *c* 6, 3 *d* $\frac{1}{4}, -\frac{1}{2}$
e $\frac{1}{2}, \frac{1}{3}$ *f* $-3, -5$ *g* $-6, 2$ *h* $0, -1$
i $0, -0{\cdot}01$ *j* $-100, -99$ *k* $9{\cdot}01, 9{\cdot}11$ *l* 0, 1

6 Which of the following sentences for integers are true?

a $1 < 3$ *b* $-5 > 3$ *c* $1 < 5$ *d* $-3 < 0$
e $-3 > 1$ *f* $-2 < -5$ *g* $-4 \leqslant -4$ *h* $0 < 7$

7 Copy each of the following pairs of numbers *in the order given*, and insert the symbol $>$ or $<$ between the members of each pair so as to make true sentences:

a $5 \ldots -7$ *b* $10 \ldots -10$ *c* $-2 \ldots -1$ *d* $-7 \ldots -100$

8 If x is a variable on the set $\{-4, -3, -2, -1, 0, 1, 2, 3, 4\}$, find the solution set of each of the following inequations:

a $x < 0$ *b* $x < -2$ *c* $x > 2$ *d* $x > -1$
e $x \geqslant 4$ *f* $x \leqslant -3$ *g* $x \geqslant 0$ *h* $x < -4$

Exercise 2B

1 Write down the following pairs of numbers *in the order given*, and insert the symbol $<$ or $>$ between the members of each pair so as to make true sentences:

a $-3 \ldots 3$ *b* $0 \ldots -2$ *c* $7 \ldots 17$ *d* $-1 \ldots -2$
e $1 \ldots -10$ *f* $-\frac{1}{2} \ldots -\frac{1}{4}$ *g* $-1{\cdot}0 \ldots 0{\cdot}1$ *h* $0{\cdot}07 \ldots -100$

2 Using the inequality symbols $<$ and then $>$, arrange each of the following triples in *ascending* order (i.e. with the *least* first and the *greatest* last), and then in *descending* order (i.e. with the *greatest* first and the *least* last):

a 3, 7, 5 *b* 4, -6, 2 *c* -2, -1, 2 *d* -100, 10, 0

3 If x is a variable on the set $\{-5, -4, -3, -2, -1, 0, 1, 2, 3, 4, 5\}$, find the solution set of each of the following inequations, and show the solution set on a number line:

a $x > 1$ *b* $x < -2$ *c* $x \geqslant 0$
d $x < 0$ *e* $x < -3$ *f* $-3 \leqslant x < 3$

4 For each of these pairs of bank balances, which is the greater sum of money?

a £80, £60 *b* £2, £(-5) *c* £(-5), £(-6)

5 Every replacement which makes the open sentence $x < 0$ a true sentence is a negative number.

a Write down five such replacements.
b Make a corresponding statement for the open sentence $x > 0$.

6 If y is a variable on the set $\{-5, -4, -3, -2, -1, 0, 1, 2, 3\}$, solve the following inequations:

a $-3 < y < 0$ *b* $-4 < y < -2$ *c* $-3 < y \leqslant 2$
d $y > 0$ and $y \leqslant 3$ *e* $y < 0$ and $y < -3$ *f* $-1 < y < 1$

---

3 *Addition of positive and negative numbers*

Two number lines can be used as a mechanical aid for adding numbers. (This is partly how a 'slide rule' works.) Cut two strips about 1 cm wide and 18 cm long from a sheet of stiff drawing paper or from a side of a cereal packet. Now draw two number lines on each of the two strips as shown in Figure 9, using a scale of 1 cm to 1 unit. Alternatively, use strips of 5-mm squared paper. We now have a 'slide rule' for addition.

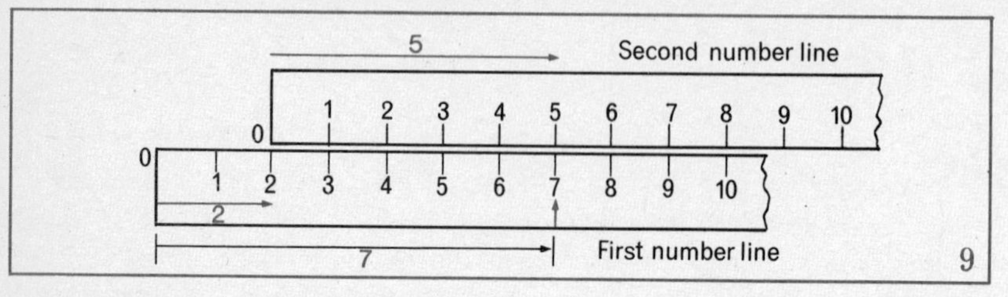

9

Suppose that we require the sum of 2+5. Figure 9 shows how you might find this. Place the zero of the second number line against the 2 on the first number line. Against 5 on the second number line we read 7 on the first number line.

We can think of a step of 2 units to the right on the first number line followed by a step of 5 units to the right on the second, so arriving at 7 units to the right on the first number line. Thus we have $2+5 = 7$.

Can you see how to obtain the answer to $7-5$ from your 'slide rule'?

Exercise 3

Use the 'slide rule' that you have made to answer the following:

1	2+3	*2*	4+2	*3*	3+6	*4*	7+2
5	8+1	*6*	5+7	*7*	3+2	*8*	6+0
9	4+9	*10*	3+8	*11*	8+5	*12*	5+8

Now make up two more number lines, running from -9 to $+9$ (see Figure 10). We can use these in the same way to add $6+(-2)$. Place zero on the second number line against 6 on the first number line. Against -2 on the second number line we read off the sum 4 on the first number line as shown in Figure 10.

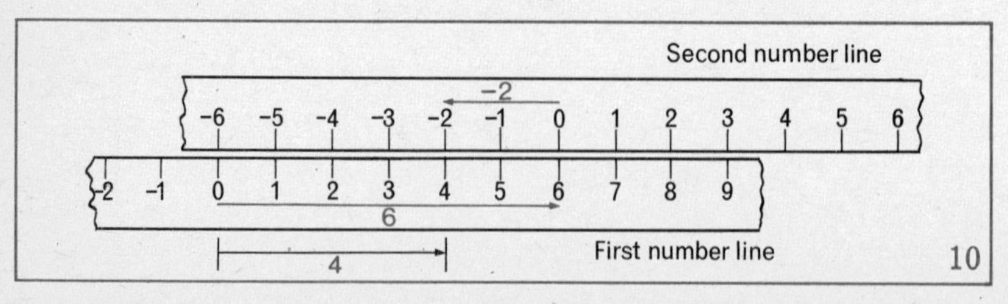

10

Just as 2 corresponds to a step of 2 units to the *right*, so -2 corresponds to a step of 2 units to the *left*. To find $6+(-2)$ we can think of a step of 6 to the right from 0, followed by a step of 2 to the left, arriving at 4 units to the right on the first number line. Thus we have $6+(-2) = 4$.

Example 1. Show on the number line $3+(-9)$.

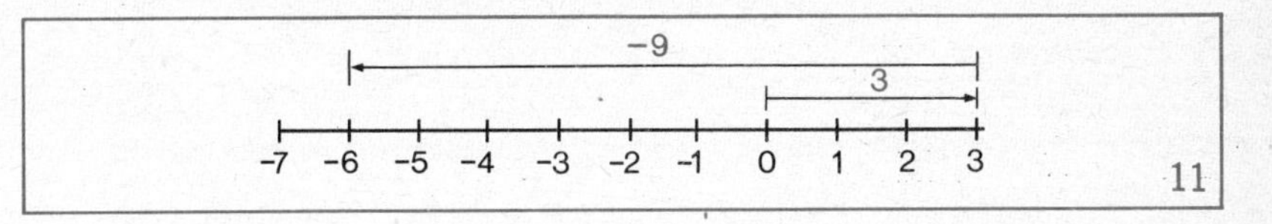

11

From 0 take a step of 3 units to the right, and follow with a step of 9 units to the left. $3+(-9) = -6$.

Example 2. Show on the number line $-4+(-2)$.

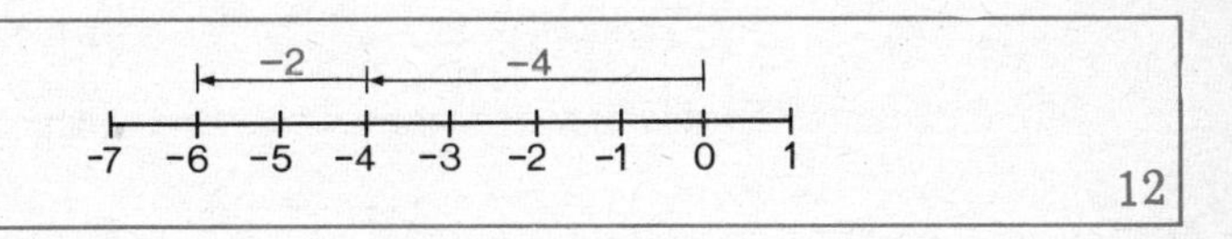

12

From 0 take a step of 4 units to the left, and follow with a step of 2 units to the left. $-4+(-2) = -6$.

Exercise 4

Use two number lines running from -9 to $+9$ (as in Figure 10) to calculate each of the following:

1 $5+3$ *2* $3+(-7)$ *3* $4+(-3)$ *4* $3+(-3)$

5 $-7+2$ *6* $-2+(-5)$ *7* $-1+(-8)$ *8* $-3+3$

9 $5+0$ *10* $6+(-6)$ *11* $-6+6$ *12* $-6+(-3)$

Use a number line as in Worked Examples 1 and 2 to illustrate each of the questions ***13*** to ***18***, and then calculate the sum in each case:

13 $5+(-3)$ *14* $3+(-7)$ *15* $-2+6$

16 $-3+3$ *17* $-4+(-2)$ *18* $9+(-6)$

Calculate each of the following, using a number line or diagram *only if you need to*:

19 $3+(-1)$ *20* $3+(-5)$ *21* $2+(-9)$ *22* $5+(-5)$

23 $-7+(-2)$ *24* $-3+(-5)$ *25* $-2+(-9)$ *26* $-5+(-5)$

27 $-1+(-14)$ *28* $-12+(-3)$ *29* $-6+14$ *30* $-34+56$

31 $-25+18$ *32* $8+(-5)+(-3)$

33 $-2+3+(-3)$ *34* $6+(-5)+5$

35 $-12+(-6)+(-2)$ *36* $-7+(-4)+11$

37 $7+(-9)+5$ *38* $-8+3+(-10)$

What replacements for x will make the following into true sentences, where x is a variable on the set of integers?

39 $x+2=0$ *40* $x+1=0$ *41* $x+(-3)=0$

42 $x+100=0$ *43* $-x+5=0$ *44* $x+(-10)=0$

45 Copy the following table of times and temperatures (degrees Celsius).

Time	6 am	9 am	12 noon	3 pm	6 pm	9 pm
Temperature	−8					

Fill in the other temperatures if the temperature rose

a 6° between 6 am and 9 am *b* 12° between 9 am and 12 noon
c 4° between noon and 3 pm *d* 16° between 6 am and 6 pm
e 6° between 6 am and 9 pm
f At what times was the temperature zero?

Exercise 4B

Find the *sums* in questions *1-12*

1 $9+(-6)$ *2* $-9+6$ *3* $-7+(-2)$

4 $-2\frac{1}{2}+\frac{1}{2}$ *5* $1\frac{1}{2}+(-\frac{1}{2})$ *6* $-2\frac{3}{4}+(-1\frac{1}{4})$

7 $2\frac{2}{3}+(-3)$ *8* $-2\frac{1}{2}+1\frac{1}{2}$ *9* $-3{\cdot}2+(-1{\cdot}5)$

10 $3{\cdot}7+(-4{\cdot}1)$ *11* $4{\cdot}3+(-6{\cdot}8)$ *12* $0{\cdot}1+(-1{\cdot}1)$

13 Copy and complete the addition table shown.

Second number

+	−3	−2	−1	0	1	2	3
−3							
−2							
−1							
0							
1							
2							
3							

First number

a About which diagonal is the table symmetrical?
b Make a list of patterns you can see in the table—where even and odd numbers occur, etc.
c What do you notice when you read a column from top to bottom?
d What do you notice when you read a row from left to right?
e List pairs of numbers which added together make zero.

14 Write down the commutative law of addition in the form $a+b = \ldots$
Check from your table in question ***13*** that $-3+(-2) = -2+(-3)$.
Is it true that $2+(-1) = -1+2$?

Addition is in fact commutative for all integers, positive and negative.

15 In your table for question ***13***, add the numbers in each of the first three columns. Write down the three sums as a sequence. Continue the sequence for four more terms.
Add the numbers in the last column, and compare with your last term.

16 Repeat question ***15***, this time adding the numbers in the rows.

17 Write down the associative law of addition in the form:

$$(a+b)+c = \ldots$$

Check whether $(-3+2)+(-1) = -3+(2+(-1))$.

Addition is in fact associative for all integers.

18 By calculating each side separately, show that:
a $(10+5)+(-12) = 10+(5+(-12))$
b $(7+(-5))+(-2) = 7+(-5+(-2))$

19 Calculate the following:

a $4+9+6$ *b* $(-4)+9+(-6)$ *c* $28+(-37)+12$

d $6+(-3\frac{1}{2})+(-2\frac{1}{2})$ *e* $-18+14+(-12)$ *f* $-0{\cdot}097+3+1{\cdot}097$

20 If x is a variable on the set of integers, solve the following equations: (The table for question ***13*** can help you if necessary.)

a $-1+x=2$ *b* $3+x=1$ *c* $x+3=-1$

d $x+(-3)=-2$ *e* $2+x=0$ *f* $x+(-3)=0$

21 Figure 13 shows two *magic squares*, in each of which the sum of the numbers in each row, and in each column, and in each diagonal, is the same. Copy the squares, and fill in the missing numbers.

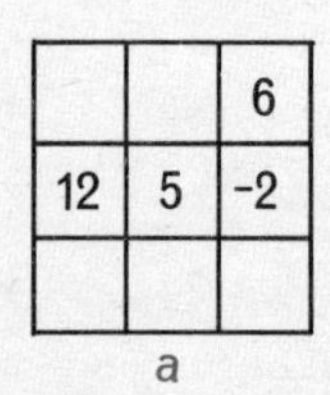

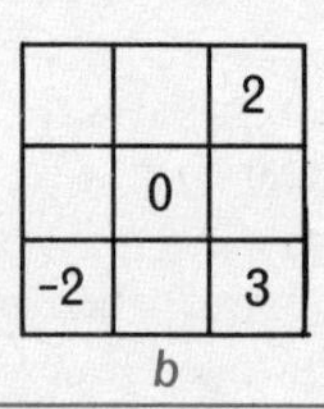

13

22 In a certain test, pupils score 2 marks for a correct answer, and score -1 mark for an incorrect answer.

a If one pupil had 76 questions correct and 19 wrong, what was his score?

b Another pupil had 50 correct and 50 wrong. What was his score?

23 In a card game the maximum possible score is 100, and negative scores are allowed. In five consecutive games, a player's score was -70, 80, -60, 50, -30. Calculate his total final score.

24 x and y are variables on the set of integers, and y is 2 more than x. Write down an equation in y and x.

If $x \in \{-5, -4, -3, -2, -1, 0, 1, 2\}$, write down the corresponding replacements for y.

---

4 The negative (or additive inverse) of a number

All positive and negative numbers fall naturally into pairs. We have 1 and -1, 2 and -2, $3\frac{1}{4}$ and $-3\frac{1}{4}$ and so on. We call each the *negative* (or *additive inverse*) of the other. For example, the negative of 1 is -1, and the negative of -1 is 1. Such relations may be shown on the number line as in Figure 14.

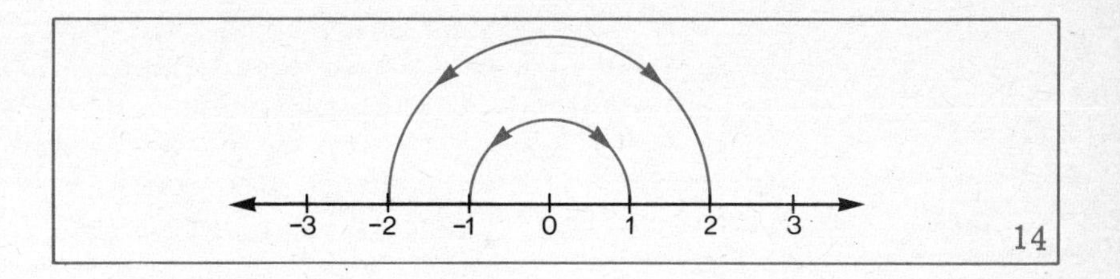

14

To each number a there corresponds an additive inverse $-a$, called the negative of a, such that $a+(-a) = 0 = -a+a$.

Notice that the negative of a need not be a negative number. If $a < 0$, then $-a > 0$; for example, $-3 < 0$, $3 > 0$.

Exercise 5

1 Write down the negative of each of the following integers:

a 3 *b* 5 *c* -6 *d* -10 *e* 10 *f* 12
g 100 *h* -100 *i* 1 *j* -9 *k* 4 *l* -4

2 On a number line mark each of the following by a dot, and the negative of each by a cross:

a 2 *b* -1 *c* -4 *d* 3·5 *e* 0 *f* $-0·5$

3 Copy and complete the following:

a $7+ \ldots = 0$ *b* $\ldots +(-2) = 0$ *c* $-9+ \ldots = 0$
d $-5+5 = \ldots$ *e* $\ldots +1·5 = 0$ *f* $8+(-8) = \ldots$

4 Solve the following equations, where x is a variable on the set of integers:

a $x+(-6) = 0$ *b* $x+8 = 0$ *c* $2x+(-10) = 0$
d $x+2+(-2) = 0$ *e* $x+7+(-3) = 0$ *f* $x+1+(-4) = 0$

5 Subtraction

When you pay 7p out of a 10p piece, you receive $(10-7)$p change, i.e. 3p. The shopkeeper probably thinks, 'What number added to 7 gives 10?' The answer is 3 (see Figure 15). Thus $10-7 = 3$.

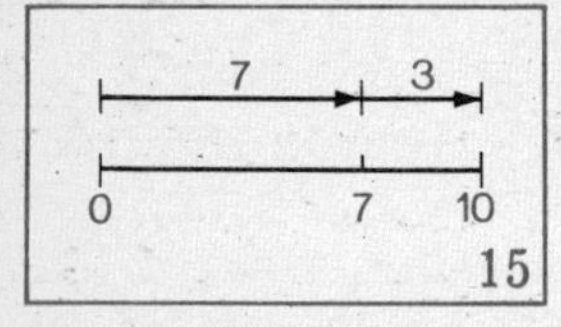

15

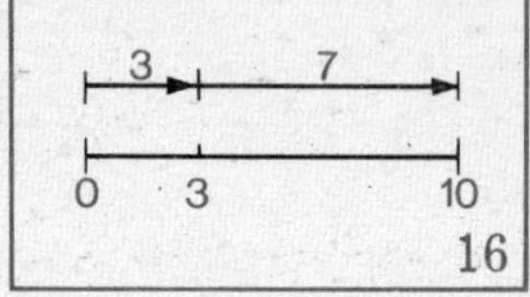

16

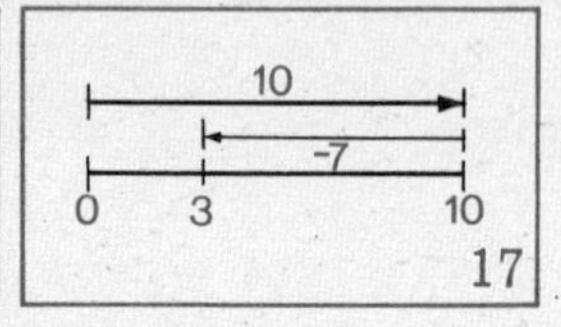

17

We could, of course, think of $(10-7)$ as the number to which 7 must be added to give 10 (see Figure 16). Again, $10-7 = 3$.

Changing the direction of the arrow in Figure 16 gives a diagram for adding -7 to 10 (see Figure 17). So $10+(-7) = 3$.

Exercise 6

Use your two number lines from Section 3, or a diagram, to calculate each of the following as in Figures 16 and 17:

1 a	$10-7$	*2 a*	$5-7$	*3 a*	$-2-3$
b	$10+(-7)$	*b*	$5+(-7)$	*b*	$-2+(-3)$
4 a	$-10-(-7)$	*5 a*	$-5-(-7)$	*6 a*	$2-(-3)$
b	$-10+7$	*b*	$-5+7$	*b*	$2+3$

Can you see that *subtracting* b from a is the same as *adding* the *negative* of b to a?

$$a-b = a+(-b)$$

Examples

(*1*) $6-(-5)$
$= 6+5$
$= 11$

(*2*) $3-8$
$= 3+(-8)$
$= -5$

(*3*) $-3x-(-2x)$
$= -3x+2x$
$= -x$

(*4*) $-7x-5x$
$= -7x+(-5x)$
$= -12x$

7 Copy and complete the following:

a $9-(+5)$ $= 9+(-5)$ $= \dots$
b $9-(-5)$ $= 9+5$ $= \dots$
c $-9-5$ $= -9+(-5)$ $= \dots$
d $-9-(-5)$ $= -9+5$ $= \dots$

8 Rewrite each of the following as an addition, and calculate the sum:

a $14-(+8)$ *b* $5-(-3)$ *c* $2-(+6)$ *d* $-3-(+3)$
e $6-(-4)$ *f* $-3-(-3)$ *g* $0-(+4)$ *h* $8-(+24)$

9 Question *7* shows the way in which *every subtraction can be thought of as an addition*, using the additive inverse. Using the method of question *7*, perform the following subtractions:

a $6-4$ *b* $3-8$ *c* $-7-9$ *d* $-5-(-10)$
e $13-(-17)$ *f* $-17-7$ *g* $8-18$ *h* $4-0$
i $0-(-4)$ *j* $0-4$ *k* $-6-(-8)$ *l* $17-9$

10 Simplify each of the following:

a $5+8-4$ *b* $3+(-7)-4$ *c* $-3-2+8$
d $2-(-5)-(-4)$ *e* $-12-6+5$ *f* $20-7-8$

11 Simplify each of the following:

a $6c-2c$ *b* $10x-7x$ *c* $3y-(-2y)$ *d* $-4h-6h$
e $-5n-(-3n)$ *f* $2p-8p$ *g* $0-(-2n)$ *h* $-5k-(-5k)$
i $x-3x$ *j* $3m^2-m^2$ *k* $-6y^2-2y^2$ *l* $10t^2-7t^2$

12 Add:

a $3p$ and $4p$ *b* $8a$ and $-2a$
c $-8a$ and $-a$ *d* x^2 and $-x^2$
e $9t$ and $-12t$ *f* $-5g$ and $-4g$
g $-m$ and $-4m$ *h* $3x$, $-7x$ and $-2x$
i $3y$, $-5y$ and y *j* a^2, $-a^2$ and a^2
k $-2w$, $-3w$ and $-4w$ *l* $5v$, $-8v$ and $9v$

13 If $p = 6$, $q = -3$ and $r = 2$, calculate:

a $p+q$ *b* $q+r$ *c* $p+q+r$ *d* $p-r$
e $p-q$ *f* $q-r$ *g* $r-p$ *h* $p-q-r$

6 Sets of numbers

We have now seen the following sets of numbers:

a The set of natural numbers, $N = \{1, 2, 3, \ldots\}$

b The set of whole numbers, $W = \{0, 1, 2, 3, \ldots\}$

c The set of integers, $Z = \{\ldots, -3, -2, -1, 0, 1, 2, 3, \ldots\}$

The set of all the numbers which can be written as *fractions* is called the *set of rational numbers*, usually denoted by Q.

For example, $\frac{1}{2}$, $-\frac{1}{3}$, $\frac{5}{1}$, $3\frac{1}{4}$, $-\frac{6}{2}$ are all rational numbers.

We saw that the positive integers $+1$, $+2$, $+3$, ... have the same properties as natural numbers and are usually written the same way as natural numbers, i.e. 1, 2, 3, ... We can think of the natural numbers as forming a subset of the set of integers.

In the same way, some rational numbers are just like integers. For example, the rational number which can be written $\frac{5}{1}$, or $\frac{10}{2}$, has the same properties as the integer $+5$, and is usually written '5'. Can you suggest further examples? We can think of the set of integers as forming a subset of the set of rational numbers.

The Venn diagram in Figure 18 illustrates the relations between the four sets. Explain what the diagram shows.

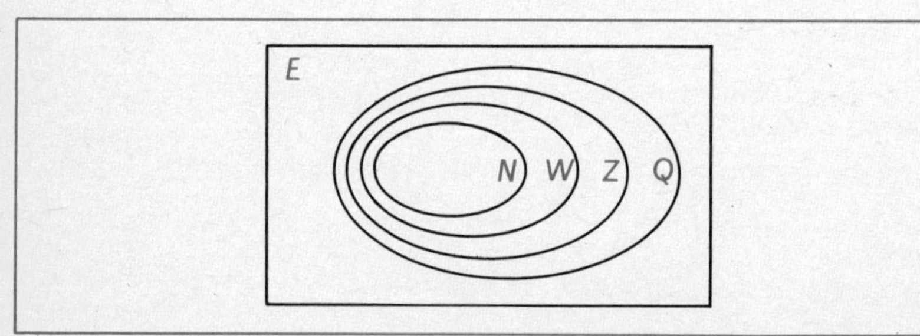

18

Exercise 7

1 Copy and complete the table, showing by a tick (√) whether each number is a member of N, W, Z, Q.

	6	−8	0·1	−5	$\frac{3}{4}$	100	0	−3·14
N								
W								
Z								
Q								

2 a Using a scale of 2 cm to represent 1 unit, draw a number line to show the integers −3, −2, −1, 0, 1, 2, 3.

b Mark and label points which show the rational numbers $\frac{1}{4}$, $\frac{3}{4}$, 1·5.

c Mark a point midway between 0 and −1. This shows the rational number $-\frac{1}{2}$.

d Mark and label points to show −1·7 and $-2\frac{3}{4}$.

3 Insert the symbol > or < between the members of each of the following pairs, *in the given order*, to make true sentences:

a $\frac{3}{4} \ldots 1{\cdot}5$ *b* $-\frac{1}{2} \ldots \frac{1}{4}$ *c* $-\frac{1}{2} \ldots -2\frac{3}{4}$
d $-1{\cdot}7 \ldots 2$ *e* $0 \ldots -\frac{1}{2}$ *f* $-2\frac{3}{4} \ldots -1$

4 Write down the negative of each of the following rational numbers:

a $\frac{1}{2}$ *b* $-\frac{3}{4}$ *c* $-1{\cdot}5$ *d* $-\frac{11}{4}$ *e* $1{\cdot}7$ *f* $-9{\cdot}81$

5 Find the solution set of each of the following equations, noting the universal sets carefully:

a $x+4 = 6, x \in W$ *b* $x+6 = 3, x \in N$ *c* $x+(-8) = 2, x \in Z$
d $2x = 5, x \in W$ *e* $2x = 5, x \in Q$ *f* $x-6 = 3, x \in Q$
g $x+4 = 0, x \in Z$ *h* $5x-1 = 3, x \in Z$ *i* $7-x = 7, x \in Z$

6 Find the solution set of each of the following inequations:

a $x-2 < 3, x \in W$ *b* $x-7 > \frac{1}{2}, x \in N$ *c* $x+2 < \frac{1}{2}, x \in Z$
d $x+x > 3, x \in Q$ *e* $3-x > 2, x \in N$ *f* $5-x < x, x \in Q$

7 Calculate the following; e.g. $\frac{3}{8} + \left(-\frac{5}{8}\right) = \frac{3+(-5)}{8} = \frac{-2}{8} = -\frac{1}{4}$

a $\frac{2}{3}+\frac{1}{3}$ *b* $\frac{4}{5}-\frac{2}{5}$ *c* $\frac{3}{8}+(-\frac{7}{8})$ *d* $\frac{1}{4}+(-\frac{3}{4})$
e $\frac{1}{2}-\frac{3}{4}$ *f* $1-\frac{1}{4}$ *g* $\frac{3}{4}-1$ *h* $\frac{3}{8}-2$
i $\frac{2}{3}-\frac{1}{2}$ *j* $\frac{2}{3}+\frac{3}{4}$ *k* $\frac{4}{5}+\frac{2}{10}$ *l* $\frac{4}{5}-\frac{4}{10}$

7 *Standard form for small numbers*

We have seen that a large positive number is *greater* than a small number, and that a large negative number is *less* than a small positive number. This is illustrated on the number line in Figure 19.

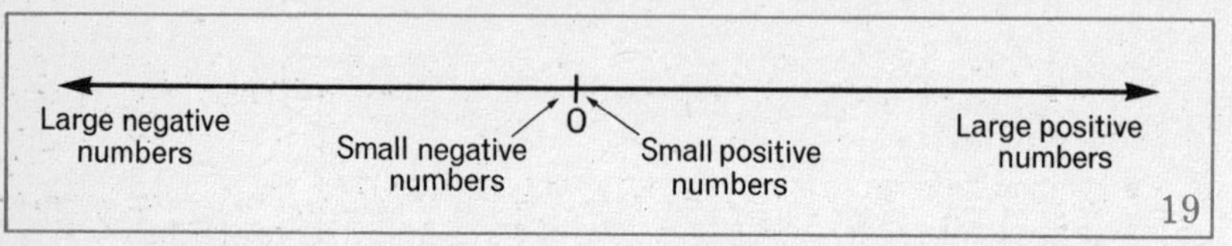

19

We can see in Arithmetic, Chapter 1, how large numbers can conveniently be expressed in *standard form*. For instance, the length known as a *light year* is approximately 9 400 000 000 000 kilometres, and this can be written as $9{\cdot}4 \times 10^{12}$ km.

With the aid of negative integers we can now extend the use of standard form to small numbers such as 0·000 000 05, the radius in millimetres of one orbit of an electron. Scientists and engineers often have to deal with very small numbers like this, and standard form is shorter to write. It also avoids the danger of confusion about the number of zeros.

* * * * *

Study this sequence of numbers: $10^5, 10^4, 10^3, 10^2, 10^1, \ldots$
Notice that:

a Each term's index is one less than that of the term before.
b Each term is one tenth of the term before it.

Continuing the sequence according to ***a*** we have
$10^5, 10^4, 10^3, 10^2, 10^1, 10^0, 10^{-1}, 10^{-2}, 10^{-3}, \ldots$
Continuing the sequence according to ***b***, we have

$$10^5, 10^4, 10^3, 10^2, 10^1, 1, \frac{1}{10^1}, \frac{1}{10^2}, \frac{1}{10^3}, \ldots$$

This suggests that:

10^0 should be taken to mean 1, 10^{-1} to mean $\frac{1}{10}$, 10^{-2} as $\frac{1}{10^2}$, 10^{-3} as $\frac{1}{10^3}$, and so on.

Without going into the details, that is in fact what is done.

It is important to observe that a number like 10^{-3} is *not* a negative number but the positive number $\frac{1}{1000}$; the index *is* a negative number.

With the help of this new notation we can now express small numbers in standard form, i.e. in the form $a \times 10^n$, where a represents a number between 1 and 10, and n is a negative integer.

Example 1. Express 0·07 in standard form.

$$0{\cdot}07 = \frac{7}{100} = \frac{7}{10^2} = 7\times\frac{1}{10^2} = 7\times 10^{-2}$$

Example 2. Express 0·0000675 in standard form.

$$0{\cdot}0000675 = \frac{6{\cdot}75}{100000} = \frac{6{\cdot}75}{10^5} = 6{\cdot}75\times 10^{-5}$$

Exercise 8

1 Continue the sequence 10^3, 10^2, 10^1, ... for seven more terms in two ways:

a so that the index of each term is one less than the term before it

b so that each term is one tenth of the term before it.

Write the two sequences carefully in lines, with corresponding terms one above the other.

2 Use your sequences in question *1* to express in the form 10^{-n}:

a $\frac{1}{10^2}$ *b* $\frac{1}{10^4}$ *c* $\frac{1}{10^5}$ *d* $\frac{1}{10}$ *e* $\frac{1}{10^3}$

Express the numbers in questions ***3–14*** in standard form, as in the Worked Examples:

3 0·3	*4* 0·47	*5* 0·05	*6* 0·0278
7 0·0006	*8* 0·00273	*9* 0·000649	*10* 0·0456
11 0·000000173	*12* 0·0045	*13* 0·7004	*14* 0·0105

* * * * *

As we saw for very large numbers, the numerical part of the index indicates the number of places the decimal point moves.

For example, $12000 (= 1{\cdot}2\times 10000) = 1{\cdot}2\times 10^4$

$$0{\cdot}00638\left(=\frac{6{\cdot}38}{1000}\right) = 6{\cdot}38\times 10^{-3} = 6{\cdot}4\times 10^{-3}$$

to 2 significant figures.

Express the numbers in questions *15–23* in standard form to 2 significant figures:

15	0·0751	*16*	0·0000456	*17*	0·123
18	0·00991	*19*	0·107	*20*	0·0000032
21	0·404	*22*	0·000782	*23*	0·000000875

Express the numbers in questions *24* and *25* in standard form:

24 The thickness of an oil film is about 0·00000005 mm.

25 The mass of a molecule of water is about

0·00000000000000000000003 gramme.

26 Work out the thickness of a page of this book, expressing your answer in standard form.

27 Use an encyclopaedia or science book to find six examples of very small numbers, and write them in standard form.

---

8 Using positive and negative numbers

Exercise 9A

1 On 5-mm squared paper take the origin near the centre of a page and draw the x- and y-axes. A tile with straight edges has corners A(1, 1), B(−2, 1), C(−2, −2), D(1, −2).

a What is the name of this shape?

b If you were to slide the tile ABCD in a direction parallel to the y-axis until CD lay on the x-axis, what would then be the coordinates of A, B, C and D?

2 Using the same coordinates for A, B, C, D as in question *1*, slide ABCD from its first position, parallel to the x-axis, until CB lies on the y-axis. Give the coordinates of the new positions of A, B, C, and D.

3 A is the point (k, 5), B(3, 1), C(7, 1), and ABC is an isosceles triangular tile with equal edges AB and AC. What is the value of k?

If the tile is moved in a direction parallel to the y-axis until A lies on the x-axis, write down the coordinates of the new positions of A, B, and C.

4 Having found k use the same coordinates for A, B, C as in question *3*, but move the tile from its original position parallel to the x-axis until A lies on the y-axis. Write down the coordinates of the new positions of A, B, and C.

5 Draw a diagram of the set of points $A = \{(-2, 1), (-2, 4), (-2, 5), (-2, -1), (-2, -3)\}$. If B is the set of all points with first coordinate -2 which of the following is true?

a $A = B$ *b* $A \subset B$ *c* $B \subset A$. If $(p, 4) \in B$, what is p?

6 The symbol $\binom{4}{2}$ gives the instructions: 'Move 4 units to the right, then move 2 units up the page'. Starting from the point $(2, -3)$, find the positions given by

a $\begin{pmatrix}4\\2\end{pmatrix}$ *b* $\begin{pmatrix}4\\-2\end{pmatrix}$ *c* $\begin{pmatrix}-2\\1\end{pmatrix}$ *d* $\begin{pmatrix}5\\-6\end{pmatrix}$

7 A ship leaves port A and sails to position B according to the instructions $\binom{4}{5}$ which means 'Sail 4 km East, then 5 km North'. It then sails from B according to the instructions $\binom{-4}{2}$, where it heaves to because of engine trouble. Using our 'code', what instructions would you give a lifeboat at A to reach the ship?

8 A police car leaving the station was ordered to a point P on the map by the instructions $\binom{-8}{4}$. At P it was ordered by radio to proceed $\binom{4}{-2}$. What is its new position relative to the police station?

9 Starting from a youth hostel a party travelled to X according to the instructions $\binom{4}{-2}$. At X one member had to return to the hostel. What instructions would take him back?

Exercise 9B

1 A is the point (3, 1), B(6, 2) and C(9, 3). A′ is the image of A in the origin, B′ is the image of B in the origin, and C′ is the image of C in the origin.

a Give the coordinates of A′, B′ and C′.

b If you continue the sequences A, B, C, ... and A′, B′, C′, ..., which of the following points would belong to them?

$(15, 5), (12, 36), (-15, -5), (-30, -10), (-30, 10), (33, 99)$

c If (p, q) belonged to one of these sequences, which of these would be true?

(1) $p = 3q$ *(2)* $p = \frac{1}{3}q$ *(3)* $p = q$ *(4)* $q = \frac{1}{3}p$

2 STUV is a tile in the shape of a rectangle. S is the point $(1, -2)$, $T(14, -2)$ and $U(14, 13)$.

a Give the coordinates of V.

b If the tile is moved parallel to the y-axis until VU lies on the x-axis, give the coordinate of S, T, U and V now.

3 Show by shading on a coordinate diagram the set P of all points with first coordinates greater than 2, and the set Q of all points with first coordinates less than -4.
Say whether each of the following is true or false:

a $A(3, 0) \in P$ *b* $B(-3, 0) \in Q$ *c* $C(-5, 5) \in Q$

d $D(5, -5) \in P$ *e* $O(0, 0) \in P \cap Q$ *f* $P \cap Q = \emptyset$

4 The symbol $\begin{pmatrix} p \\ q \end{pmatrix}$ gives the instructions: 'Move p units to the right, then move q units up the page.' Starting from the origin O, carry out the following instructions:

a $\binom{4}{5}$; this takes you to A. *b* From A, $\binom{4}{-6}$; this takes you to B.
Is it true to say that $\binom{4}{5}$ followed by $\binom{4}{-6}$ took you from O to B.?
Give one instruction for $\binom{4}{5}$ followed by $\binom{4}{-6}$.

5 Give one instruction for:

a $\binom{-2}{3}$ followed by $\binom{3}{-4}$ *b* $\binom{-2}{6}$ followed by $\binom{-1}{-5}$.
Plot the moves on squared paper.

6 Make up a more complicated example, checking the answer by plotting the moves on squared paper.

7 From O, X is reached according to the instructions $\begin{pmatrix} a \\ b \end{pmatrix}$. What instruction would take you from X to O?

8 Draw axes X′OX and Y′OY, with OX and OY in the usual positions.
If you started at the origin and the instructions $\begin{pmatrix} p \\ q \end{pmatrix}$ took the point into the YOX′ quarter of the page, which of the following would be true?

a $p > 0$ and $q > 0$ *b* $p > 0$ and $q < 0$

c $p < 0$ and $q < 0$ *d* $p < 0$ and $q > 0$

Repeat the question for the XOY′ quarter of the page.

Summary

1 *a* *Positive integers*: $+1, +2, +3, \ldots$; or $1, 2, 3, \ldots$
A positive integer a is *greater* than zero; $a > 0$

b *Negative integers*: $\ldots, -3, -2, -1$.
A negative integer b is *less* than zero; $b < 0$

c *Zero* is neither positive nor negative.

d The set of integers $Z = \{\ldots, -3, -2, -1, 0, 1, 2, 3, \ldots\}$

e Q is the set of *rational numbers*

$\frac{3}{4}$, 2·4, $-\frac{1}{2}$, $\frac{10}{5}$ are rational numbers.

2 *Addition* of numbers:

a $5+3 = 8$ *b* $5+(-3) = 2$

c $-5+3 = -2$ *d* $-5+(-3) = -8$

3 Addition of numbers is:
(i) commutative $a+b = b+a$
(ii) associative $(a+b)+c = a+(b+c)$

4 *Negatives* (or *additive inverses*)
a and $(-a)$ are *negatives* of each other, and are such that

$$a+(-a) = (-a)+a = 0$$

5 *Subtraction*: To subtract b from a, *ADD* the negative of b to a.

$$a-b = a+(-b)$$

e.g. $$4-9 = 4+(-9) = -5$$

Thus, every subtraction may be thought of as an addition.

6 *Standard form* $a \times 10^n$, where $1 \leqslant a < 10$, and n is a positive or negative integer. Examples:

a $342 = 3{\cdot}42 \times 100 = 3{\cdot}42 \times 10^2$

b $0{\cdot}045 = \dfrac{4{\cdot}5}{100} = \dfrac{4{\cdot}5}{10^2} = 4{\cdot}5 \times 10^{-2}$

Distributive Law

1 *Expressing a product of factors as a sum or a difference of terms*

Exercise 1

1 Copy and complete the following table:

a	b	c	$(b+c)$	$a\times(b+c)$	ab	ac	$ab+ac$
3	2	4	6	18	6	12	18
2	5	3	.	.	.	.	.
4	3	6	.	.	.	.	.
0	3	8	.	.	.	.	.
8	1	5	.	.	.	.	.
10	5	7	.	.	.	.	.
4	$\frac{1}{2}$	$\frac{1}{4}$	.	.	.	.	.

By comparing the numbers in the fifth and eighth columns do you see that in each case $a(b+c) = ab+ac$?

2 Copy and complete the following table:

a	b	c	$(b-c)$	$a\times(b-c)$	ab	ac	$ab-ac$
5	4	2	2	10	20	10	10
3	5	2	.	.	.	.	.
1	8	3	.	.	.	.	.
0	7	5	.	.	.	.	.
4	9	2	.	.	.	.	.
5	10	6	.	.	.	.	.
3	$\frac{1}{2}$	$\frac{1}{4}$	.	.	.	.	.

Do you find that $a(b-c) = ab-ac$ in each case?

Question *1* above illustrates the distributive law of multiplication over addition, which we met in Book 1. This law may be shown as follows

$$a(b+c) = ab+ac$$

Note that $a(b+c)$ means 'Add b and c and then multiply the sum by a', and $ab+ac$ means 'Multiply a by b, a by c, and add the products'.

We also have:

$$a(b-c) = ab-ac$$

The distributive law can be used to express certain products of factors as sums or differences of terms.

Example 1. $3(2p+3q+r) = 6p+9q+3r$

Example 2. $4(2a - 3b) = 8a-12b$

Example 3. $m(m-2a) = m^2-2am$

Exercise 2

Express each of the following as a sum or difference of terms:

1 $2(a+1)$ *2* $7(c+2)$ *3* $5(1+p)$ *4* $8(m+n)$

5 $1(a+b)$ *6* $2(n+8)$ *7* $3(x^2+1)$ *8* $4(p-2q)$

9 $5(2-2x)$ *10* $7(3u+4v)$ *11* $6(2x-y)$ *12* $3(3c+2d)$

13 $3(y^3-1)$ *14* $p(4+q)$ *15* $a(5h+2k)$ *16* $10(m-5n)$

17 $a(a+b)$ *18* $2h(l+b)$ *19* $4a(a-b)$ *20* $x(2-3x)$

21 $2x(x+y)$ *22* $p(2p-3q)$ *23* $ab(c+2)$ *24* $rs(1-r)$

25 $3(a+b+c)$ *26* $2(a-b+c)$ *27* $a(a+b+c)$

28 $x(x+2y-z)$ *29* $3(2p-3q+4r)$ *30* $xy(1-x+y)$

Example. Express $\frac{1}{2}(2a+6b)$ as a sum of terms.

$$\tfrac{1}{2}(2a+6b) = (\tfrac{1}{2}\times 2a)+(\tfrac{1}{2}\times 6b) = a+3b$$

Exercise 2B

Express each of the following (questions *1–15*) as a sum or a difference of terms:

1 $5(3p+2q)$ *2* $2p(2x-y)$ *3* $3a(3a+5)$

4 $r^2(r-s)$ *5* $ab(a+b)$ *6* $x^2(1-7x)$

7 $2(xy+yz+zx)$ *8* $a(2a+3b+1)$ *9* $2p(p+q-2r)$

10 $\frac{1}{2}(m+n)$ *11* $\frac{1}{2}(4x+8y)$ *12* $\frac{1}{4}(4a-12b)$

13 $\frac{3}{4}(12x+4y)$ *14* $0{\cdot}4(5h-10)$ *15* $1{\cdot}2(5u+7v)$

16 Here is an arithmetic system containing a set of five whole numbers {0, 1, 2, 3, 4}. For this system, we define two operations $\oplus$ and $\otimes$ as follows:

$a \oplus b$ means 'Divide the sum of a and b by 5 and write down the remainder', e.g. $2 \oplus 4 = 1$.

$a \otimes b$ means 'Divide the product of a and b by 5 and write down the remainder', e.g. $2 \otimes 4 = 3$.

a Now copy and complete the 'addition' and 'multiplication' tables below:

$\oplus$	0	1	2	3	4
0	0	1	2	3	4
1	.	.	.	4	0
2	.	.	.	.	.
3	.	.	.	.	.
4	.	.	.	.	.

$\otimes$	0	1	2	3	4
0	0	0	0	0	0
1	.	.	.	3	4
2	.	.	.	1	.
3	.	.	.	.	.
4	.	.	.	.	.

b Copy and complete the following table using the results from the table in ***a***:

a	b	c	$b\oplus c$	$a\otimes(b\oplus c)$	$a\otimes b$	$a\otimes c$	$(a\otimes b)\oplus(a\otimes c)$
2	1	3	4	3	2	1	3
4	2	1	.	.	.	.	.
1	4	3	.	.	.	.	.
3	3	3	.	.	.	.	.
4	2	3	.	.	.	.	.

c From an inspection of your entries in columns 5 and 8, is the 'multiplication' $\otimes$ distributive over the 'addition' $\oplus$?

Note.—The above arithmetic system resembles 'clock arithmetic'. It is an example of 'arithmetic modulo 5', since numbers are replaced by their remainders on division by 5.

2 *Expressing a sum or a difference of terms as a product of factors*

Interchanging the sides of the equality $a(b+c) = ab+ac$ gives

$$ab+ac = a(b+c)$$

Notice that a is a *common factor* of the terms ab and ac. Whenever two terms of an expression have a common factor we can use the distributive law to express a sum (or a difference) of terms as a product of factors. This is the opposite of what we were doing in Section 1.

Example 1. $(3\times 5)+(3\times 8) = 3(5+8) = 3\times 13 = 39$
Example 2. $4x-20y = 4x-(4\times 5y) = 4(x-5y)$
Example 3. $2a^2+6ab = (2a\times a)+(2a\times 3b) = 2a(a+3b)$

Exercise 3A

1 Copy the following and insert the missing terms in the brackets:

a $2p+2q = 2(......)$ *b* $2p+10 = 2(......)$ *c* $3c+9 = 3(......)$
d $7x-7y = 7(......)$ *e* $4m-6 = 2(......)$ *f* $8a+12 = 4(......)$
g $5n-10m = 5(......)$ *h* $2a+ab = a(......)$ *i* $x^2-x = x(......)$

2 Express each of the following as a product of factors:

a $2x+4$ *b* $3a+3b$ *c* $4c+12d$ *d* $3y+6z$
e $4p+12$ *f* $ax+ay$ *g* $4x-32$ *h* $7m-49$
i $16t-24$ *j* $pq-ps$ *k* $pq-qr$ *l* $15x+20xy$
m $6a+12b$ *n* x^2+x *o* y^2-y *p* $xy+yz$

3 Underline the common factor in each of the following. Then form a product of factors, and calculate the value of the product as in Worked Example 1.

a $(7\times 11)+(7\times 9)$ *b* $(14\times 13)+(16\times 13)$
c $(38\times 14)+(14\times 12)$ *d* $(15\times 23)-(15\times 19)$
e $(16\times 16)-(16\times 11)$ *f* $(\frac{3}{4}\times 27)-(23\times \frac{3}{4})$

4 Express the right-hand side of each of the following as a product of factors, and then carry out the calculation asked.

a $s = 3u+3v$. Find s when $u = 10$ and $v = 160$.
b $Q = 15x+15y$. Find Q when $x = 17{\cdot}5$ and $y = 2{\cdot}5$.
c $A = 2ah+2bh$. Find A when $a = 15$, $b = 10$ and $h = 4$.

5 Figure 1 shows the plan of a corridor of uniform width a metres; b and c are in metres also. The dotted line divides the plan into two rectangular parts.

a Find a formula for the area A m² of the whole corridor, and express the formula in factorized form.
b Calculate the area if $a = 2$, $b = 5$, $c = 8$.

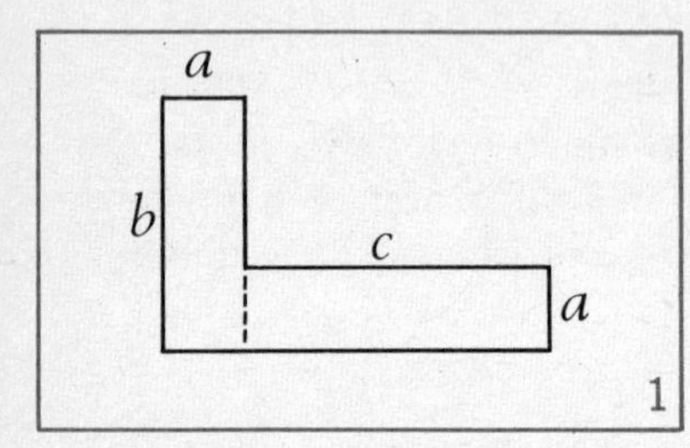

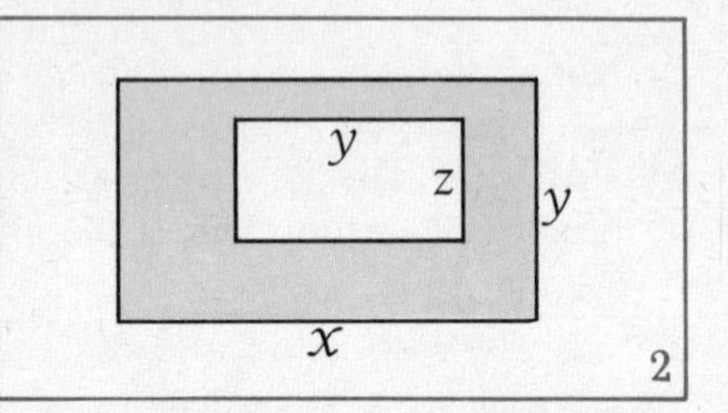

6 Figure 2 shows a rectangle of length x mm and breadth y mm, from which a rectangle of length y mm and breadth z mm is removed, so forming a frame.

a Find a formula for the area A mm² of the frame, and express the formula in factorized form.
b Calculate A if $x = 175$, $y = 80$, $z = 55$.

Exercise 3B

1 Express each of the following as a product of factors:

a $5x+15y$ *b* $8y+18$ *c* $3x^2+6$ *d* p^2+pr
e a^2-2ab *f* p^2-pq *g* x^2y+x *h* $15x^2+5x$
i $9p^2-15pq$ *j* $ax+ay+az$ *k* $4p+6q-8r$ *l* a^3+a^2+a
m $x-x^3$ *n* x^2y+x^2z *o* $2ab+4a^2b^2$

2 Write down the common factor in each of the following. Then use factors to calculate the answer mentally if you can.

a $(3\times 25)+(3\times 75)$ *b* $(0{\cdot}5\times 11)+(2{\cdot}5\times 11)$
c $(29\times 14)-(4\times 29)$ *d* $13^2+(13\times 17)$
e $(3{\cdot}4\times 2{\cdot}7)+(3{\cdot}4\times 2{\cdot}3)$ *f* $(360\times \frac{1}{2})-(360\times \frac{1}{4})$

3 Express the right-hand side of each of the following as a product of factors, and then carry out the calculation asked.

a $s = 20t-5t^2$. Find s when $t = 2{\cdot}4$.

b $I = mv - mu$. Find I when $m = 4$, $v = 50$ and $u = 10$.

c $A = \frac{1}{2}ah + \frac{1}{2}bh$. Find A, given that $h = 6$, $a = 16$ and $b = 9$.

4 Complete the following:

a $\frac{1}{2}(8a+4b) = (\frac{1}{2} \times \ldots) + (\frac{1}{2} \times \ldots) = \ldots + \ldots$ (See Worked Example on page 51)

b $\dfrac{8a+4b}{2} = \dfrac{4(\ldots+\ldots)}{2} = 2(\ldots+\ldots) = \ldots+\ldots$

What do you notice about your results?

5 Simplify each of the following in the two ways shown in question ***4***:

a $\dfrac{2x+2y}{2}$ *b* $\dfrac{6p+6q}{3}$ *c* $\dfrac{6a+9b}{3}$ *d* $\dfrac{4m-8n}{4}$

e $\dfrac{9x-15}{3}$ *f* $\dfrac{10c-15d}{5}$ *g* $\dfrac{3a+6b}{3}$ *h* $\dfrac{20x-20y}{10}$

i $\dfrac{5c+5d}{5}$ *j* $\dfrac{x^2+xy}{x}$ *k* $\dfrac{4t^2+6t}{2t}$ *l* $\dfrac{6p^2-24p}{6p}$

6 Find a formula for the perimeter P metres of the shape in Figure 1. Express the formula in factorized form.

7 With the help of factors, calculate each of the following:

a $\dfrac{(8 \times 11)+(8 \times 19)}{15}$ *b* $\dfrac{(1{\cdot}5 \times 13)+(2{\cdot}5 \times 13)}{4}$

c $\dfrac{(49 \times 18)-(18 \times 24)}{45}$ *d* $\dfrac{45^2-(45 \times 30)}{75}$

8 Simplify:

a $\dfrac{5a+10b}{5}$ *b* $\dfrac{6a-12b}{2}$ *c* $\dfrac{a^2-4a}{a}$

d $\dfrac{x^3+x^2+x}{x}$ *e* $\dfrac{y^3-y^2}{y}$ *f* $\dfrac{2z^3+3z^2+4z}{z}$

3 Collecting like terms

In the expression $4x+3y+2x+y$, $4x$ and $2x$ are *like terms*, and $3y$ and y are *like terms*. We can collect like terms together as shown in the following examples.

Examples

1. $3a+5a = (3+5)a = 8a$ using the distributive law; a is a common factor.
2. $5xy-2xy = (5-2)xy = 3xy$
3. $3m+4m+m = (3+4+1)m = 8m$
4. $7x+3y$ cannot be simplified, as $7x$ and $3y$ are unlike terms.

Exercise 4

1 Copy and complete the following:

a $2m+3m = (2+\ldots)m = \ldots$ *b* $5a+4a = (\ldots+4)a = \ldots$
c $8y-5y = (\ldots-5)y = \ldots$ *d* $12p-7p = (12-\ldots)p = \ldots$
e $6x+x = (\ldots+1)x = \ldots$ *f* $9n-n = (\ldots-\ldots)n = \ldots$

Simplify the following, by inspection if you can:

2 $4a+3a$ *3* $3c+c$ *4* $8m+9m$
5 $2y+4y$ *6* $x+9x$ *7* $7h+6h$
8 $10p+13p$ *9* $5w-3w$ *10* $8n-2n$
11 $9a-5a$ *12* $8c-c$ *13* $7k-7k$
14 $12d-5d$ *15* $\frac{1}{2}b+\frac{1}{2}b$ *16* $\frac{3}{4}p-\frac{1}{4}p$
17 $2{\cdot}6x+3{\cdot}4x$ *18* $7{\cdot}5q+3{\cdot}5q$ *19* $5{\cdot}9m-4{\cdot}9m$
20 $2ab+4ab$ *21* $7x^2+3x^2$ *22* $8xy-7xy$
23 $4pq+10qp$ *24* $14cd-8dc$ *25* $a+2a+3a$

Simplify *where possible* (i.e. where there are like terms):

26 $6a+7a$ *27* $2y+4y$ *28* $3p+4q$
29 $10x-x$ *30* $2x^2+y$ *31* $6a-7b$
32 $5pq+2qp$ *33* $7yx-3xy$ *34* $2a+3a+4a$
35 $8c+3c-2c$ *36* $6m-2m+m$ *37* $12r+3r+s$
38 $5n+4n-3$ *39* $7y+5y-y$ *40* $13a^2-3$
41 $6x+2y+4x$

Example. Simplify $3(3a+2b)+2(2a-2b)$

$3(3a+2b)+2(2a-2b)$
$= 9a+6b+4a-4b$ (distributive law)
$= 9a+4a+6b-4b$ (commutative law of addition)
$= 13a+2b$ (distributive law)

Exercise 4B

Simplify as far as possible:

1	$4x+1(x+3)$	*2*	$4x+2(x+3)$	*3*	$3(2x-1)+2x$
4	$7m+2(4m+5)$	*5*	$3(8m-2n)+10n$	*6*	$a+(a+b)$
7	$x+(2x+3x)$	*8*	$5y+2(3-y)$	*9*	$5(a-3k)+16k$
10	$x(y+3)-xy$	*11*	$x(x+1)+x^2$	*12*	$3(p^2+2q)-3p^2$
13	$3x+2(y-x)$	*14*	$a(a-1)+a$	*15*	$2x(x+4)+3x^2$

16	$2(a+2)+3(a+1)$	*17*	$3(5a+2)+2(a-3)$
18	$3(a-1)+2(a+2)$	*19*	$4(2x+7)+7(x-4)$
20	$6(3c+4d)+3(2c-5d)$	*21*	$2(p+3)+3(5p-2)$
22	$2a(a^2+1)+a^2(2a-1)$	*23*	$3(a+2b+3c)+2(a+b+c)$
24	$5(x+y+4z)+2(x+y-3z)$	*25*	$2(a-3b-3c)+3(a+2b+2c)$
26	$4a(3+b)+3b(3+a)+5c(1+b)$	*27*	$5a(a-b)+3b(2-a)+8ab$

28 If $P = a^2+2b$ and $Q = b^2-2a$, express $aP+bQ$ in its simplest form.

4 Application of the distributive law to the solution of equations and inequations

Example 1. Solve the equation $2x+3x = 15$, x being a variable on the set of natural numbers.

$$\begin{aligned} 2x+3x &= 15 \\ \Leftrightarrow \quad 5x &= 15 \text{ (The symbol } \Leftrightarrow \text{ means 'is equivalent to'.)} \\ \Leftrightarrow \quad x &= 3 \end{aligned}$$

Example 2. Find the solution set of $y+2(y-1) > 3$, $y \in W$.

$$\begin{aligned} y+2(y-1) &> 3 \\ \Leftrightarrow \quad y+2y-2 &> 3 \\ \Leftrightarrow \quad 3y-2 &> 3 \\ \Leftrightarrow \quad 3y &> 5 \\ \Leftrightarrow \quad y &> \tfrac{5}{3} \end{aligned}$$

Hence the solution is {2, 3, 4, ...}.

Exercise 5

Solve the following equations, the variables being on the set of natural numbers:

1	$3x+2x = 15$	*2*	$4y+y = 35$	*3*	$7z-4z = 18$
4	$11p-8p = 3$	*5*	$2t+5t = 63$	*6*	$9x+6x = 45$
7	$9x+2x-8x = 45$	*8*	$9y-5y+2y = 72$	*9*	$7m+2m+3m = 24$
10	$3p+2p+5p = 20$	*11*	$8q+2q-6q = 12$	*12*	$9r-7r-r = 6$
13	$3(x+2) = 9$	*14*	$2(x+5) = 20$	*15*	$2(x-1) = 4$
16	$5(y+1) = 10$	*17*	$4(y-2) = 12$	*18*	$6(y+3) = 24$
19	$5(x+2)-2x = 13$	*20*	$3(x-2)+4x = 8$	*21*	$4(y+1)-3y = 6$
22	$2(y+5)-y = 19$	*23*	$z+2(z+3) = 12$	*24*	$6(z-1)+z = 1$

Find the solution set of each of the following inequations, the variables being on the set of natural numbers:

25	$3x+5x < 24$	*26*	$4x-x < 15$	*27*	$5x+2x > 22$
28	$x+2x < 10$	*29*	$3y+4y > 10$	*30*	$8x-3x < 30$
31	$9v+5v-2v < 50$	*32*	$m+m < 6$	*33*	$2z+z > 10$
34	$3y+2(y+1) \geqslant 12$	*35*	$5p+3(2+p) \leqslant 8$	*36*	$4(x+2)+7 \leqslant 31$
37	$5(y-1)+3y \geqslant 3$				

Exercise 5B

1 Solve the following, where x is a variable on the set $\{0, 1, 2, 3, 4, 5\}$:

a $7x+2x+3x = 24$ *b* $4x-2x+x > 2$ *c* $x+x < 5$
d $\frac{1}{2}x+\frac{1}{2}x > 2$ *e* $\frac{3}{4}x+\frac{1}{4}x = 0$ *f* $\frac{3}{4}x-\frac{1}{2}x = 1$

Find the solution set of each of the following, the variables being on the set of whole numbers:

2	$7x+4x+x = 60$	*3*	$21x+x-4x = 54$	*4*	$11p-p-7p > 10$
5	$5(v+3)+2v > 30$	*6*	$8t+2(1+t) < 50$	*7*	$2x+(x-3) < 7$

8	$5(n-2)+5n = 60$	*9*	$2(n+1)+3(2n+3) \geqslant 18$
10	$3(2n+1)+2(n+6) \leqslant 47$	*11*	$(n-1)+3(2n+1) < 18$
12	$7(x+1)+3(x+2) > 23$	*13*	$\frac{1}{2}(8x+4) = 10$
14	$\frac{1}{3}(12x+15) < 17$	*15*	$\frac{1}{2}(8x+6) > 7$

Find the solution sets of the following, x being a variable on the set $\{0, 1, 3, 5\}$:

16 $3(7x-5)-2x = 4$ *17* $x-\frac{1}{4}x < 2$ *18* $\frac{1}{4}x-\frac{1}{5}x < \frac{1}{4}$

19 $0{\cdot}4x-0{\cdot}1x < 1$ *20* $\frac{1}{2}x+\frac{3}{4}x = 5$ *21* $\frac{1}{2}(4x+10) > 10$

5 *Illustrations of the distributive law*

Example. n represents a whole number. From n subtract 4, and then multiply the difference by 2. If the value of this expression is 20, find n.

From the given instructions we have: n; $n-4$; $2(n-4)$; $2(n-4) = 20$

$$\begin{aligned} 2(n-4) &= 20 \\ \Leftrightarrow \quad 2n-8 &= 20 \\ \Leftrightarrow \quad 2n &= 28 \\ \Leftrightarrow \quad n &= 14 \end{aligned} \qquad \textit{or} \qquad \begin{aligned} 2(n-4) &= 20 \\ \Leftrightarrow \quad n-4 &= 10 \\ \Leftrightarrow \quad n &= 14 \end{aligned}$$

Exercise 6A

1 Form an equation and find the whole number in each of the following:

Representing a whole number	First instruction	Second instruction	Value of expression
p	Add 2	Multiply by 3	12
q	Subtract 1	Multiply by 2	10
r	Add 10	Multiply by 4	60
s	Subtract 6	Multiply by $\frac{1}{2}$	5

2 To x add 2, multiply the sum by 2, then subtract 4. Find a simple expression for the result.
If x represents a whole number, is the result even or odd?

3 From k subtract 1, then multiply the difference by 3. If the value of this expression is 72, find k.

4 n is a variable on the set of natural numbers. To n add 5, and then multiply your answer by 3. If the result is greater than 33, form an inequation in n and hence find its solution set.

5 The equality $(3\times 2)+(3\times 4) = 3(2+4)$ may be illustrated by the 'dot pattern' in Figure 3 on page 60:

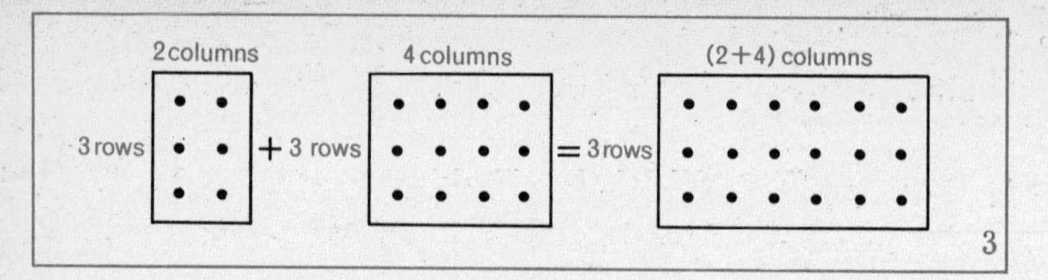

Make a 'dot pattern' to illustrate:

a $(3\times 1)+(3\times 2) = 3(1+2)$ *b* $(3\times 3)+(3\times 4) = 3(3+4)$

6 Figure 4 shows an L shape in which all the angles are right angles, and all the units are centimetres. Check (by drawing in lines if necessary) that each of the following expressions indicates a way of finding the area of the shape in cm^2:

a $(3\times 7)+(3\times 5)+(3\times 3)$ *b* $(3\times 7)+(3\times 8)$

State two more expressions for finding the area, and then calculate the area.

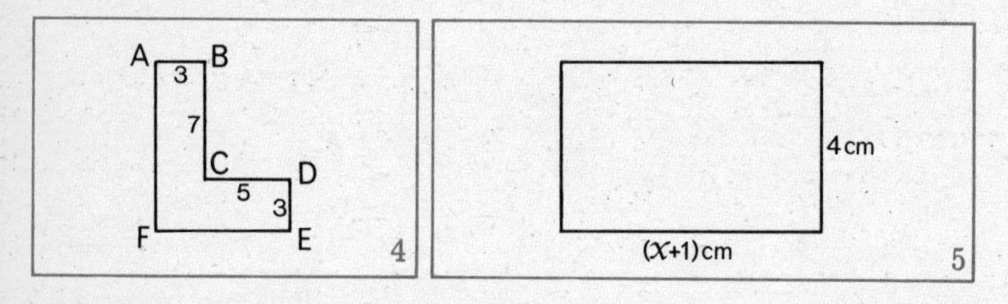

7 Figure 5 shows a rectangle with length $(x + 1)$ cm and breadth 4 cm. Find expressions for:

a its perimeter in cm *b* its area in cm^2.

If the perimeter exceeds 12 cm, what can you say about x?
What can you say about the area?

8 If n is a whole number, $11n = (10+1)n = 10n+n$.
Hence, to multiply a number by 11, 'multiply the number by 10, and then add the number to the product'. Use this rule to calculate:

a 7×11 *b* 35×11 *c* 86×11 *d* 987×11

9 Noting that $9n = (10-1)n$, make up a rule for multiplying a whole number by 9, using question ***8*** as a guide. Test your rule by working out the following, and check your answers by multiplying out in the usual way:

a 36×9 *b* 9×74 *c* 460×9 *d* 538×9

Try to invent a rule for multiplying a whole number by 99, and check it by calculating 99×57.

10 An operation * means 'multiply the first number by itself, and then add the second number'. Thus, $5 * a = 25+a$.

In a similar way, write down expressions for:

a $3 * x$ *b* $3 * y$ *c* $x * 3$ *d* $y * 3$

e $3 * (x+y)$, where + has its usual meaning.

(*1*) Is operation * commutative (i.e. is $3 * x = x * 3$, etc.)?

(*2*) Does $3 * (x+y) = (3 * x)+(3 * y)$? What does this suggest?

Exercise 6B

1 The length of a rectangle is $(x+5)$ cm, and its breadth is 6 cm.

a Write down an expression in x for the area of the rectangle.

b If the area is 90 cm², form an equation, find x, and then give the length of the rectangle.

2 Make a sketch of a cuboid and mark the length, breadth and height x cm, y cm and z cm respectively. Find formulae for:

a the total length L cm of its edges in terms of x, y, z

b the total surface area, A cm².

3 If $2n$ pence and $(n+5)$ pence together total £2, form an equation in n and solve it.

Hence write the value of $2n$ pence, and of $(n+5)$ pence in £s.

4 x belongs to the set of even natural numbers. From x subtract 2 and multiply your answer by 4. If the value of this expression is less than 40, form an inequation in x and find its solution set.

5 If in the L shape in Figure 4, $AB = p$ cm, $DE = p$ cm, $BC = q$ cm and $CD = r$ cm, show by drawing sketches (with any necessary lines added) how each of the following expressions for the area is obtained (in cm²):

a $p(q+p)+pr$ *b* $p(r+p)+pq$

c $pq+pr+p^2$ *d* $p(p+q+r)$

Use the distributive law to show that all the expressions are equivalent.

6 n is a whole number.

a Write down a number 1 more than n.

b Write down a number 1 less than n.

c If the sum of these three consecutive numbers is 144, find n and hence state the three numbers in ascending order.

7 The smallest angle of a certain triangle is $x°$, and the largest angle is four times the smallest angle. The remaining angle is 30° larger than the smallest angle.

a Write down an expression for the sum of the angles of the triangle in terms of x.

b Form an equation in x and solve it.

c State the size of each of the angles of the triangle. Check your answers by finding the sum of the angles of the triangle.

8 x boys, and two more girls than boys, went on a short bus excursion; each paid 18 pence.

a Write down an expression for the total bus fares, and simplify it as far as possible.

b The total fares were £2·16. Form an equation in x, and solve it.

c How many were in the bus party?

9 A collection of coins consisted of $3x$ five-pence coins, $(x+2)$ ten-pence coins and $(x-6)$ fifty-pence coins.

a Find in its simplest form an expression for the amount of the collection in pence.

b If the collection amounted to £3·20, form an equation in x and solve it.

c If each person contributed one coin only, find the number of contributors.

10 A rectangle has length $(x+9)$ cm and breadth x cm. Obtain expressions for:

a its perimeter in cm *b* its area in cm^2.

If the perimeter exceeds 30 cm, but is less than 42 cm, find inequations in x, and simplify these.

Hence complete the inequation $\ldots < x < \ldots$

Summary

1 *Products to sums or differences: multiplying out.*

(i) $a(b+c) = ab+ac$
(ii) $a(b-c) = ab-ac$
(iii) $\frac{1}{a}(b+c) = \frac{b}{a}+\frac{c}{a}$

2 *Sums or differences to products: common factors.*

(i) $ab+ac = a(b+c)$
(ii) $ab-ac = a(b-c)$
a is a *common factor* of ab and ac.

3 *Collecting like terms.*

In $4x+3y+2x+y$, $4x$ and $2x$ are like terms, and $3y$ and y are like terms.

$$\begin{aligned} &4x+3y+2x+y \\ =\ &4x+2x+3y+y \\ =\ &6x+4y \end{aligned}$$

Revision Exercises

Revision Exercises on Chapter 1
Replacements and Formulae

Revision Exercise 1A

1 Find the value of each of the following when $x = 4$:

a $x+10$ *b* $12x$ *c* $5x+15$
d $20-4x$ *e* $2x^2$ *f* x^2-8

2 Copy and complete the following table for the given replacements of a and b:

a	b	$2a+4$	$3ab$	$4a-3b$	$2ab+5$	a^2+b^2
2	1	8	6	5	9	5
3	0					
4	3					
$\frac{1}{2}$	$\frac{1}{3}$					
3	3					

3 If x is a variable on the set {0, 1, 2, 3, 4, 5}, and $n = 3x+1$, find the set of all possible values of n.

4 a If $Q = xy+25$, find Q when $x = 8$ and $y = 7$.
b If $A = 20-mn$, find A when $m = 6$ and $n = 2$.

5 The perimeter P cm of a square is given by the formula $P = 12a$.
a Write down the length of a side of the square in terms of a.
b Write down the area of the square.

6 State whether each of the following is true or false:

a $x+4 = 15$, when $x = 9$ *b* $4p-3 = 9+p$, when $p = 4$
c $6y+1 = 4$, when $y = \frac{1}{2}$ *d* $2q-5 = 70$, when $q = 7$

7 Figure 1 shows a 3×3 tiling of squares, each of side x cm. Find a formula for the area A cm^2 of the octagon (with 8 sides) outlined.
Use this formula to calculate the area when $x = 3$.

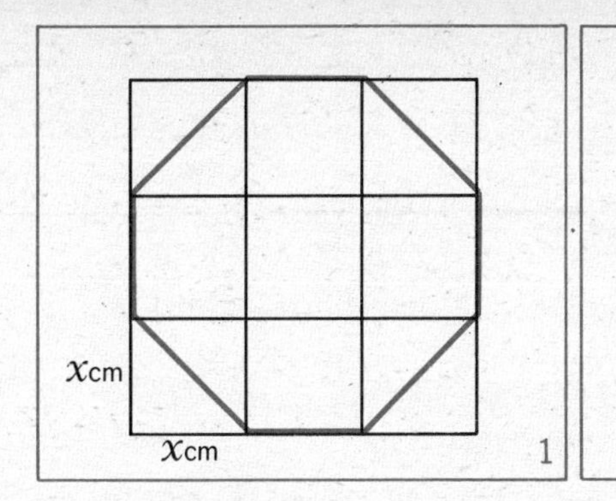

1

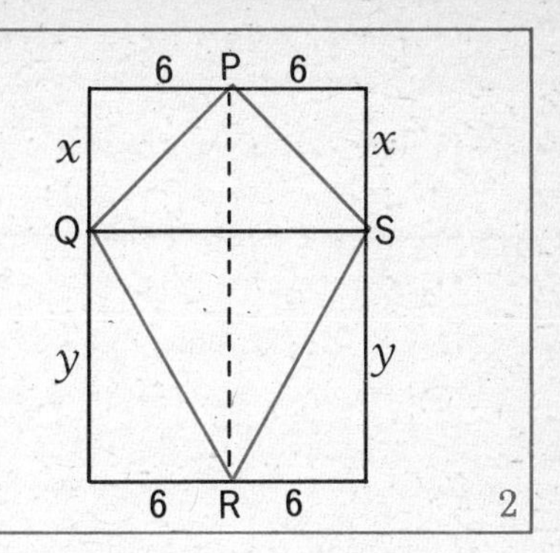

2

8 Figure 2 shows a 'kite' PQRS inside a rectangle. The dimensions are in centimetres.

a Write down expressions for the areas of triangles PQS and QRS in terms of x and y.

b State the formula for the area A cm^2 of the kite.

c Find A when $x = 6$ and $y = 8$.

9 An aircraft flies at 1050 km/h.

a Write down how far it flies in 2 hours, 3 hours, 5 hours, x hours.

b If it flies D km in x hours, state the formula for D in terms of x.

c Calculate D when $x = 3\frac{1}{2}$.

10 If x is a variable on the set {1, 2, 3, 4, 5, 6} and $y = 2x+3$, copy and complete this table:

x	1	2	3	4	5	6
y						

Revision Exercise 1B

1 If $p = 5$, $q = 2$, and $r = 3$, find the values of:

a $4p+3q+2r$ *b* $pq+qr+rp$ *c* $p^2+q^2+r^2$

d pqr *e* $p+qr$ *f* p^2-2qr

2 If $v = 5$ and $w = 4$, find the values of:

a v^2 *b* v^2+w^2 *c* $(v+w)^2$ *d* v^2-w^2 *e* $(v-w)^2$

3 Given that $p = 10$ and $q = 5$, find the values of:

a $\frac{p}{q}$ *b* $\frac{q}{p}$ *c* $\frac{p+q}{3}$ *d* $\frac{100}{pq}$

4 A formula relating the number of faces, corners, and edges of a solid with plane faces is $F+C = E+2$. Use this formula to find:

a the number of edges of a snub cube, which has 38 faces and 24 corners

b the number of faces of an octahedron, which has 6 corners and 12 edges.

5 a Is the open sentence $12(x-2) = 2x + 25$ changed into a true sentence when x is replaced by 5?

b When $a = 2$, $b = 3$, and $c = 4$, find the value of $3a+2b+4c$. Does $3(2a+b)+(b+c)$ have the same value?

6 Draw a sketch of a cuboid. Mark the length x cm, the breadth y cm, and the height also y cm.

a Find a formula for the sum S cm of all the edges of the cuboid.

b Find a formula for the area A cm^2 of all the faces.

c Write down a formula for the volume V cm^3 of the cuboid.

d Calculate each of these quantities if $x = 12$ and $y = 5$.

7 A bus company advertises a coach trip at £0·50 per head. The estimated expenses to the bus company are £9.

a Complete the following table:

Number of passengers	Income	Expenditure	Profit
20	£10	£9	£1
30	...	£9	...
35	...	£9	...
40	...	£9	...
n	...	£9	...

b If the profit for n passengers is denoted by £P, find a formula for P, and test it by putting $n = 30$.

c Find n if the bus company makes neither a profit nor a loss.

8 Here is a mathematical system whose elements are 0, 1 and 2. Addition and multiplication in this system are defined by the following tables:

+	0	1	2
0	0	1	2
1	1	2	0
2	2	0	1

×	0	1	2
0	0	0	0
1	0	1	2
2	0	2	1

If x and y are variables on the set $\{0, 1, 2\}$, find the value of:

a $2x+y$, when $x = 1$ and $y = 2$
b $2x+2y$, when $x = 2$ and $y = 1$
c $xy+2y$, when $x = 2$ and $y = 2$

9 Figure 3 shows two rectangles, each of length x cm and breadth y cm, placed end to end.

a Write down a formula for the perimeter P cm of the large rectangle formed.

b Find a formula for the perimeter P cm in each case if the following number of x cm by y cm rectangles are placed end to end:

(*1*) 3 (*2*) 4 (*3*) 5 (*4*) 6 (*5*) n

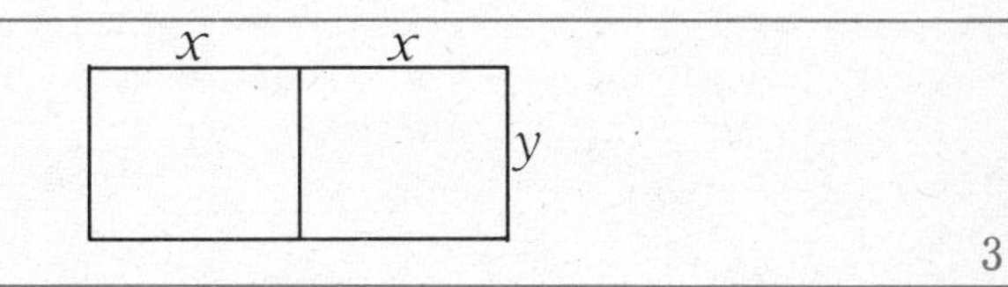

3

Revision Exercises on Chapter 2
Inequalities and Inequations

Revision Exercise 2A

1 Write the symbol for each of the following:

a 'is greater than' *b* 'is equal to'
c 'is less than' *d* 'is less than or equal to'

2 Copy and insert one of the signs $>$, $=$, $<$ between each of the following pairs of numbers to give a true sentence:

a 3 ... 9 *b* 3^2 ... 9 *c* $\frac{1}{2}$ ... 1
d 1 ... 0 *e* 5·80 ... 5·81 *f* $\frac{1}{5}$ ... $\frac{1}{10}$

3 Which of these sentences are true?

a $4 > 2$ *b* $9 \leqslant 9$ *c* $0 < 2$
d $1^2 = 1^3$ *e* $8 \geqslant 7$ *f* $8 \leqslant 7$

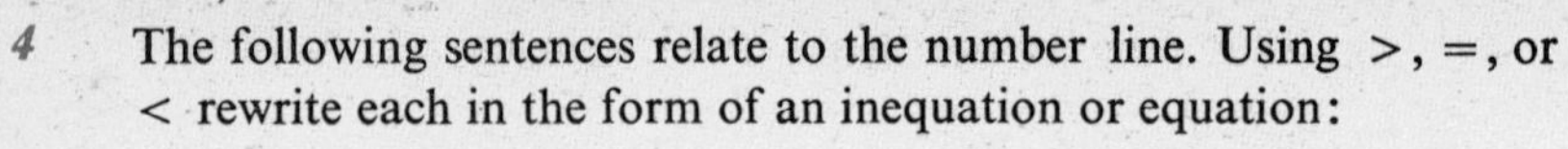

4 The following sentences relate to the number line. Using $>$, $=$, or $<$ rewrite each in the form of an inequation or equation:

a p lies to the left of 5. *b* q lies to the right of 12.
c r is between 4 and 10. *d* n is between 15 and 17.

5 Using the appropriate inequality sign, rewrite each of the following:

a 8 added to x is less than 25.
b 5 times y is greater than 30.
c n is not equal to zero.
d 9 subtracted from p is less than or equal to 15.

6 Show on number lines the following subsets of the set of whole numbers:

a $\{3, 4, 5\}$ *b* $\{2, 3, 5, 7\}$ *c* $\{0, 3, 6, 9\}$ *d* $\{0\}$

7 Using the inequality symbols $>$ and then $<$, show the order of each of the following in *two* ways:

a 8, 3, 5 *b* 1, 10, 6 *c* $\frac{1}{3}, \frac{1}{4}, \frac{1}{2}$

8 If x is a variable on the set $\{0, 1, 2, \ldots, 10\}$, find the solution set of these inequations:

a $x > 7$ *b* $x < 1$ *c* $2x < 14$ *d* $4x \geqslant 12$ *e* $3x > 30$

9 If $x \in \{1, 3, 5, 7, 9\}$, solve each of the following and show the solutions on separate number lines:

a $x+4>9$ *b* $x-7 < 1$ *c* $7-x<7$
d $2x > 11$ *e* $2x+1 < 7$ *f* $2x-1 > 8$

10 In each of the following write down an inequation containing the variable indicated so as to express the idea given in the sentence:

a The number (n) of matches in a box is not more than 50.
b The time (t minutes) for which some packet soups must be simmered is not less than 5 minutes.
c On many continental roads, the speed (v km/h) of a vehicle must not exceed 110 km/h.
d The mean radius (R km) of the earth lies between 6300 km and 6400 km.

11 Given that $x \in \{0, 1, 2, 3, 4, 5, 6\}$, list the members of this set which are solutions of each of the following, and illustrate on a number line where possible:

a $x+2 < 6$ *b* $x+5 > 12$ *c* $x+4 > 4$
d $x+4 \leqslant 4$ *e* $2x > 9$ *f* $8-x < 5$

12 The perimeter of a square is less than 24 cm and each side is x cm long. Write down an equation in x, and simplify it if you can.

Revision Exercise 2B

1 $5 > x$ and $1 < x$ can be combined into a single sentence $1 < x < 5$ or $5 > x > 1$. Combine each of the following in a similar way.

a $x > 5$ and $x < 10$ *b* $3 > x$ and $x > 0$
c $y = 4$ or $y > 4$ *d* $x \geqslant 8$ and $x < 10$

2 Each of the following graphs shows the possible positions of a whole number n on a number line. Using the inequality symbol $>$ or $<$, write down the corresponding inequations.

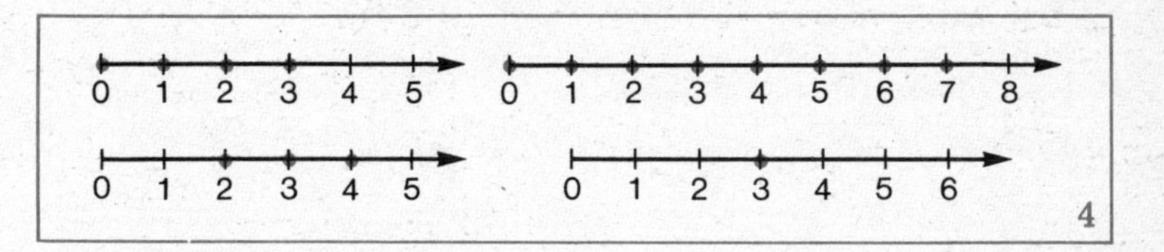

3 If x is a variable on the set {0, 2, 4, 6}, find the solution set of each of the following inequations:

a $2x+5 > 10$ *b* $3x-6 < 8$ *c* $9x > 2x+22$

4 Given that $p \in \{1, 2, 3, 4, 5\}$, find the solution sets of the following, and show them on a number line:

a $p+11 > 14$ *b* $8-p < 6$ *c* $p^2+6 > 5p$
d $p^2 < p+10$ *e* $1 \leqslant p < 4$ *f* $3p-12 > 0$

5 From the set of whole numbers, find the solution sets of:

a $3m < 20$ *b* $2+5x < 12$ *c* $2y+1 \geqslant 9$
d $3 < t < 8$ *e* $1 \leqslant h \leqslant 5$ *f* $3w-2 > 10$

6 From the set of natural numbers give the solution set of each of the following:

a $3x-2 \leqslant 7$ *b* $2y-1 > 10$ *c* $4z > 5+z$
d $1+9x < 10$ *e* $10+x < 3x$ *f* $12-m \geqslant 4+m$

7 a The perimeter of a square is less than 16 cm and each side is x cm long. Write down an inequation in x.

b The length of a rectangle is y cm and its breadth is $(5-y)$ cm, y

being a variable on the set of natural numbers. Find the set of replacements of y, and hence obtain the set of values the area may take.

8 In each of the following sentences, write down an inequation containing the variable indicated so as to express the given information in mathematical form:

a The petrol tank in a motor car could hold 50 litres of petrol; the tank contained x litres of petrol.

b The minimum weight of a certain box of biscuits is 1·6 kg, and W kg is the weight of a box of biscuits.

c The minimum flying speed of an aircraft is 190 km/h; the maximum speed is 1200 km/h. The speed of the aircraft is S km/h.

d The cost (£C) of a holiday in Europe can be as little as £50.

9 The maximum safe load for Mr Brown's car is 410 kg. On holiday, he and his passengers together weigh 250 kg and Mr Brown decides to restrict the weight of luggage, w kg, so as to be under the maximum safe load.
From this information, form an inequation in w and then express it in its simplest form.

10 One of the equal angles of an isosceles triangle is $x°$.

a Express the size of the third angle in terms of x.

b If the third angle is greater than 30°, form an inequation in x and complete the sentence $x <$ ....

c If the third angle is also less than 40°, form another inequation in x and complete the sentence $x >$ ....

d Hence make up a sentence of the form $a < x < b$, where a and b are whole numbers.

11a Express in the form of an inequation:
'14 added to x is greater than x subtracted from 20'.

b If x is a variable on the set $\{1, 2, 3, \ldots, 20\}$, find the solution set of this inequation.

12 Let S_r denote the sum of the first r terms of the *series* of numbers $2+4+6+8+10+12+\ldots$. Then

$S_1 = 2 = 1\times 2$; $S_2 = 2+4 = 6 = 2\times 3$;
$S_3 = 2+4+6 = 12 = 3\times 4$;
$S_4 = 2+4+6+8 = 20 = 4\times 5$, and so on.

Study the above pattern and then answer the following:

a Express S_5 and S_6 as a product of two factors.

b Find the greatest number r such that $S_r < 100$.
c Find the least number r such that $S_r > 400$.
d Obtain a formula for S_r as a product of two factors, and test your formula by putting $r = 4$.

Revision Exercises on Chapter 3
Negative Numbers

Revision Exercise 3A

1 a If $+$£10 stands for a gain of £10, write down:
(*1*) a gain of £25 (*2*) a loss of £30 (*3*) a loss of £p

b If $+5$ km is used to denote a journey of 5 km east, write down
(*1*) a journey of 80 km east (*2*) a journey of 125 km west
(*3*) a journey of x km east followed by a journey of y km west.

2 Copy and complete the following by placing one of the symbols $>$, $=$, or $<$, between each of the pairs to make a true sentence:

a $6 \ldots 4$ *b* $0 \ldots -1$ *c* $-2 \ldots (-2)$ *d* $-3 \ldots 3$

3 Write down in simpler form:

a $(+5)$ *b* (-7) *c* $n+(-3)$ *d* $(-x)+1$
e -0 *f* $m+(-m)$ *g* $(-m)+(-m)$ *h* $0-1$

4 Which of the following are true?

a $5 > 1$ *b* $-5 > 1$ *c* $-5 < -1$ *d* $-5 < 0$
e $-5 \leqslant 5$ *f* $-5 < 5$ *g* $-(-2) = 2$ *h* $7-3 = 3-7$

5 Use number lines to illustrate each of the following:

a $2+5 = 7$ *b* $-3+(-4) = -7$ *c* $8+(-3) = 5$
d $4+(-7) = -3$ *e* $5-3 = 2$ *f* $5-9 = -4$

6 Simplify:

a $-4+(-9)$ *b* $15+(-8)$ *c* $7+(-13)$ *d* $14-9$
e $-20-12$ *f* $6-34$ *g* $0-4$ *h* $0-(-4)$
i $-15-(-8)$ *j* $10a-3a$ *k* $2a-9a$ *l* $24m-(-12m)$

7 If $p = 8$, $q = -5$ and $r = -3$, calculate:

a $p+q$ *b* $p-q$ *c* $q-p$ *d* $q+r$
e $p-r$ *f* $p+q+r$ *g* $p+q-r$ *h* $p-q+r$

8 If x is a variable on the set $\{-4, -3, -2, -1, 0, 1, 2\}$, solve the following:

a $x > -2$ *b* $x \leqslant 1$ *c* $x < 0$
d $x+(-3) = 0$ *e* $-1+x = 1$ *f* $x+x = -4$

9 In a game, a player counts all even numbers scored as being positive and all odd numbers as being negative. What is his total score if his individual scores are 5, 4, 6, 3, 2, 4, 1, 3, 6, 5?

10 Calculate:

a $\frac{14}{5}-\frac{4}{5}$ *b* $1-\frac{3}{4}$ *c* $\frac{1}{4}-\frac{1}{8}$
d $-\frac{1}{2}+2$ *e* $1{\cdot}50-0{\cdot}75$ *f* $-0{\cdot}50-0{\cdot}25$

11a Show on a number line the set $\{4, -2, 7, -5\}$, marking each point with a dot.

b What is the average of the integers 4, -2, 7 and -5?

c Indicate with an arrow the point on the number line which shows the average.

12 Let N, W, Z, and Q have the meanings given on page 42. Find the solution set of each of the following:

a $x+(-4) = 2, x \in Z$ *b* $4+n < 7, n \in W$
c $y+3 = 1, y \in N$ *d* $1-x > 0, x \in Z$
e $2z-1 = -2, z \in Q$ *f* $2m-3 = 0, m \in Z$

13 Take an origin O near the centre of your page of squared paper and draw OX and OY as axes. On OX mark $-3, -2, -1, 0, 1, 2, 3$, and on OY mark $-6, -5, -4, \ldots, 4, 5, 6$.

Plot the points A$(-3, 6)$, B$(-2, 4)$, C$(-1, 2)$, O$(0, 0)$. Write down the coordinates of three more points in this sequence.

14 Express each of the following in standard form, to 2 significant figures:

a 0·474 *b* 0·0638 *c* 0·000001258 *d* 2310

Revision Exercise 3B

1 Insert the correct symbol $>$, $=$, or $<$, between the members of each of the following pairs so as to make true sentences:

a $3 \ldots 5$ *b* $-3 \ldots -5$ *c* $\frac{1}{3} \ldots 0{\cdot}333$ *d* $-\frac{1}{2} \ldots (-\frac{1}{2})$
e $a \ldots -(-a)$ *f* $-0{\cdot}2 \ldots -0{\cdot}7$ *g* $-2000 \ldots 2$ *h* $0{\cdot}08 \ldots 0{\cdot}1$

2 The table shows certain replacements for x. Copy and complete the entries:

x	$x+3$	$x-5$	$-x+4$	$-x-2$
5	8	0	-1	-7
4	.	.	.	.
2	.	.	.	.
0	.	.	.	.
-1	.	.	.	.
-2	.	.	.	.
-9	.	.	.	.

Can you see a way to check your answers in the third and fourth columns?

3 If x is a variable on the set $\{-5, -4, -3, -2, -1, 0, 1, 2, 3, 4, 5\}$, find the solution set of each of the following:

a $x < -2$ *b* $x > 0$ *c* $x > 4$ *d* $x < -5$ *e* $-2 < x < 2$

4 Given that x is a variable on the set of integers, solve:

a $x+6 = 8$ *b* $x+(-5) = -9$ *c* $x+(-3) = 2$
d $x+4 = -2$ *e* $7+(-x) = 3$ *f* $-8+(-x) = -17$
g $x-7 = 1$ *h* $x+5 = 3$

5 Simplify:

a $12-15$ *b* $-19+14$ *c* $1{\cdot}4-2{\cdot}8$ *d* $0-\frac{3}{4}$

6 In an examination a pupil gains three marks for every correct answer and scores -2 for every wrong answer. If the pupil has 33 answers correct and 17 wrong, what is his total mark?

7 Simplify:

a $3p+5p-2p$ *b* $9m-4m-3m$ *c* $6y-10y+y$
d $-4x+5x+4x$ *e* $-t-2t-3t$ *f* $n-n+n+n-n$
g $7p-3p+p+2p$ *h* $-2a^2+5a^2-7a^2$ *i* $-4k-3k+k+6k$

8 Write down the resulting temperature if the temperature:

a changes by $+15°$ from $-4°$ *b* changes by $-15°$ from $-4°$.

9 Find three consecutive integers whose sum is -27.

10 Copy and complete the addition table for the '5 minute clock' shown in Figure 5 (page 74), remembering that the first number is taken from the left-hand column and the second number from the top row.

a Here $3+2 = 0 = 2+3$, so 2 is the additive inverse of 3. Find the additive inverses of 2, 4 and 1.

+	0	1	2	3	4
0					
1					
2					
3					
4					

0 1 2 3 4

5

b Interpreting -2 as a move of two places anti-clockwise, express the additive inverse of 2, 4 and 1 in a second way. Equate these to your answers for ***a***.

11 P = {all points whose first coordinates are greater than -5 and less than -3}
Q = {all points whose second coordinates are greater than -2 and less than 0}.

a Show in a diagram, and colour, the areas representing the sets (or part of the sets) P, Q, and $P \cap Q$.

b State whether each of the following is true or false:

(*1*) $A(-3\frac{1}{2}, 21) \in P$ (*2*) $B(-4, -1) \in P \cap Q$
(*3*) $C(-\frac{1}{2}, -\frac{1}{2}) \in Q$ (*4*) $D(-1, -4) \in P \cap Q$
(*5*) $E(-13, -7) \in P$ (*6*) $F(2, \frac{1}{2}) \in Q$

12 Express each of the following in standard form to 2 significant figures:

a 0·956 *b* 0·001 140 *c* 3280 *d* 0·000001 762

13a What shape is the quadrilateral A(2, 1), B(5, 5), C(1, 8), D(-2, 4)?

b Show in a diagram the image of each point in the x-axis.

c Give the coordinates of these images, and state the shape of the figure formed by joining them together.

14 A game is played on the number line according to the rule 'Integer n goes to integer $14-n$'. For example, 10 goes to 4.

a At which number should you start so that one move will take you four units to the *left* of your starting point?

b Where should you start so that *two* moves take you back to your starting point?

Revision Exercises on Chapter 4
The Distributive Law

Revision Exercise 4A

1 Which of the following are equal?
$3a+3b$, $3a\times 3b$, $3(a+b)$, $3ab$, $3a+b$, $a\times 3b$

2 Simplify:

a $4n+9n$ *b* $3x^2+5x^2$ *c* $2p+2p+2p$
d $\frac{1}{2}ab+\frac{1}{2}ba$ *e* $10z-9z$ *f* $4m+m+n$

3 Figures 6 and 7 each show rectangles divided into two parts. The unit of length in each case is 1 centimetre. Find in *two* ways, formulae for the area A cm^2 of each rectangle.

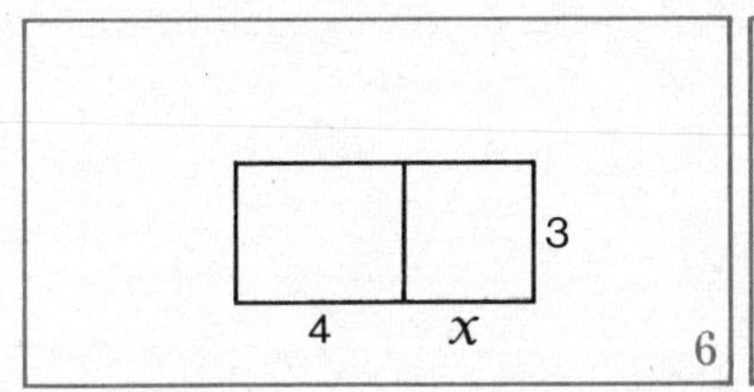

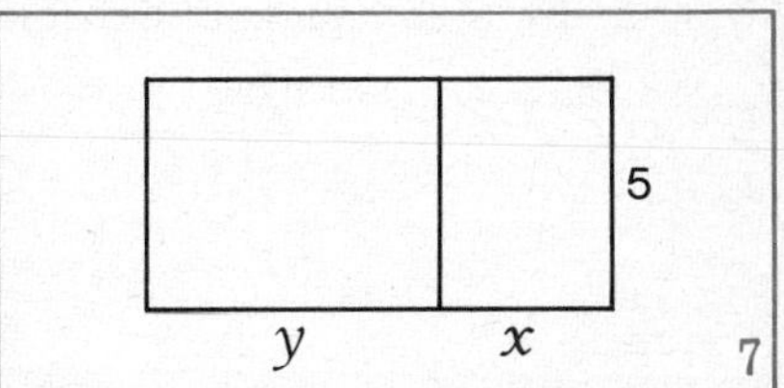

4 Express each of the following as a sum or difference of terms:

a $4(c+9)$ *b* $3(c-2)$ *c* $c(c+9)$ *d* $c(c-2)$
e $c(a+3)$ *f* $c(a-3)$ *g* $3(2a+7)$ *h* $3(4x-y)$

5 A loose-leaf notebook costs p pence and a refill costs r pence.

a Find the total cost of a notebook and two refills. Hence write down the cost of 5 notebooks, each with two refills, leaving your answer in factorized form.

b Find the total cost of 5 notebooks, each with two refills, in another way and check that the two answers are equivalent.

c Construct a formula for the total cost P pence of x notebooks, each with two refills.

6 Factorize the following:

a $5x+30$ *b* $6a+15$ *c* $10m-25n$ *d* $3x+xy$
e $5y^2+30y$ *f* c^2-c *g* $12pq+20q^2$ *h* $4x^2+4x-2$

7 Find the solution set of each of the following, x being a variable on the set of natural numbers:

a $5x+3x = 24$ *b* $3(x+4) = 21$ *c* $9x-5x < 20$
d $2x+3x < 18$ *e* $2x+(x-4) = 20$ *f* $2(x-1)+3x = 18$
g $x+3(x+2) > 30$ *h* $3(x+2)-2x < 7$ *i* $3x+\frac{1}{2}(4x+6) \leqslant 23$

8 Simplify:

a $2a+5a+3b$ *b* $9x-3y+5x$ *c* $5p+3q-3p+6q$
d $7(x+y)-7x$ *e* $4m+3(m-n)$ *f* $5(u+2v)+8(2u-v)$

9 *a* How many pence are there in £P and P pence?
b What must be added to $5m$ to make $24m$?
c Add $8x$ and $9x$ and take $13x$ from the sum.
d Two sides of a triangle are $8k$ and $17k$ cm respectively, and the perimeter is $40k$ cm. What is the length of the third side?
e The base of an isosceles triangle is x cm long and each of the other two sides is three times the length of the base. Find the perimeter of the triangle.

10 Write down an equation in x for each of the following, and hence find x:

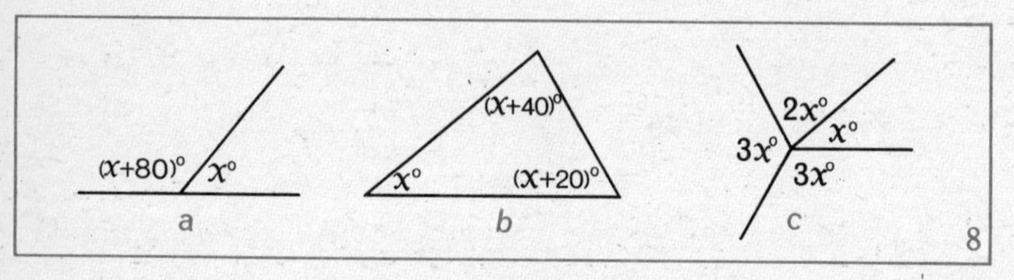

Revision Exercise 4B

1 Expand (i.e. *multiply out*) each of the following, and simplify where possible:

a $3(2x+2y+5z)$ *b* $2(p+3q)+3(p-2q)$
c $xy(1+x)+x^2(1-y)$ *d* $x(x^2+x+1)+x^2(x-1)$

2 Factorize:

a $5a^2+10a$ *b* $6mn+15n^2$ *c* x^2y+xy^2
d $4x^3+4x^2+6x$ *e* $a^2+2ab-ac$ *f* $4x^2-6xy+8xz$

3 *a* If n is an even number, write down the next two *even* numbers greater than n.
b Given that the sum of the three even numbers is equal to 180, find n.

4 Which of the following are equal?

$\frac{1}{2}a+b,\ \frac{1}{2}(a+b),\ a+\frac{1}{2}b,\ \frac{a}{2}+\frac{b}{2},\ \frac{1}{2}a\times\frac{1}{2}b.$

5 a A torch costs x pence and a battery y pence. Each torch requires 3 batteries to operate it. Write down the cost of a torch and its 3 batteries.

b Find a formula for the cost c pence of n torches, each supplied with 3 batteries. State the formula in two ways:
(*1*) with brackets (*2*) without brackets.

c Use each formula to find c when $n = 6$, $x = 95$ and $y = 5$. Hence express the cost in £s. Which formula was easier to use?

6 a Calculate the value of $5x^2+5x$ when $x = 19$.

b Factorize $5x^2+5x$, and hence calculate its value when $x = 19$.

c Which method did you prefer for the calculation?

7 Simplify the following:

a $\frac{2a+2b}{2}$ *b* $\frac{5x-10y}{5}$ *c* $\frac{4x+6y}{2}$ *d* $\frac{9a-12b}{3}$

8 Figure 9 shows a wire network, the unit of length being 1 centimetre.

a Find an expression for the length of wire needed to make the network.

b Given that the total length of the wire is 52 cm and that $x = 2y$, find x and y.

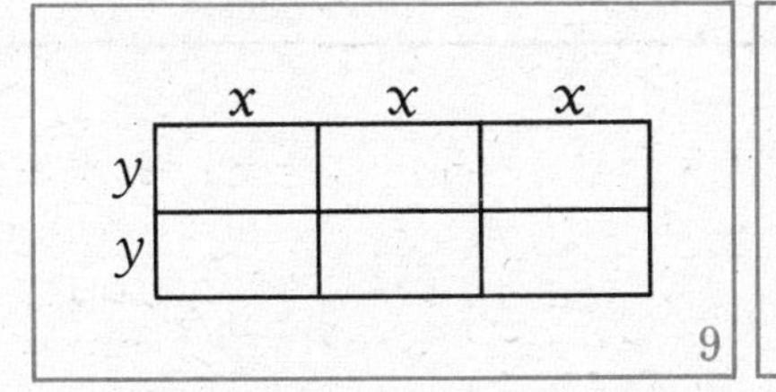

9

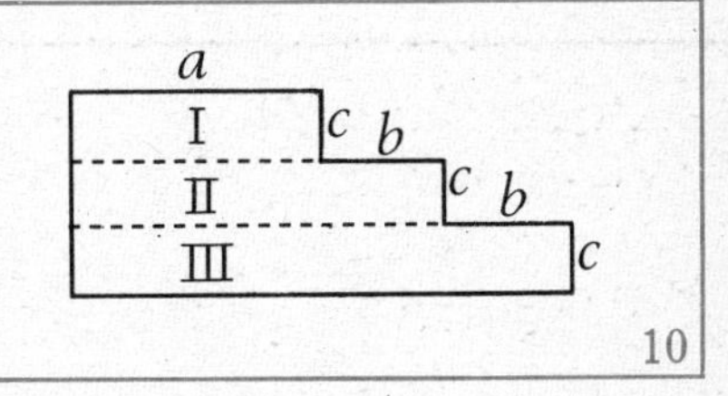

10

9 Figure 10 shows a stepped plate, lengths being measured in centimetres. Find the areas of regions I, II, and III.

Hence obtain a formula for the area A cm² of the plate. Simplify the formula as much as you can, and state the result in factorized form.

10 Find the solution set of each of the following, the variables being on the set of whole numbers:

a $5x+4+3x = 36$ *b* $5y+2y-3 < 22$ *c* $\frac{1}{2}n+\frac{1}{3}n = 5$

d $3(x+2)+2 = 32$ *e* $2(x-3)+3(x+2) = 20$

f $3(2x+1)+(x+1) < 46$ *g* $8(p+5)+4(p+1) > 128$

11 A boy cycled home from a friend's house at a speed of 14 km/h for $3x$ hours. Due to a puncture, he had to walk the remaining distance which took him x hours at a speed of 6 km/h. If the total distance travelled was 24 km, form an equation in x and solve it. What distance did he cycle?

12a Factorize $ac+ad$ and $bc+bd$.
Hence show that $ac+ad+bc+bd = (a+b)(c+d)$.

b Using a similar method, factorize $xp+xq+2p+2q$.

Cumulative Revision Section (Books 1 and 2)

Book 1 Chapter Summaries

Chapter 1 An introduction to sets

1	A *set* is a collection of clearly defined objects.	$A = \{a, b, c\}$
2	Each object in a set is a *member* or *element* of that set, and *belongs* to the set.	$a \in A, p \notin A$
3	*Equal sets* have exactly the same members.	$B = \{c, a, b\}, A = B$
4	The *empty set* is the set with no members.	$\emptyset$ *or* $\{\ \}$
5	A *universal set* is the set of all elements being discussed.	E('Entirety')
6	A set B is a *subset* of a set A if every element of B is a member of A.	$B \subset A, A \subset A$ $\emptyset \subset A, \emptyset \subset E$
7	The *intersection* of two sets A and B is the set of elements which are members of A and are also members of B.	$A \cap B$

8 *Venn diagrams*:

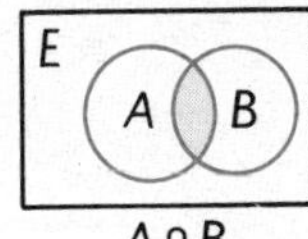

A∩B

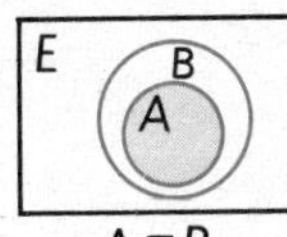

A⊂B

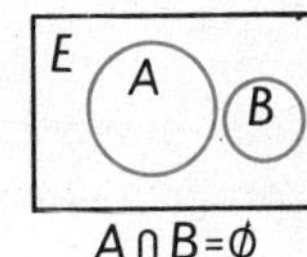

A∩B=∅

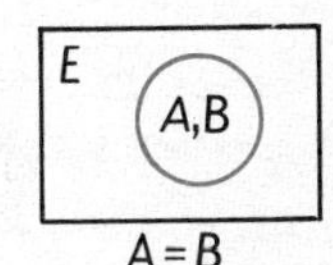

A=B

Chapter 2 Mathematical sentences

1 A *sentence* may be true or false.

2 An *open sentence* is a sentence containing a variable or variables, e.g. '12 is divisible by x'; '$x+y = 5$'.

3 A *variable* is a symbol which can be replaced by members of a given set.

4 The *solution set* of an open sentence is the set of replacements of the variable which give a true sentence.
Each member of the solution set is a *solution* of the open sentence.

5 An *equation* is an open sentence containing the verb 'is equal to'.

6 Mathematical sentences can be represented by *graphs*.

Chapter 3 Multiplication

1 *Multiplication is commutative*
e.g. $3 \times a = a \times 3$, and is written $3a$.

2 *Multiplication is associative*
e.g. $(3 \times a) \times b = 3 \times (a \times b) = 3ab$.

3 *Multiplication is distributive over addition*
e.g. $a(b+c) = ab+ac$.

4 *The coefficient in a term is the numerical factor in the term*
e.g. the coefficient of $3a$ is 3.

5 *The value* of $3x^2$ when $x = 2$ is $3 \times 2 \times 2 = 12$.

6 $a \times a = a^2$, and $a \times a \times a = a^3$.
Also $a+a = 2a$, $a+a+a = 3a$.

Cumulative Revision Exercises

Exercise A

1 $S = \{1, 2, 3, 4, 5, 6\}$.

a Describe the set S in words. How many members has S?
b List the set A of odd numbers in S.
c List the set B of prime numbers in S.
d List $A \cap B$.

2 Draw a rectangle PQRS with its diagonals PR and QS. List the set V of vertices (or corners) of the rectangle, the set S of its sides, and the set T of triangles which have one vertex at O, where PR and QS cross.

3 Which of the following sets are empty?

a The solution set of the equation $x+2 = 2$ where $x \in W$.
b The set of integers between 2·1 and 2·9.
c The set of even prime numbers. *d* $\{0\}$.

4 Write down pairs of equal sets from the following:
$A = \{-1, -3, 0\}$, $H = \{1, 2\}$, $P = \{12, 21\}$, $K = \{a, b, c\}$,
$F = \{u, v, w\}$, $G = \{0, -1, -3\}$, $B = \{c, a, b\}$, $S = \{1, 0, 2\}$

5 In this question, the universal set $E = \{1, 2, 3, \ldots, 10\}$.

a Construct Venn diagrams to show the following subsets of E:
(*1*) $A = \{1, 2, 3, 4, 8\}$, $B = \{1, 2, 3, 5, 6, 7\}$.
(*2*) $P = \{4, 5, 6\}$, $Q = \{4, 5, 6, 7, 8, 9\}$.
(*3*) $S = \{1, 2, 3, 4\}$, $T = \{5, 6, 7\}$.

b List the sets $A \cap B$, $P \cap Q$, $S \cap T$.

c Which of these is false? $T \subset Q$, $S \subset A$, $P \subset Q$, $B \subset Q$.

6 Find a replacement which makes each sentence below true, and also a replacement which makes each sentence false:

a $x+4 = 12$ *b* $y-5 = 5$ *c* p is less than 10
d f is a factor of 48 *e* g is greater than -5

7 The variables in the open sentences below are on the set $S = \{0, 1, 2, 3, 4, 5\}$. Find the solution set of each sentence.

a $x+3 = 7$ *b* $x+3 < 7$ *c* $m-1 = 2$ *d* y^2 is in S
e $p \neq 3$ *f* $n+4 > 10$ *g* $0 < z < 5$ *h* c is even

8 Solve the following equations, the variables being on the set of whole numbers:

a $x+14 = 25$ *b* $y-17 = 13$ *c* $4 \times y = 56$
d $z \times 3 = 33$ *e* $19 = p+11$ *f* $23 = 40-x$
g $15-x = 9$ *h* $y+y = 24$ *i* $5(x+3) = 30$
j $4(t-3) = 12$ *k* $3x+4x = 63$ *l* $\frac{1}{2}p = 30$

9 Given $p = 2, q = 1, r = 0$, find the values of:

a $3pqr$ *b* $2p+2q+2r$ *c* $pq+qr+rp$
d $p^2+q^2+r^2$ *e* $5p^2$ *f* $(p+q+r)^2$

10 Write the following in as short a form as possible:
$a \times b$, $3 \times c$, $p \times 8$, $m \times m$, $2p \times q$, $3p \times 4q$, $a \times b \times c$

11 Express each of the following in its shortest form:

a Add p to q. *b* Subtract x from y. *c* Multiply m by n.
d Four times the product of a and b.
e Add b to a, and multiply the sum by 10.

12*a* $p+3 = q$, and $p = 15$; find q.
b $c+d = 45$, and $d = 24$; find c.
c $x-y = 20$, and $x = 38$; find y.

d $s = 500t$, and $t = 2{\cdot}5$; find s. *e* $pv = 480$, and $v = 24$; find p.
f $x+2y = 27$, and $x = 15$; find y.

13 Using algebraic notation, write each of the following in the form of an equation, and then solve the equation.

a x metres of power cable are joined to a cable of length 200 metres. The total length is now 450 metres. Find x.

b The sides of a triangle are $4p$ cm, $3p$ cm and $5p$ cm long. The perimeter of the triangle is 84 cm. Find p.

14a A bus fare is 15p. Find the cost of 10 journeys.
b A bus fare is x pence. Find the formula for the cost C pence of $2n$ journeys.

15 Figure 1 shows four frames made of strips joined together. Copy and complete the table, where S is the number of strips and J the number of joins. State a formula connecting S and J. Hence calculate J when $S = 23$.

S	$S+3$	J	$2J$
3	.	3	6
.	.	5	.
11	.	.	.
.	.	9	.

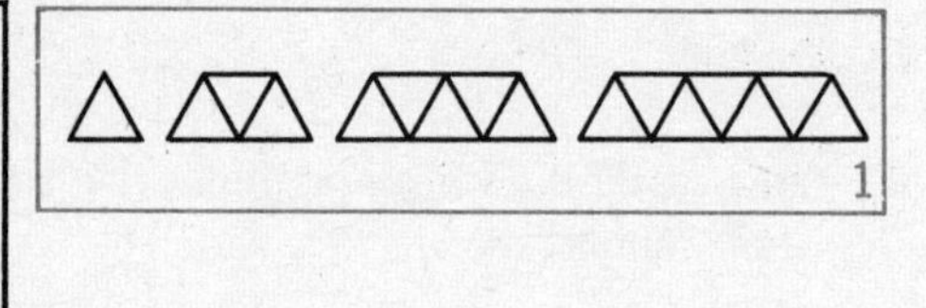

16 Complete the following:

a $12w = 3\times\ldots$ *b* $24c = 4c\times\ldots$ *c* $15pq = 5p\times\ldots$
d $5a\times 7 = \ldots$ *e* $4\times 9m = \ldots$ *f* $8m\times 9n = \ldots$
g $t\times t = \ldots$ *h* $3p\times p\times p = \ldots$

17 Write down a sequence of three consecutive numbers of which the least is n. Find n if the sum of the three numbers is 225.

18 Given that x is a variable on the set {0, 1, 2, 3, 4}, calculate all the possible values of each of the following:

a $4x+9$ *b* $12-4x$ *c* $2x^2$ *d* $3(x+5)$

19 x is a variable on the set {0, 1, 2, 3, 4, . . ., 10}. Find the solution set of each of the following inequations, and show these on a number line.

a $x+5 < 10$ *b* $x-4 > 3$ *c* $x+3 > 10$ *d* $2x+3 < 9$
e $2x-4 > 11$ *f* $4x+5x < 45$ *g* $9x-3x > 25$ *h* $1 < x < 8$

20 Find the sum of:

a 14 and 23 *b* 32 and -12 *c* 1 and -20
d $-12a$ and $12a$ *e* $-15y$ and $-23y$ *f* $-18z$ and $13z$

21 Which of these sentences is false?

a $9 > 4$ *b* $-1 < 0$ $-7 > 4$ *d* $-8 < -5$

22 Calculate the following:

a $8+(-7)$ *b* $-5+10$ *c* $-13+5$
d $4a+(-11a)$ *e* $-9c+(-14c)$ *f* $8m+(-8m)$
g $-6p+6p+(-4p)$ *h* $-3h+(-4h)+(-5h)$

23a A(a, 7), B(3, 2) and C(9, 2) are the vertices of a triangular tile ABC with equal edges AB and AC. Find a.

b If the tile is moved parallel to the y-axis so that A lies on the x-axis, write down the new coordinates of A, B, and C.

c If the tile is moved from its original position parallel to the x-axis so that A lies on the y-axis, write down the new coordinates of A, B, and C.

24 Solve the following equations, the variables being on the set of integers:

a $x+(-2) = 8$ *b* $y+4 = 15$ *c* $z+10 = 3$
d $t+(-10) = -20$ *e* $x+5 = -2$ *f* $10 = 7-x$

25 Express as a sum or difference of terms:

a $3(x+2)$ *b* $5(2x+4)$ *c* $5(4a+3b)$
d $p(2q+r)$ *e* $5(a+b+2c)$ *f* $3(3x-8)$
g $(2a-7)c$ *h* $10(2a+b)+9a$

26 Factorize each of the following:

a $5p+5q$ *b* $4n+8$ *c* $ab+a$
d a^2-4a *e* $6p-21q$ *f* x^2-8x

27 Simplify the following, giving your results in factorized form:

a $2p+2(p+q)$ *b* $5(2m+3)+15m$
c $2(x+5)+3(x+10)$ *d* $x(x+4)+x(x+6)$

28 If $x = 2a+b$ and $y = a+2b$, express each of the following in terms of a and b:

a $x+y$ *b* $2x+y$ *c* $x+2y$ *d* $3x+4y$

29 Factorize the right-hand side of the formula $s = ut+5t^2$, and hence calculate s when $u = 35$ and $t = 3$.

30 Construct a rectangular diagram to show that

$$x(a+b+c) = xa+xb+xc.$$

Exercise B

1 State whether each of the following is true or false.

a $3 \in \{1, 2, 3\}$ *b* $2 \subset \{1, 2, 3\}$ *c* $A \subset A$
d $\{p, q, r\} = \{a, c, b\}$ *e* $\emptyset \subset \{p, q\}$ *f* $A \cap \emptyset = A$

2 List the sets
A = {prime factors of 210}, B = {odd numbers less than 10},
C = {common factors of 30 and 105}. Write out in full:

a $A \cap B$ *b* $B \cap C$ *c* $A \cap C$ *d* $A \cap B \cap C$
From your results, give examples of *(1)* subsets, and *(2)* equal sets.

3 P = {all odd integers}, Q = {all even integers},
R = {all prime numbers}.

Give a universal set E, and find $P \cap Q$ and $Q \cap R$.

4 Write down the set S of numbers on a clockface, and list these subsets:

a each number is odd *b* each member is a multiple of 3
c twice each number is in S *d* the square of each number is in S.
Write down the intersections of subsets ***a*** and ***b***, and ***c*** and ***d***.

5 Find the solution set of each of the following inequations, the variables being on the set {1, 2, 3, 4, 5, 6}:

a $x+7 < 12$ *b* $y+6 > 10$ *c* $2z-1 \leqslant 9$

d $n^2 < 15$ *e* $x+\frac{1}{x} > 2$ *f* $p^2+4 \leqslant 5p$

6 The variables in this question are on the set $S = \{1, 2, 3, 4, 5\}$.
Find replacements, if you can, which make true sentences from each of the following open sentences:

a $x+y = 7$ *b* $x+y = 12$ *c* $x-y = y$
d $xy = y$ *e* $x+y > xy$ *f* $x^2+6 \geqslant 5x$

7 The relation between t°C (Celsius) and T K (Kelvin) is given by the equation $T = t+273$. Calculate:

a T when $t = 100$, and when $t = -100$
b t when $T = 500$, and when $T = 0$.

8 The distance d km travelled by a rocket in t hours at an average speed of v km/h is given by the formula $d = vt$. Calculate:

a d when $v = 7000$ and $t = \frac{3}{4}$ *b* v when $d = 10000$ and $t = 4$.

9 A piece of wire 75 cm long is bent to form three sides of a rectangle of lengths h cm, w cm and h cm.

a Write down an equation in h and w.

b Find w given that $h = 10$, and find h given that $w = 45$.

10a What integer added to -28 gives zero?

b Give the next three numbers in the sequence 10, 7, 4, 1, ...

c What is the rise in temperature from -5°C to 8°C?

d Write down the result of the following: $-5a + 7a + (-8a)$.

e Given that $p = 5$ and $q = -4$, calculate the value of $3p + 2q$.

11 If $p = -2$, $q = 7$ and $r = -5$, calculate:

a $p+q$ *b* $p+(-q)$ *c* $p+r$ *d* $q+(-r)$ *e* $q+r$

12 Assuming that x is a variable on the set of integers, solve the following equations:

a $x+4 = 1$ *b* $x+(-4) = 2$ *c* $x+(-10) = -3$

d $x+15 = 0$ *e* $-x = 14$ *f* $-x = -8$

g $-5+x = -5$ *h* $-5+x = 0$

13 Figure 2 shows a *magic square* in which the sum of the numbers in each row, and in each column, and in each diagonal, is the same. Can you fill in the missing numbers?

		2
	0	
−2		3

2

$x+(-y)$	$x+y+(-z)$	$x+z$
$x+y+z$	x	$x+(-y)+(-z)$
$x+(-z)$	$x+(-y)+z$	$x+y$

3

14 Figure 3 gives a plan for making magic squares. x, y, z represent whole numbers. Add up each row, column, and diagonal in turn. What answer do you obtain? Replace x, y, z by integers and add up the rows, columns and diagonals.

15 In a certain test, pupils score 2 marks for a correct answer and score -1 mark for an incorrect answer. If one pupil had 38 questions correct and 12 wrong, what was his score? Another pupil had 25 correct and 25 wrong; what was his score?

16 Find the solution set of each of the following inequations, given that the variables are on the set of whole numbers:

a $2x+7 < 25$ *b* $5x-10 \leqslant 22$ *c* $18-3x > 2$

17 Copy and complete each of the following by placing one of the symbols $<$, $=$, $>$ between the pairs:

a $8 \ldots 6$ *b* $-2 \ldots 5$ *c* $3 \ldots -3$
d $-7 \ldots -4$ *e* $(-5+0) \ldots (-5)$ *f* $4\times 0 \ldots (-1)$

18 A square plate of side 10 cm has two rectangular pieces x cm by 2 cm removed from it.

a Find a formula for the area A cm² of the remaining part.
b If $A < 80$, write down an inequation in x, and hence show that $x > 5$.
c Obtain another inequation in x by considering the original length of the plate, and hence complete the *compound* inequation $5 < x \leqslant \ldots$

19 Simplify:

a $5x^2+6x^2-x^2$ *b* $4(m-3)+3(7+m)$ *c* $12\left(\frac{a}{2}+\frac{b}{3}+\frac{c}{6}\right)$
d $\frac{1}{2}(8a-10b)$ *e* $a(a+3)+3a(a+4)$ *f* $10ab-7ab+5ba+8c$

20 Copy and complete this multiplication table:

$\times$	a	$3b$	$2c$
$2a$	$2a^2$	$6ab$	$4ac$
b	.	.	.
$4c$	.	.	.

Check the accuracy of your entries by considering the replacements $a = 2$, $b = 1$ and $c = 3$.

21 Factorize:

a $4n+6$ *b* $2ah+2bh+2ch$ *c* $6p^2+4p$

22 All the pupils in a class of 35 take French or German or both. 29 take French, and 15 take German.

a If x pupils take French *and* German, how many pupils take (*1*) French only (*2*) German only?
b By first forming an equation in x, find how many take French and German.
c Illustrate the problem by means of a Venn diagram.

23 Two operations $\triangle$ and $*$ are defined on a certain number system. $\triangle$ is distributive over $*$ and $*$ is distributive over $\triangle$. Given that a, b, c are numbers belonging to this system, complete the following:

a $a \triangle (b * c) = \ldots\ldots$ *b* $a * (b \triangle c) = \ldots\ldots$

24 A variable x is defined on the set

$$X = \{-5, -4, -3, -2, -1, 0, 1, 2, 3, 4, 5\}.$$

Find the solution set of each of the following inequations:

a x is less than zero *b* $-4 < x < 4$.

Write down the intersection of these solution sets.

25 Using squared paper, draw a set of axes with origin O. Mark the points A(5, 0), B(3, 4), C(−3, 4), and join AO, BO, CO. Produce AO, BO, CO their own lengths to A′, B′, C′ so that OA′ = AO, etc.

a Write down the coordinates of A′, B′, C′.

b What name is given to the figure obtained by joining the points A, B, C, A′, B′, C′, A? Can you find the area of this figure?

26 If $x \odot y$ means 'Add 2 to the sum of x and y', calculate $x \odot y$ in the following cases:

a $x = 5, y = -7$ *b* $x = -4, y = -1$ *c* $x = 2a, y = -5a$

Find three different replacements for x and y such that $x \odot y = 0$.

Geometry

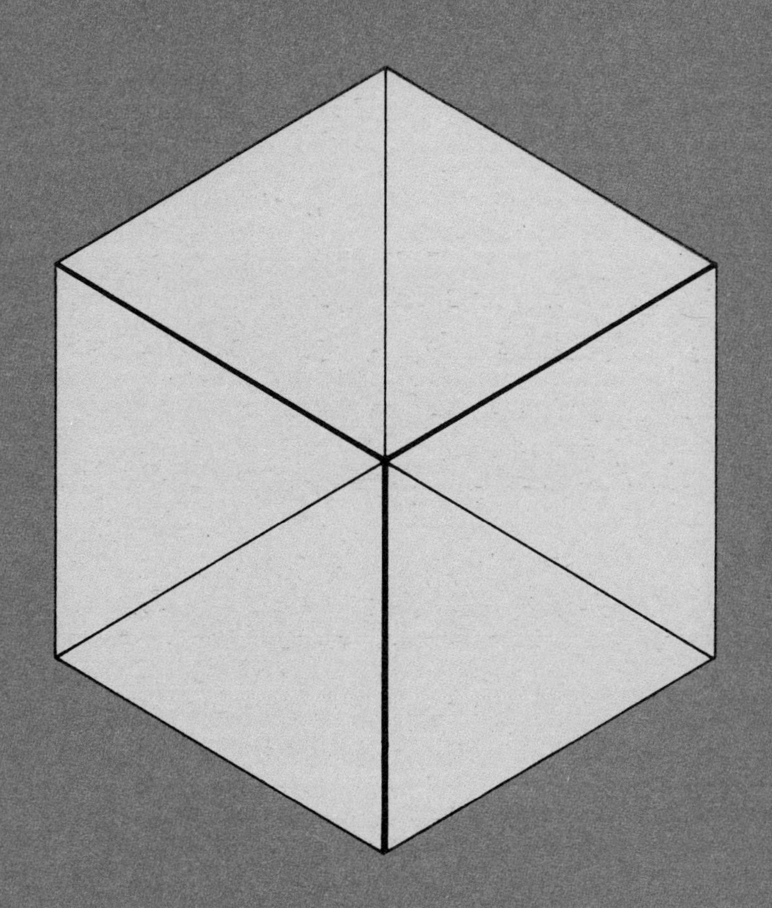

Geometry

Rectangle and Square

1 *Looking back, and looking ahead*

Revision Exercise 1 Tilings

1 Figure 1 shows a tiling of squares. Make a tracing of this, and by moving the tracing in different directions extend the tiling over your tracing paper.

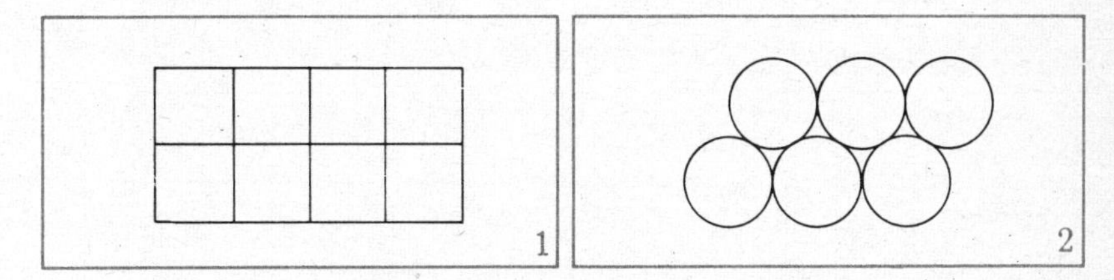

2 Figure 2 shows a tiling of circles. Make a tracing of this, and by moving your tracing check that the circles are all exactly the same. They are said to be *congruent*.
As in question *1*, extend the tiling over your tracing paper.

3 Repeat question *2* for Figures 3 and 4.

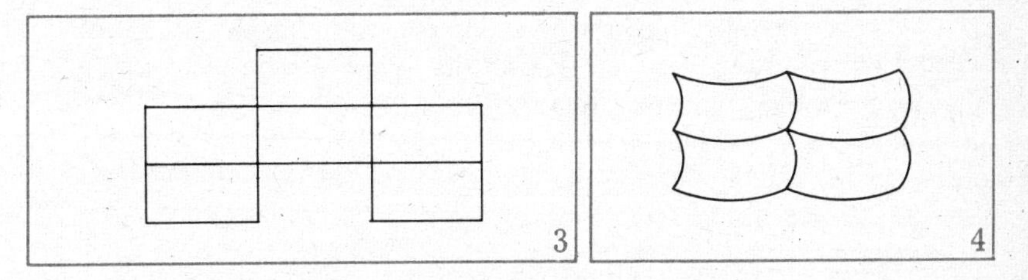

Revision Exercise 2 Parallel lines

In Book 1 we met straight lines which kept the same distance apart. They were called *parallel lines*.

1 Which of the pairs of lines in Figure 5 do you think are parallel?

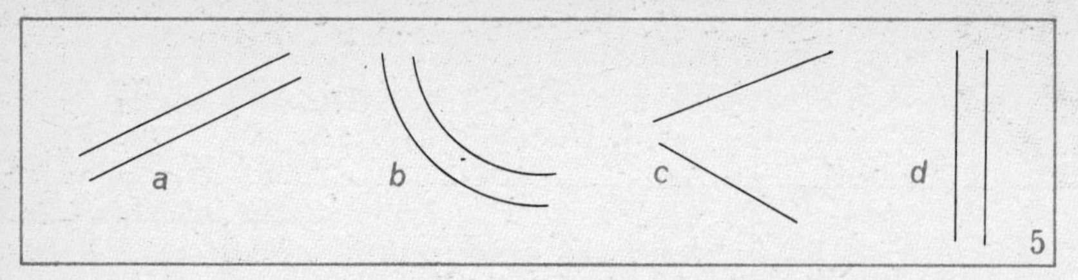

2 Pick out sets of parallel lines in Figures 1, 2, 3 and 4 where possible.

3 List parallel lines:

a in the classroom *b* outside *c* at home

4 In Figure 6, CD is parallel to XY, and AB is perpendicular to XY.

a Make a tracing of the figure.

b Continue AB on your tracing until it meets XY.

c By sliding your tracing, check that all the angles on it are right angles.

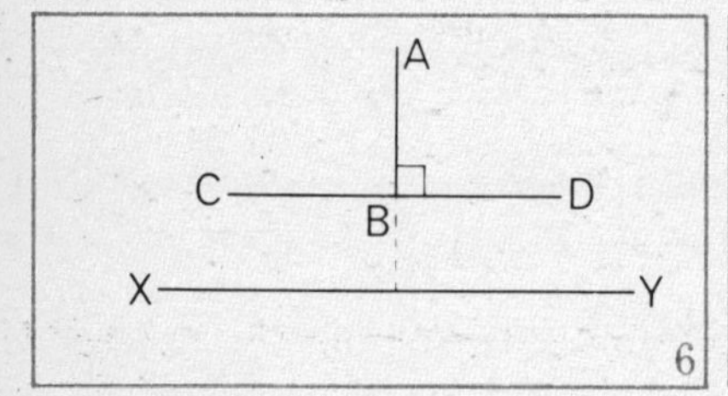

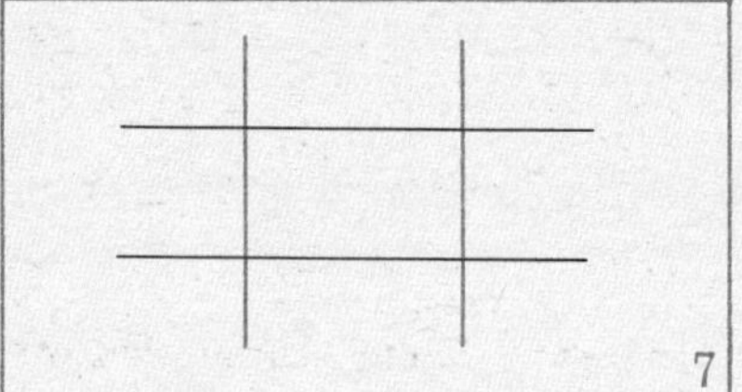

5 In Figure 7 the pair of black parallel lines cuts the pair of green parallel lines at right angles.

a Make a tracing of the figure.

b Move the tracing along the black lines a distance equal to that between the green lines, and extend the figure to the right and left.

c Repeat ***b*** for movements along the green lines.
You now have a *rectangular tiling*.

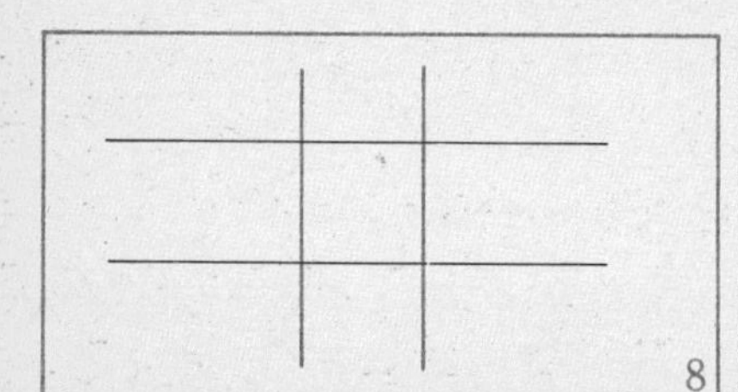

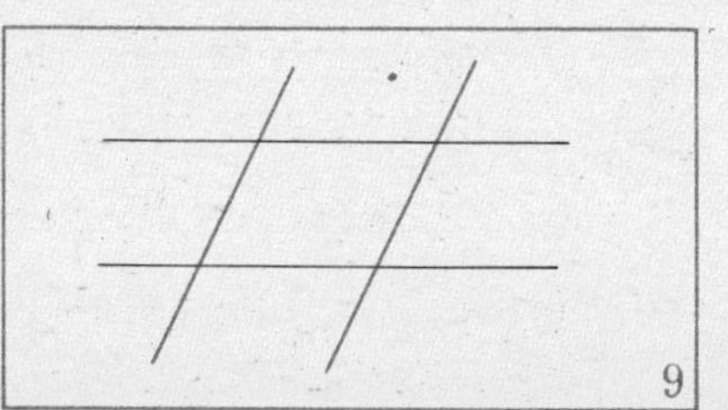

6 Repeat question ***5*** for Figures 8 and 9, to obtain a *square tiling* and a *parallelogram tiling*.

Revision Exercise 3 Half turns

1 Make a tracing of Figure 10. Rotate the tracing about O until the two shapes change places.
This movement is called a *half turn*.

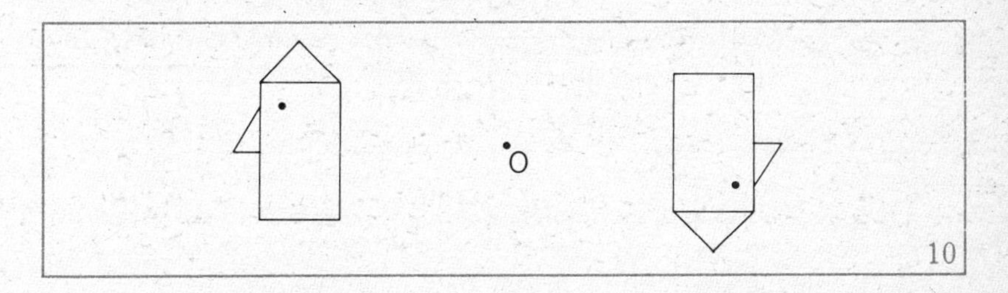

10

2 Trace the shapes shown in Figures 11 and 12. Then rotate your tracings through a half turn about O. What happens in each case?

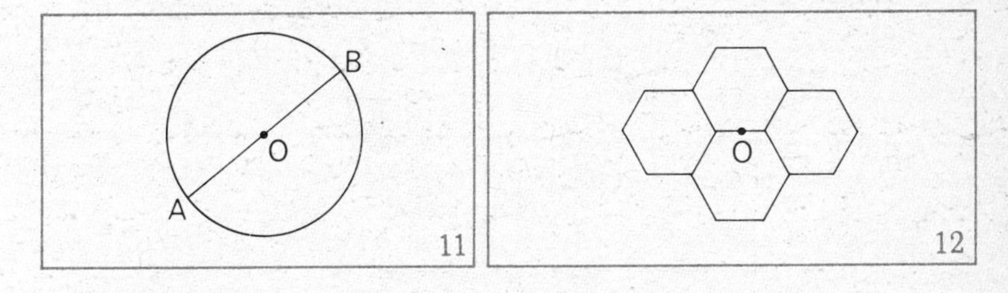

11 12

3 a If you are facing north, and you make a half turn, in what direction will you be facing?

b Repeat ***a***, starting from east, then from north east.

4 In Figure 13, P and Q are points on two parallel lines. O is the midpoint of PQ.
What happens to P, Q and the parallel lines if the figure is given a half turn about O?

5 In Figure 14, R and S are points on two parallel lines, and X is the midpoint of RS. If the figure is given a half turn about X, what happens to:

a X *b* R and S *c* XR and XS *d* the parallel lines?

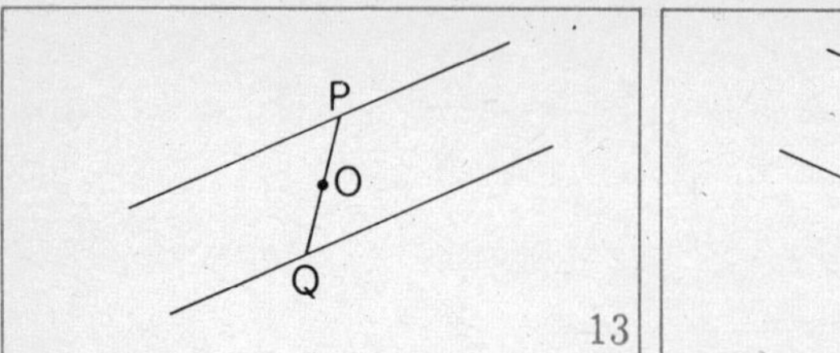

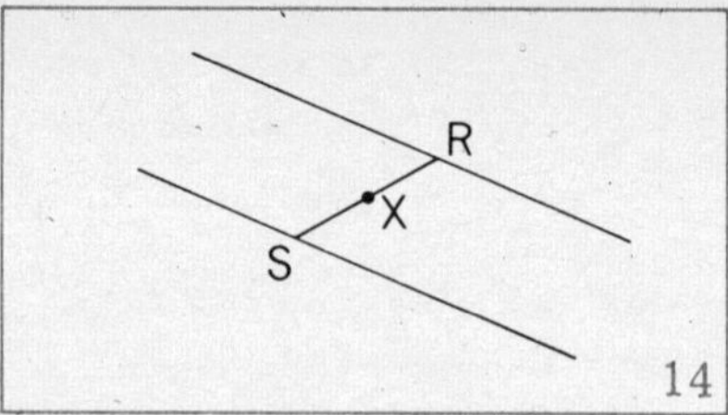

2 Fitting shapes

Figure 15 shows another version of the child's posting toy described in the section on 'Angle as shape' in Book 1. The idea is to replace the shapes in the spaces from which they have been cut in as many different ways as possible. Each of the shapes has four straight sides.

You can turn the shapes round, or turn them over. A card cut-out, or tracing, of each shape might help you in Exercise 4.

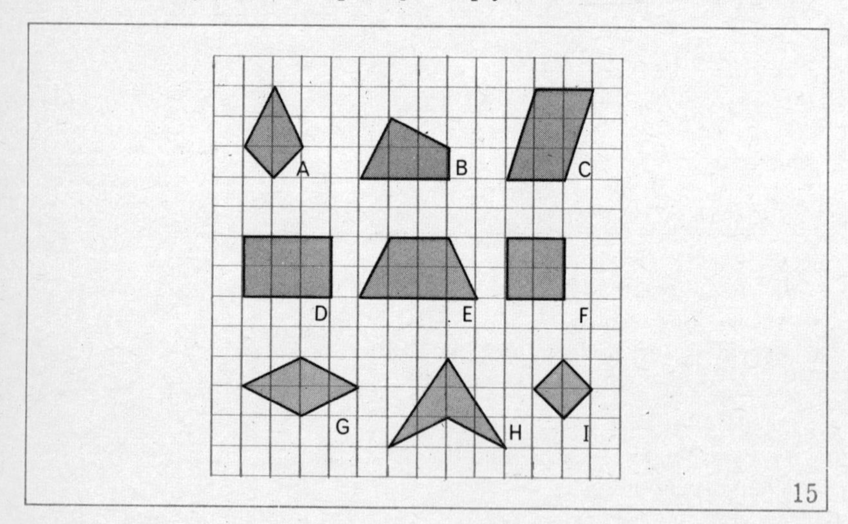

Exercise 4

1 Copy the table below, and fill in the total number of ways in which each shape in Figure 15 can be fitted into its space.

Shape	A	B	C	D	E	F	G	H	I
Number of ways of fitting									

2 Which shape will fit in only one way?

3 Which shapes will fit in exactly two ways?

4 Which shapes will fit in exactly four ways?

5 Which shapes will fit in more than four ways?

3 The rectangle

In Exercise 4 you should have found two shapes, D and G, that fit their outlines in exactly four ways. In the case of shape D each corner changes position from one fitting to another; this does not happen with shape G. Check the fittings of D and G again to make sure that you see what is meant. Shape D is called a *rectangle*.

From the work above, and from experience earlier in the course, we decide to make the following assumptions, or *axioms*, about all rectangles.

Axiom 1

A rectangular tile can be fitted into its outline in the four ways shown in Figure 16. Note that each corner changes position from one fitting to another.

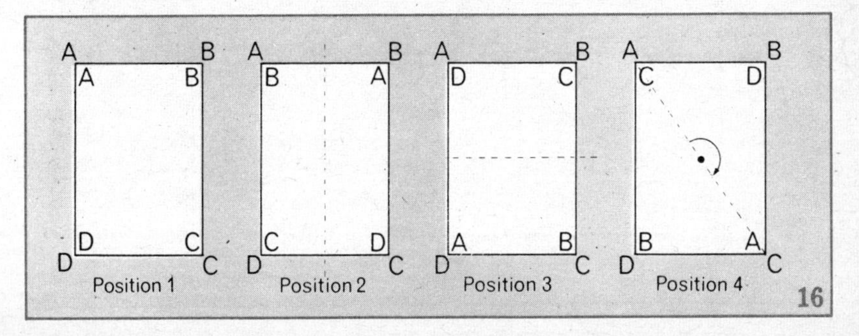

Axiom 2

Congruent rectangular tiles can be fitted together, side by side, so as to cover a plane (flat) surface without leaving any gaps, as shown in Figure 17.

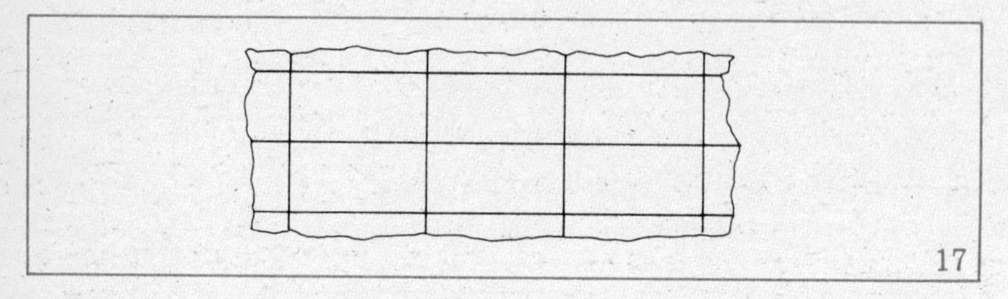

17

*From these **assumptions** we are now going to **deduce** certain properties of the rectangle. This kind of thinking, from an assumption to a deduction, is very important in mathematics.*

Example 1

In Figure 18, Position 1, a rectangular tile ABCD coincides with its outline ABCD as shown.

Assumption.—The tile can be picked up and turned over about PQ so that it fits its outline as shown in Position 2.

What can we deduce about AD and BC?

A goes to B, which can be written $A \rightarrow B$.
D goes to C, which can be written $D \rightarrow C$.
AD goes to BC, which can be written $AD \rightarrow BC$.

We deduce that $AD = BC$.

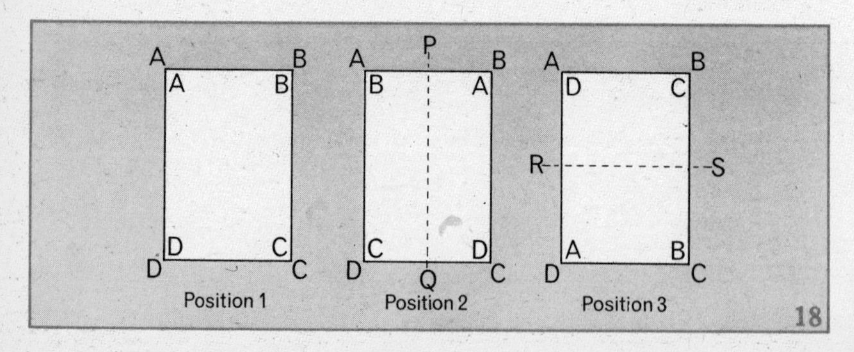

18

Example 2

Assumption.—The tile can be picked up and turned over about RS so that it fits its outline as shown in Position 3.

What can we deduce about AB and DC?

A goes to D, i.e. A → D.
B goes to C, i.e. B → C.
AB goes to DC, i.e. AB → DC.

We deduce that AB = DC.

The opposite sides of a rectangle are equal.

Exercise 5

1 In Figure 19, PQRS is a rectangle.

a Name two pairs of equal sides.
b What are the lengths of PS and SR?

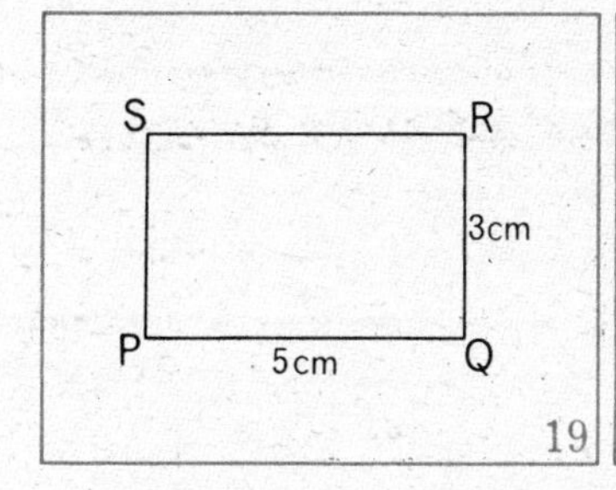

19

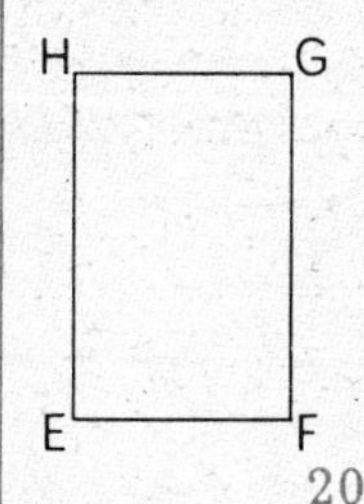

20

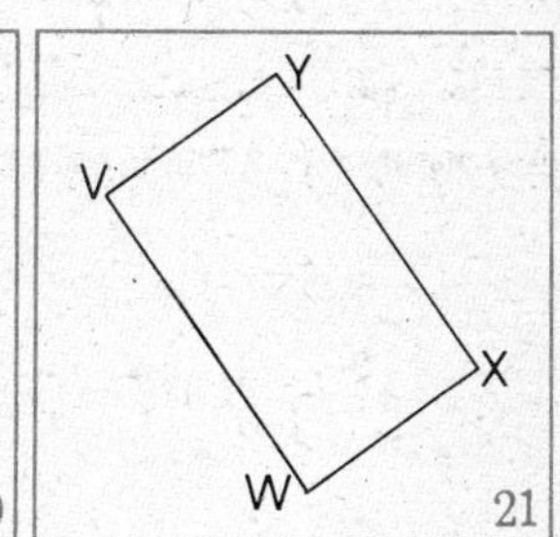

21

2 In Figure 20, EFGH is a rectangle.

a Which side is equal in length to (*1*) EF (*2*) FG?
b If EF = 2·8 cm and EH = 4·5 cm, give the lengths of the other sides.

3 a In rectangle VWXY in Figure 21 name two pairs of equal sides.
b Two of the sides are 8 mm and 17 mm long. List the names and lengths of all the sides.

4 a Make a list of rectangular objects inside and outside the classroom.
b Which of these rectangular objects are made so as to fit as they move?

5 Copy Figure 18, but replace the letters A, B, C, D by K, L, M, N. Using the methods of Worked Examples 1 and 2, show that KN = LM and KL = NM.

6 Suppose you had to make a model of a rectangular window frame using four 'Meccano' strips for the sides. What would have to be true about the lengths of the strips?

* * * * *

Example 3

Assumption.—The rectangular tile shown in Figure 22 can be turned over about PQ to fit its outline as in Position 2.

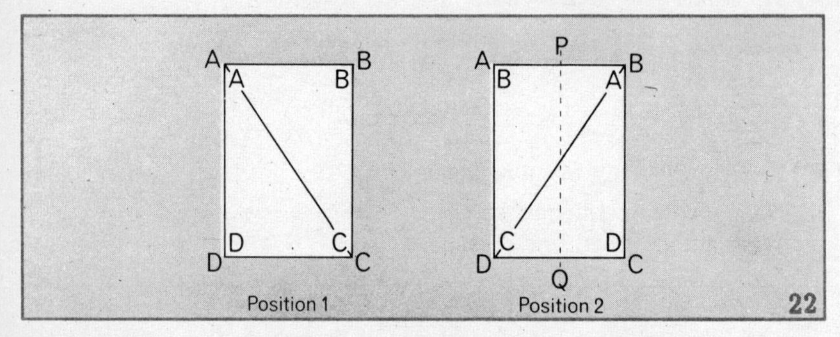

What can we deduce about the *diagonals* AC and BD?

A → B, and C → D, so that AC → BD.

We deduce that AC = BD.

The diagonals of a rectangle are equal.

Exercise 6

1 a Sketch the rectangles shown in Figures 19, 20 and 21.
b Draw in their diagonals.
c Copy and complete:
Diagonal QS = diagonal ...; diagonal EG = diagonal ...;
diagonal ... = diagonal VX.

2 a Sketch a rectangle, and name it STUV.
b What are the names of its diagonals?
c List three pairs of equal lines in your drawing.

* * * * *

Example 4

Assumption.—The rectangular tile in Figure 23 can be given a half turn so that it fits its outline as in Figure 23 (iv).

Can we deduce that if the half turn is about O, O is the point of intersection of the diagonals?

Figure 23 (ii) shows us that A → C, and O remains on AC.

Figure 23 (iii) shows us that B → D, and O remains on BD.

We deduce that O is the point of intersection of the new positions of AC and BD, and is at the same point as before.

After a half turn of the rectangle, the point of intersection of the diagonals is unchanged.

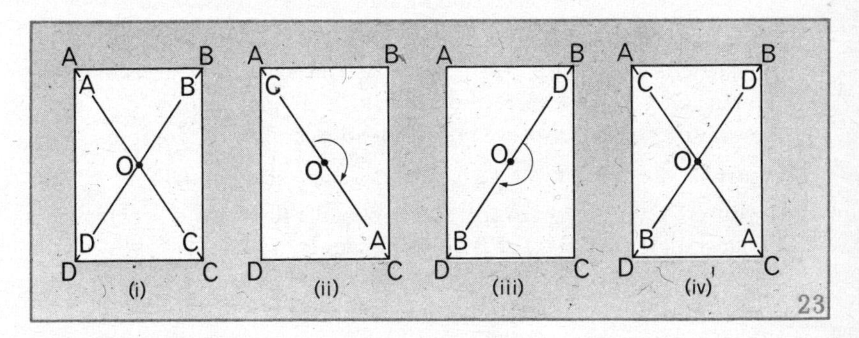

23

Further deductions about the diagonals.

After a half turn from (i) to (iv),

O → O, A → C, OA → OC.

We deduce that OA = OC.

In the same way, OB = OD.

The diagonals of a rectangle bisect each other.

Exercise 7

1 Use the method described above to deduce that OB = OD.

2 Explain how you would turn the rectangle over to fit in one of the four possible ways in order to deduce that OA = OB and OC = OD. Write out the deduction.

3 Write down a statement about the lengths of OA, OB, OC and OD.

4 a Sketch a rectangle PQRS, with its diagonals intersecting at O.

b Name four equal lines in your drawing.

c Name the two pairs of equal sides.

5 Repeat question *4* for a rectangle TUVW.

6 a Sketch a rectangle EFGH, with its diagonals intersecting at Z.
b Name the two diagonals. Are their lengths equal?
c If EZ = 2 cm, write down the length of three other lines in the figure.
d Name one line equal in length to EF, another equal to EH, and another equal to EG.

7 a Draw a line PQ 10 cm long, and mark its midpoint X.
b Draw another line AB 10 cm long having the same midpoint X.
c Join AP, PB, BQ and QA. What kind of shape is APBQ?

8 If two equal rods are bolted at their centres, and their ends are joined by elastic, what shapes are formed by the elastic as the rods are rotated about their centres? (Some sketches may help.)

9 How many rectangles can be drawn with diagonals 10 cm long?

10 Figure 24 shows a bicycle wheel with spokes to the centre O.
a If you drew in the necessary lines, would ALFE be a rectangle?
b Would ALGF, ACGI, LIEB be rectangles?
c Name a rectangle which has LJ as one side; LH as one side; LC as one side; LA as one side.

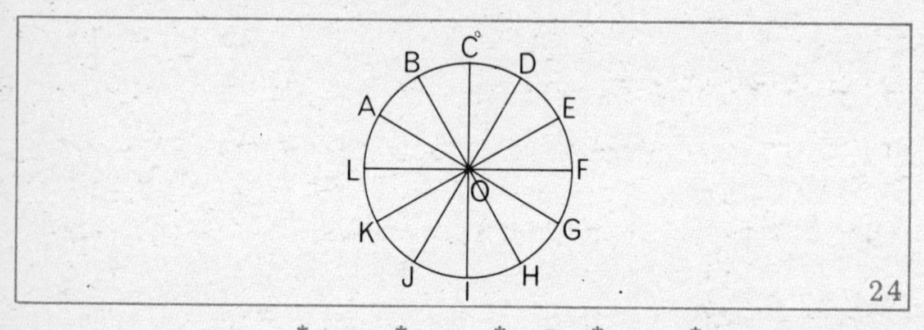

24

* * * * *

Example 5

Assumption.—The rectangular tile in Figure 25 can be turned over about PQ to fit its outline as shown in Position 2.

What can we deduce about angles A and B, and angles C and D?

Angle A ⟶ angle B
So angle A = angle B

Angle D ⟶ angle C.
So angle D = angle C.

In the same way, from Position 3, *we can deduce that*:

angle A = angle D, and angle B = angle C.

Combining these results, angle A = angle B = angle C = angle D.

All the angles of a rectangle are equal.

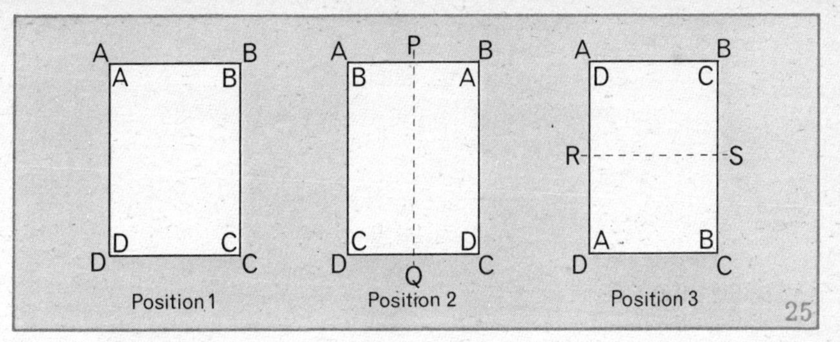

25

Example 6

Assumption.—Congruent rectangular tiles can be fitted together side by side so as to cover a plane (flat) surface without leaving any gaps.

What can we deduce about the size of the angles of a rectangle?

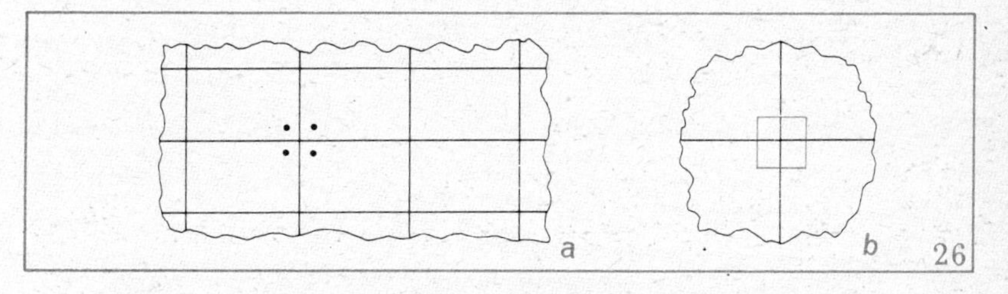

26

Since all the angles of a rectangle are equal, the four angles marked with dots in Figure 26*a* are equal. Figure 26*b* reminds us that if four equal angles fit round a point without leaving gaps, then each is a right angle.

All the angles of a rectangle are right angles.

Example 7

Assumption.—Any of the rectangular tiles in Figure 26*a* can slide along a row of the tiles, or up and down a column.

What can we deduce about the directions of the sides of the rectangle?

The opposite sides of the rectangle remain the same distance apart, and so must be parallel.

The opposite sides of a rectangle are parallel.

* * * * *

Summary—A rectangle:
fits its outline in four ways;
has its opposite sides equal and parallel;
has equal diagonals, which bisect each other;
has four right angles.

Exercise 8

1 a Name all the right angles in Figures 19, 20 and 21 (page 97).
b Name all the pairs of parallel lines in Figures 19, 20 and 21.

2 a Draw and name two rectangles on squared paper.
b List the names of the right angles, and the parallel sides.

3 What is the sum in degrees of the four angles of a rectangle?

4 The markings in Figure 27 show the properties of the rectangle that we have deduced so far.

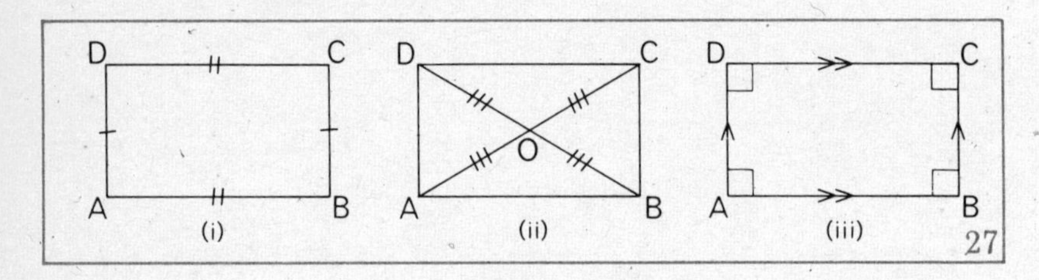

27

a Copy Figure 27, and list the sets of equal lines and angles marked.
b State the two axioms about the rectangle that enabled us to deduce all these properties.

5 Copy Figure 27 (ii), and mark in:
a two pairs of vertically opposite angles (mark them v and w)
b four pairs of complementary angles (mark them c and d)

6 Draw a rectangle ABCD, with its diagonals crossing at O, as in Figure 27 (ii). Mark $\angle AOB = 100°$ and $\angle OAB = 40°$.
By thinking of the 'fitting' properties of the rectangle, and the angle properties that you know, fill in the sizes of as many angles as you can.

7 Repeat question *6* if $\angle AOB = 110°$ and $\angle OAB = 35°$.

8 Repeat question *6* if $\angle OCD = 28°$ and $\angle BOC = 56°$.

4 Coordinates

Exercise 9A

1 a On a sheet of squared paper draw the rectangle with vertices A(1, 2), B(5, 2), C(5, 8), D(1, 8).

b List the lengths of its sides.

c Draw the diagonals, and write down the coordinates of their point of intersection.

2 Repeat question ***1*** (on the same diagram) for the rectangle E(7, 0), F(15, 0), G(15, 5), H(7, 5).

3 a Plot the points R(2, 2), S(12, 2), T(12, 8), U(2, 8), and join them in all possible ways.

b What kind of shape is RSTU?

c Name a line equal in length to RS, another equal to RU, and another equal to RT.

d Calculate the area of RSTU.

e Write down the coordinates of the point of intersection of the diagonals.

4 A is the point (5, 2), and C is (12, 11).

a Draw a rectangle ABCD with its sides AB and DC parallel to the x-axis, and its sides AD and BC parallel to the y-axis.

b Write down the coordinates of B and D.

c Calculate the area of the rectangle.

5 Repeat question ***4*** for the given points A(15, 5), and C(25, 10).

6 a K is the point (3, 1), L is (8, 1), M is (8, 3), and KLMN is a rectangle. What are the coordinates of N?

b In the same diagram, NMPQ is a rectangle with the same area as KLMN. Draw the rectangle, and write down the coordinates of P and Q.

7 a A is (2, 0), B is (6, 0), C is (6, 5), and ABCD is a rectangle. What are the coordinates of D?

b AC is joined, and produced its own length to F. What are the coordinates of F?

c Rectangle AEFG is drawn, with AE on the x-axis. Write down the coordinates of E and G.

d What is the ratio of the area of rectangle ABCD to the area of rectangle AEFG?

Exercise 9B

1 a On squared paper draw the rectangle with vertices A(0, 3), B(8, 3), C(8, 7), D(0, 7).

b Find the area of the rectangle, and the point of intersection of its diagonals.

2 P is the point (3, 3), Q is (7, 3), and M is (5, 6). Draw the rectangle PQRS whose diagonals intersect at M. What are the coordinates of R and S?

3 a Draw the rectangle A(3, 4), B(12, b), C(c, 10), D(p, q) on squared paper with its sides AB and DC parallel to the x-axis, and AD and BC parallel to the y-axis; b, c, p, q represent certain numbers.

b What are the numbers represented by b, c, p and q?

c If the sides of the rectangle are all produced to cut the x-axis at K and L, and the y-axis at M and N, give the coordinates of K, L, M and N.

4 PQRS is a rectangle. P is the point (0, 4), Q is (4, 0), and the mid-point of RS is T(10, 10),

a Draw the rectangle on squared paper, and give the coordinates of R and S.

b Write down the coordinates of the point where the diagonals intersect.

c If RU is drawn perpendicular to the x-axis, and RV is drawn perpendicular to the y-axis, give the coordinates of U and V.

5 A is the point (3, 2), and C is (9, 6).

a Draw the rectangle ABCD with its sides AB and DC parallel to the x-axis and AD and BC parallel to the y-axis.

b Produce AB its own length to E, and complete the rectangle BEFC. Write down the coordinates of F.

c Draw rectangles CFGH and CHKD congruent to BEFC.

d Give the area of the rectangle AEGK, and the point of intersection of its diagonals.

e What are the coordinates of the next three tile corners in the sequence A, C, G, ...?

6 *a* P is the point (10, 10). Draw four congruent rectangles, each of which has one vertex at P and its sides parallel to the x- and y-axes. Give the coordinates of all their vertices.

b Draw four more rectangles congruent to the first four, each of which also has one vertex at P and its sides parallel to the axes. Give the coordinates of all their vertices.

5 The Square

In Figure 15 on page 94 you should have found two examples, F and I, of a shape that fits its outline in eight ways. This shape is called a *square*.

Axiom

A square tile can be fitted into its outline in the eight ways shown in Figure 28.

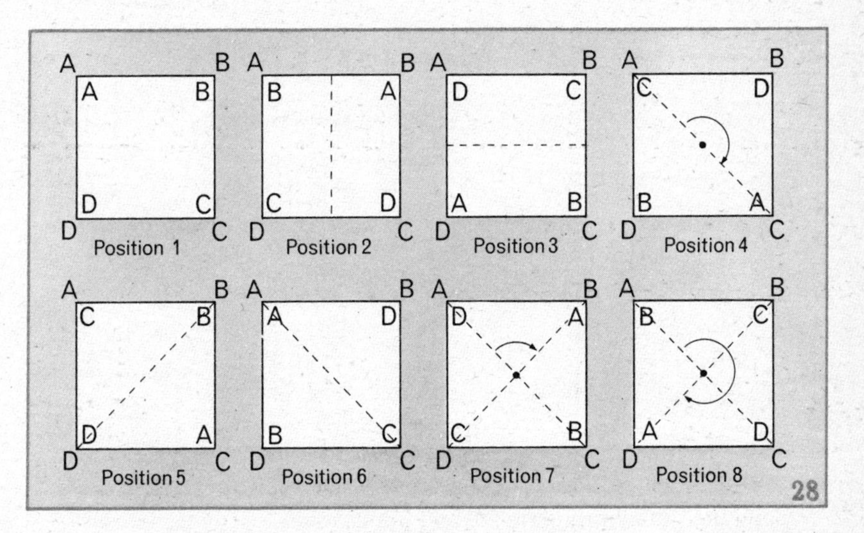

Positions 1, 2, 3 and 4 are the same as for a rectangle. Positions 5, 6, 7 and 8 are not shared with the rectangle, so the square is a special kind of rectangle.

Notice that Positions 7, then 4, then 8, can be obtained from Position 1 by giving the square quarter turns about its centre.

Exercise 10

1 Using the assumptions which allow the square to fit its outline when it is turned over about the diagonals BD and AC, deduce that:

a all the sides of a square are equal
b BD bisects the angles of the square at B and D
c AC bisects the angles of the square at A and C.

2 Using the 'quarter turn' assumptions about the square fitting its outline, deduce that the diagonals of a square intersect at right angles at O.

3 Show that the diagonals of a square are equal.

4 Show that the diagonals of a square bisect each other.

* * * * *

Figure 29 shows the properties of a square; they can be recalled by thinking of the ways in which a square fits its outline.

Exercise 11

1 Copy Figure 29, and mark in all the properties shown.

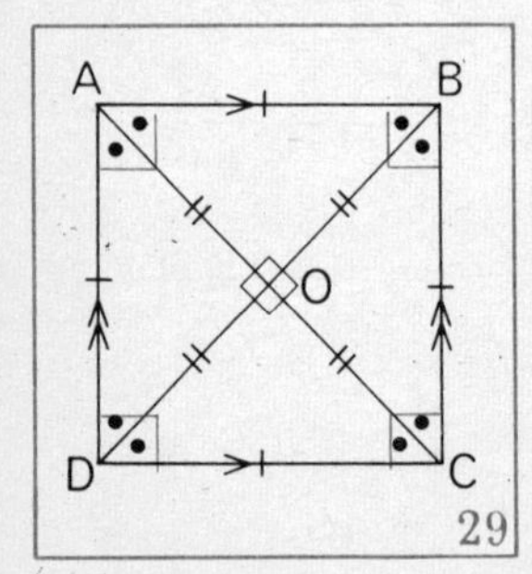

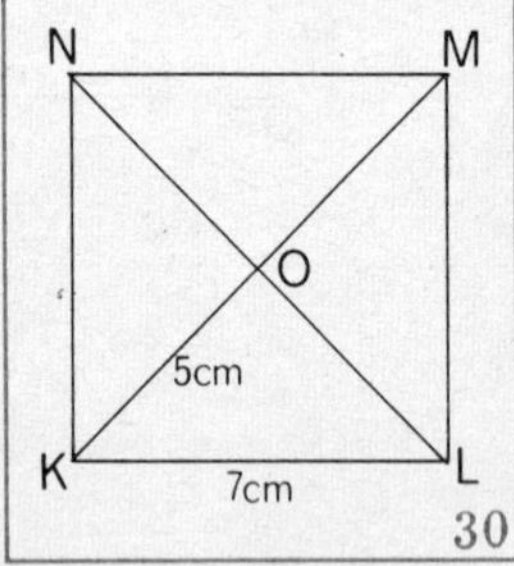

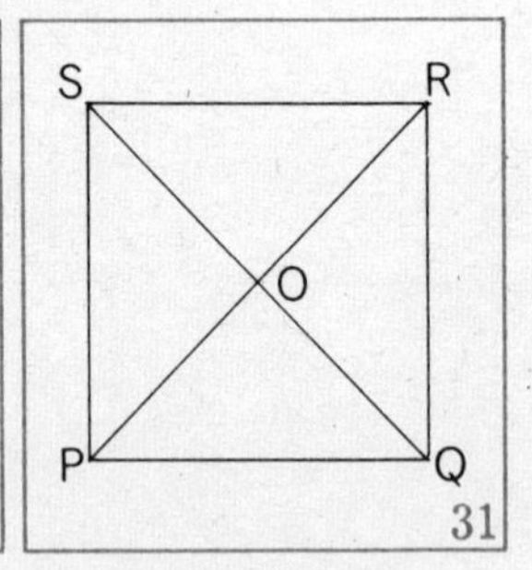

2 Copy the square in Figure 30, and fill in:

a the lengths of all the lines *b* the sizes of all the angles.

3 List the following from Figure 31.

a A set of four equal lines.
b Another set of four equal lines.

c A set of four right angles.
d Another set of four right angles.

How many angles of 45° are there in the diagram?

4 On squared paper plot the points P(2, 2), Q(8, 2), R(8, 8), S(2, 8).

a Is PQRS a square? Is it a rectangle?
b Write down the coordinates of the point of intersection of the diagonals.

5 On squared paper draw a square with diagonals 8 cm long. (Draw the diagonals first, then complete the square.)

6 A is the point (4, 1) and B is (10, 1). Complete the square ABCD on squared paper.

Write down the coordinates of C and D, and of the point where the diagonals intersect.

7 In your diagram for question *6* draw squares BCFE, DCHK and CFGH which are congruent to ABCD.

What are the coordinates of G, and of the next three tile corners in the sequence A, C, G, ...?

8 a Draw the square A(3, 2), B(15, 2), C(15, 14), D(3, 14) on squared paper.

b Mark E, F, G, H, the midpoints of AB, BC, CD, DA, and fill in their coordinates.

c Join EF, FG, GH, HE. Mark K, L, M, N, the midpoints of EF, FG, GH, HE, and fill in their coordinates. Can you make any more 'squares within squares'?

Exercise 12

Say whether each of the following is true or false. It will help if you think of the ways in which rectangles and squares fit their outlines.

1 A rectangle fits its outline in four ways.

2 All the angles of a rectangle are right angles.

3 All the sides of a rectangle are equal.

4 The diagonals of a rectangle are equal.

5 All the sides of a rectangle are parallel.

6 A square fits its outline in eight ways.

7 Each angle in a square is 90° in size.

8 All the sides of a square are equal.

9 The diagonals of a square bisect the angles of the square.

10 The set of all squares is a subset of the set of all rectangles.

6 Axes of symmetry

An *axis of symmetry*, or a *line of symmetry*, is a line about which a shape can be turned so as to fit exactly into its outline.

The axes of symmetry are shown by dotted lines in Figure 32.

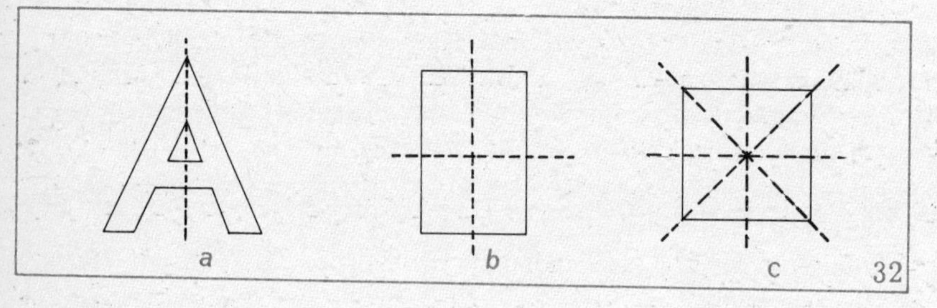

Exercise 13

1 Draw larger copies of the letters A, T, Y, H, O, and show their axes of symmetry in colour.

2 Draw four more capital letters that have at least one axis of symmetry, and show the axes in colour.

3 Draw a rectangle on squared paper, and show its axes of symmetry.

4 Draw a square on squared paper, and show its axes of symmetry.

5 On squared paper draw another shape with straight sides which has an axis of symmetry.

6 Repeat question **5** for shapes with more than one axis of symmetry.

7 A(2, 0), B(10, 0), C(10, 14), D(2, 14) is a rectangle. Which of the following is true for all points on one of its axes of symmetry?

a $x = 6$ *b* $x = 7$ *c* $y = 6$ *d* $y = 7$

8 O(0, 0), A(6, 0), B(6, 6), C(0, 6) is a square. Which of the following is true for all points on one of its axes of symmetry?

a $x = 0$ *b* $x = 3$ *c* $y = 3$ *d* $y = 6$ *e* $x = y$

7 Drawing rectangles and squares

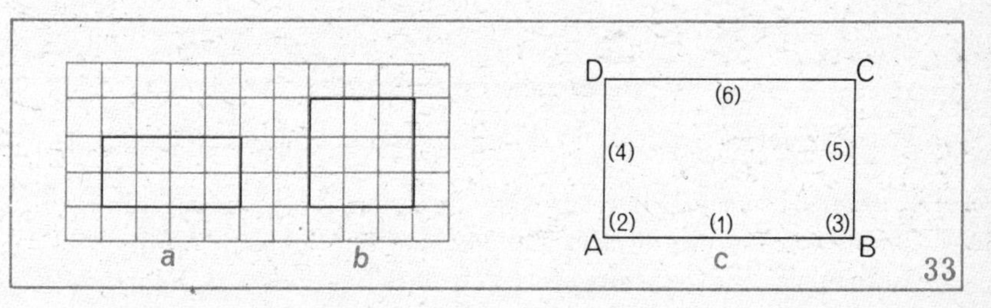

Rectangles and squares can be drawn easily on squared paper using the grid lines, as shown in Figure 33*a* and *b*.

A rectangle with length 5 cm and breadth 3 cm may be constructed on plain paper as shown in Figure 33*c*, as follows:

(*1*) Draw the base AB 5 cm long.

(*2*), (*3*) Using a protractor or setsquare, make angles of 90° at A and B.

(*4*), (*5*) Using a ruler or compasses make AD and BC each 3 cm long.

(*6*) Join DC.

Exercise 14

1 Draw the following rectangles and squares on 5-mm squared paper, and calculate their areas:

Length (cm)	4	5	7	6	4·5
Breadth (cm)	3	5	1	2·5	4·5

2 Construct the following rectangles and squares on plain paper, as described above:

Length (cm)	6	5	8	10	7·5
Breadth (cm)	5	2	8	4·5	7·5

Which are squares?

3 Construct a square with diagonals 10 cm long. Do this on squared paper, and then on plain paper; which is easier?

4 a Construct a rectangle with diagonals 10 cm long.

b How many different shapes of rectangle can you draw with diagonals 10 cm long?

c If in addition the acute angle between the diagonals is 40° draw the rectangle.

8 The cube and cuboid

In Book 1 we saw that the *vertical* direction at a point is given by a 'plumb line' as shown in Figure 34*a*, and that a *horizontal* direction can be checked by means of a 'spirit level' as shown in Figure 34*b*.

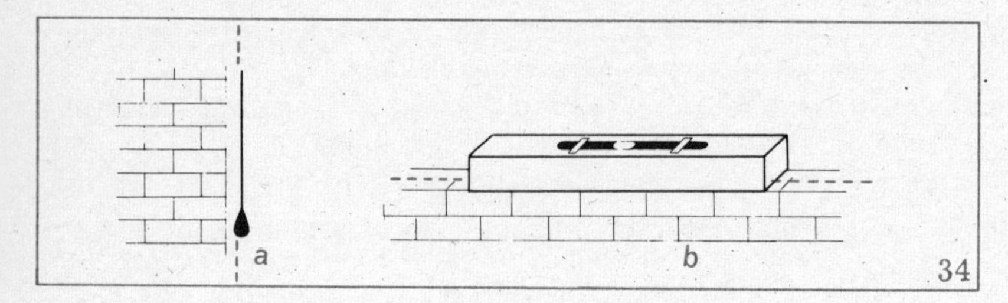

Exercise 15

1 *Either*: Use a plumb line and a spirit level to find objects in the classroom containing vertical or horizontal edges and surfaces.
Or: List objects in the classroom which contain vertical or horizontal edges and surfaces.

2 Figure 35 shows a cuboid ABCDEFGH with its base ABCD in a horizontal plane, and its sides in vertical planes. In the figure, name:

a eight horizontal lines *b* four vertical lines

c two rectangles in horizontal planes

d four rectangles in vertical planes

e three sets of four parallel lines.

3 Repeat question *2* for the cube in Figure 36, replacing 'rectangles' by 'squares'.

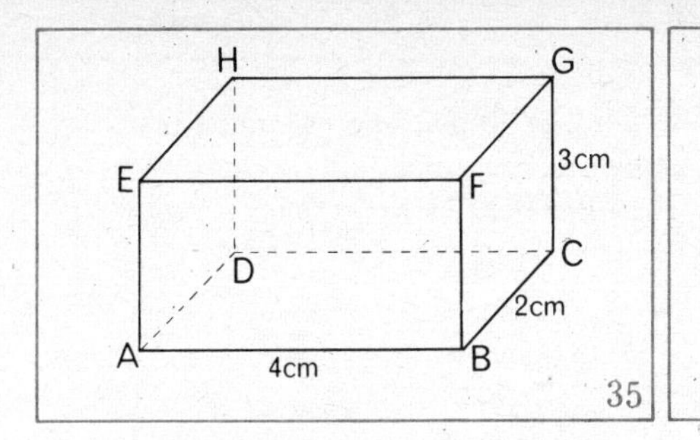

35

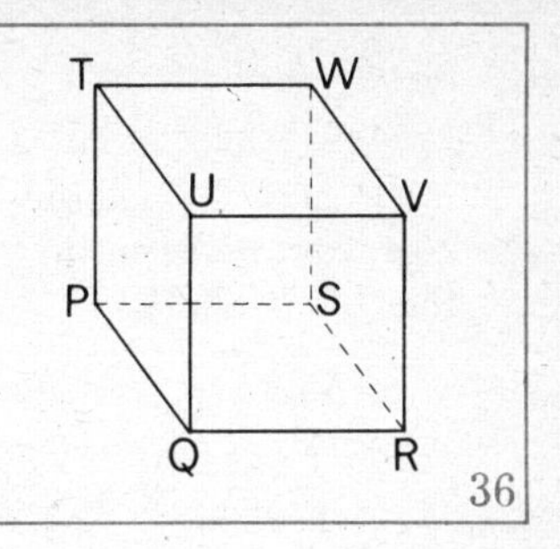

36

4 The cuboid in Figure 35 could be filled with cards congruent to rectangle ABCD.

a Would they be laid horizontally or vertically?

b What would be the length and breadth of each card?

5 Repeat question ***4*** for cards congruent to rectangle BCGF.

6 Repeat question ***4*** for cards congruent to rectangle ABFE.

7 Suppose we fit a card into the shaded position BCHE shown in Figure 37*a*. Into what other position in the cuboid could the card be fitted?

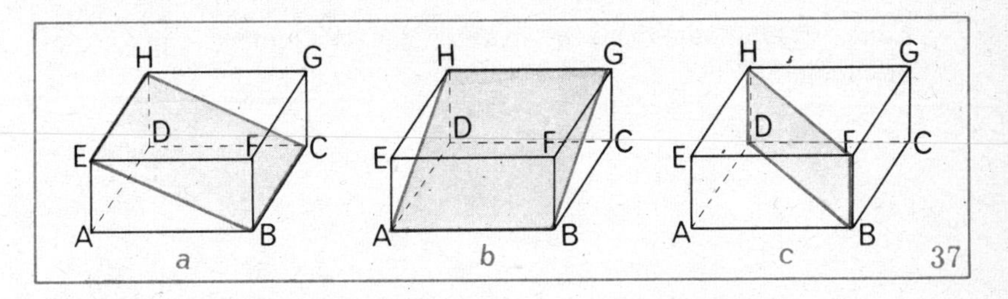

37

8 Repeat question *7* for Figure 37*b*.

9 Repeat question *7* for Figure 37*c*.

10 Each of the shaded planes in Figure 37 is called a *diagonal plane*.

a How many diagonal planes are there in a cuboid?

b How many diagonal planes are there in a cube? What can you say about these planes?

11 Sketch some cubes and cuboids with their diagonal planes, on isometric-ruled paper if possible, then on squared paper, then on plain paper.

12 Construct a cube or cuboid from its net, and make some diagonal planes to fit.

13a Make a skeleton cube using wire, or 'pipe-cleaners' and straws, or other suitable material, as shown in Figure 38.

b Use wire or thread to join H to B. HB is called a *space diagonal* of the cube? Why?

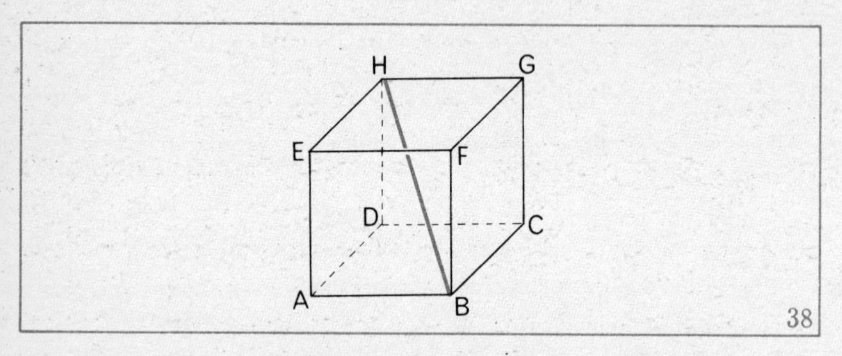

38

14a Name three more space diagonals in the cube in Figure 38.

b What can you say about the lengths of the space diagonals?

15 Sketch a skeleton cuboid, and draw its space diagonals in colour.

16 Can you think out where the space diagonals all intersect in a cube or cuboid?

Summary

1 *The rectangle*

Two axioms

1. A rectangle fits its outline in the four ways shown on page 95.

2. Congruent rectangles can be fitted together to cover the plane exactly.

Deductions

a The opposite sides are equal and parallel.
b The diagonals are equal, and bisect each other.
c All the angles are right angles.

2 *The square*

The square is a special kind of rectangle, and fits its outline in eight ways.

Deductions

The square has all the properties of a rectangle, and in addition:

a All the sides are equal.
b The diagonals bisect the angles, and bisect each other at right angles.

3 *Parallel lines*

Parallel lines are straight lines which keep the same distance apart, and so never meet.

Triangles

1 Introduction

This chapter is about triangles of all shapes and sizes. Figure 1 shows a set of triangles. Which of these triangles have:

a one angle a right angle ***b*** one angle an obtuse angle
c two equal sides ***d*** three equal sides?

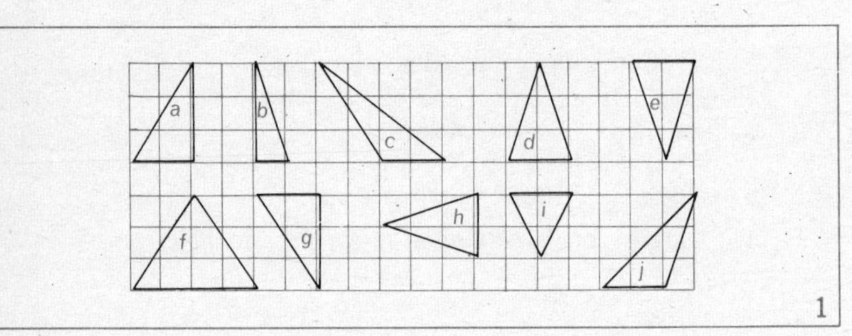

1

2 Right-angled triangles

Figure 2 shows a rectangle ABCD, with its diagonal AC and midpoint O.

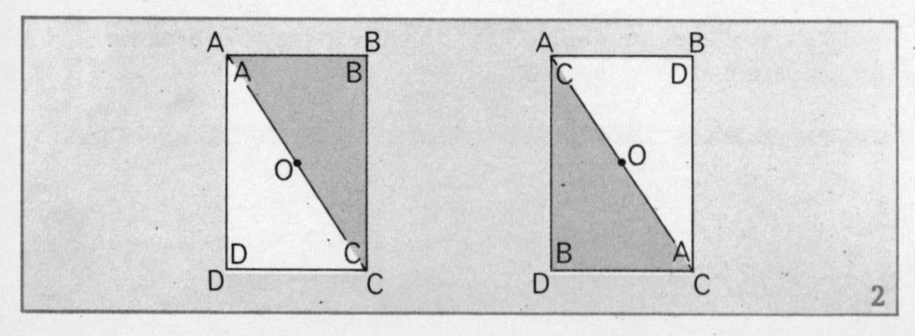

2

Under a half turn about O, triangle ABC goes to triangle CDA
i.e. $\triangle ABC \rightarrow \triangle CDA$

So triangles ABC and CDA have exactly *the same size and shape* i.e. they are *congruent.*

Also, the triangles are *right-angled,* as they each have one right angle.

Exercise 1A

1 Draw a rectangle on squared paper. Cut it out, and then cut it in half by cutting along a diagonal.

You now have two congruent right-angled triangles. Fit one on to the other.

2 a Draw a square ABCD, and draw its diagonal AC. Name the two congruent right-angled triangles in the square.

b Draw the diagonal BD, and name two more congruent right-angled triangles.

3 a In Figure 3, what fraction of the area of the rectangle is the area of the triangle?

b Calculate, in terms of the small squares:
(*1*) the area of the rectangle (*2*) the area of the triangle.

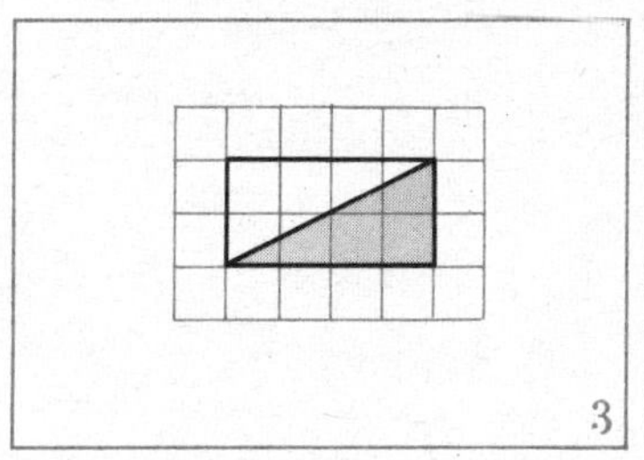
3

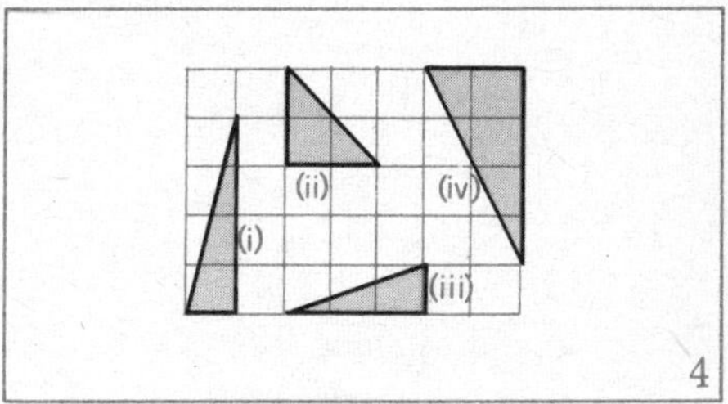

4

4 a Copy the right-angled triangles in Figure 4 on to squared paper.

b As in Figure 3, draw the rectangle round each triangle.

c Calculate the area of each rectangle, and of each triangle.

5 a Plot the points O(0, 0), A(6, 0), B(6, 8) on squared paper.

b Join the points to make a right-angled triangle.

c Complete the rectangle round it, and calculate the area of the triangle.

6 a Draw right-angled triangles on a sheet of squared paper from the following sets of points:
(*1*) {(0, 0), (4, 0), (4, 5)} (*2*) {(0, 5), (0, 10), (5, 10)}
(*3*) {(5, 0), (12, 0), (12, 12)}.

b Calculate the area of each triangle in square units.

Exercise 1B

1 a Name two congruent right-angled triangles in Figure 5.

b Copy the diagram, and draw the diagonal AC. Name two more congruent right-angled triangles in the drawing.

2 Name four congruent right-angled triangles in the square in Figure 6.

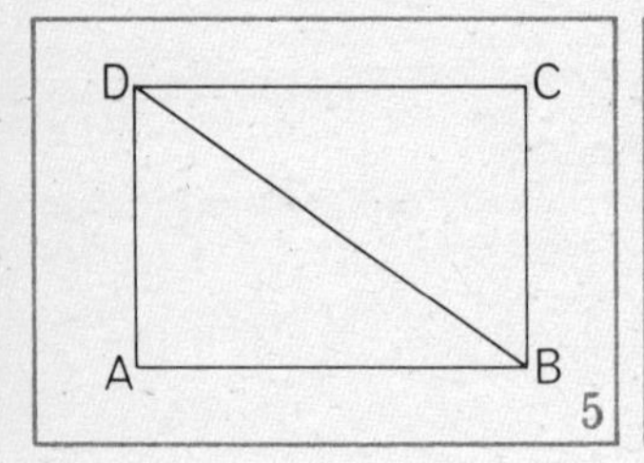

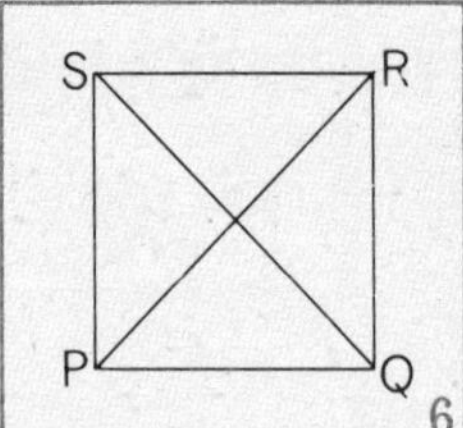

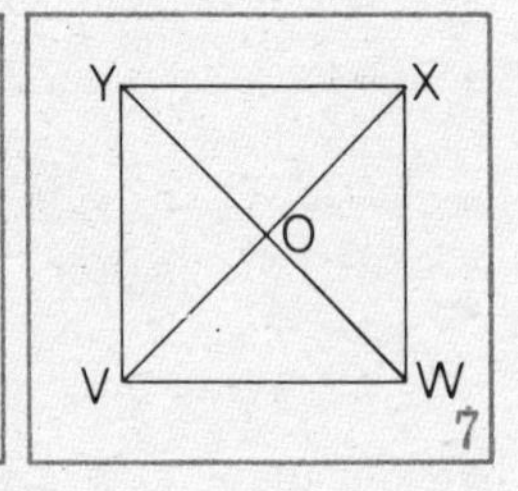

3 In Figure 7, VWXY is a square, and the diagonals intersect at O.

a What do you know about the size of the angles of the triangles at O?

b Name four congruent right-angled triangles in the square, each with one corner, or vertex, at O.

4 a In Figure 5, what fraction of the area of the rectangle is the area of triangle ABD?

b If AB = 6 cm and BC = 4 cm, calculate the areas of rectangle ABCD and triangle ABD.

5 In Figure 6, PQ = 8 m. Calculate the areas of square PQRS and triangle PQR.

6 In Figure 7, VW = 10 cm. Calculate the areas of square VWXY, triangle VWX, and triangle VOW.

7 Calculate the areas of the right-angled triangles in Figure 8. Copy them on to squared paper, with a rectangle round each if you wish. The units are centimetres.

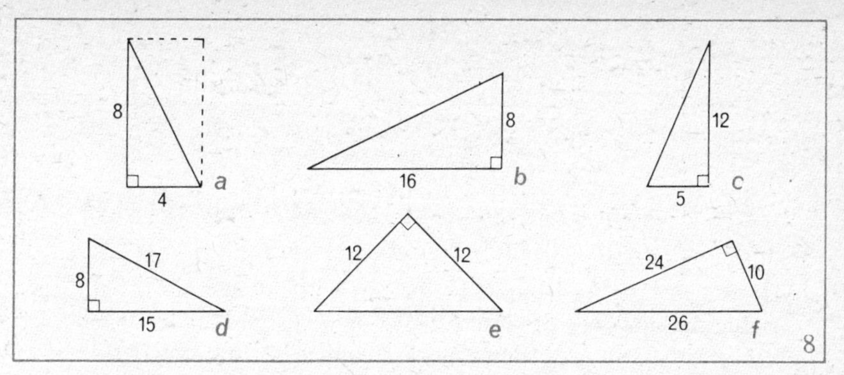

8 In the rectangular tiling in Figure 9, in which certain diagonals have been drawn, give the area of each shaded part in terms of the area of one of the small rectangles (e.g. $\frac{1}{2}$ rectangle, $1\frac{1}{2}$ rectangles).

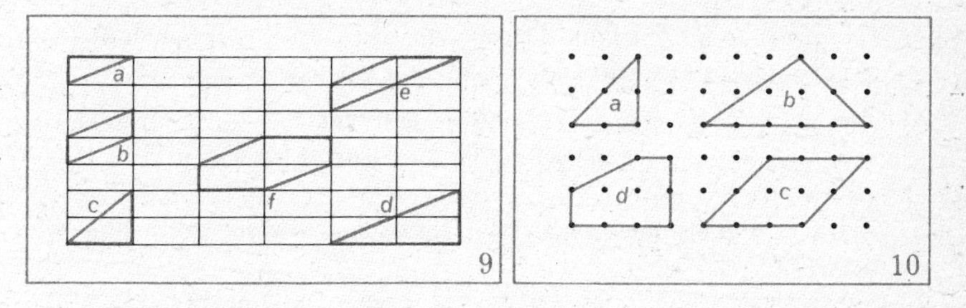

9 Figure 10 shows a square array of nails on a board. Elastic bands have been stretched round some nails to make the shapes shown.

Find the area of each shape in terms of the area of one small square by dividing the shapes up into areas you know. Copy the shapes on to squared paper if you wish.

10 Plot these sets of points on a sheet of squared paper, and then calculate the areas of the shapes formed by joining the points in each set.

a {A(10, 0), B(13, 0), C(13, 8)} {D(10, 10), E(10, 15), F(15, 15)}

c {K(2, 8), L(7, 9), M(5, 13), N(0, 12)}

d {P(1, 2), Q(6, 2), R(6, 4), S(3, 6)}

3 Isosceles triangles

Exercise 2

Do either question 1 or question 2. Question 1 involves cutting out shapes, and question 2 involves drawing the shapes.

1 Draw a rectangle on squared paper, or on card, and cut it out. Cut along one diagonal to obtain two congruent right-angled triangles.

Place the triangles alongside each other so that equal sides fit together.

Try to make shapes which have four sides; draw the outline of each shape. You should be able to find four different shapes with four sides.

Now try to make triangles from your two shapes, and sketch them. You should find two different triangular shapes.

2 On squared paper draw six separate congruent right-angled triangles.

Alongside each triangle draw another one congruent to it so that equal sides fit together.

Try to make shapes which have four sides; you should be able to draw four of these.

Now try to make shapes with only three sides; you should be able to find two such triangles.

3 Why were you able to get some shapes with only three sides in questions ***1*** and ***2***?

The triangles you made from two congruent right-angled triangles (half-rectangles) are called *isosceles* triangles (from Greek words meaning 'equal' and 'legs').

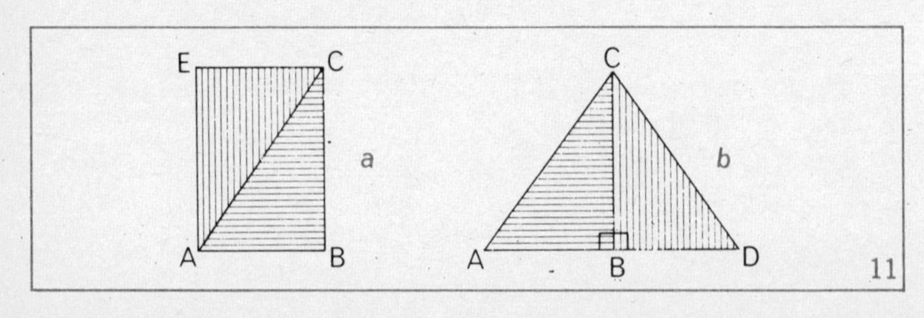

11

Figure 11*b* shows an isosceles triangle ACD composed of the two congruent right-angled triangles in the rectangle in Figure 11*a*.

4 Copy Figure 11 into your notebook. Explain how to form Figure *b* from *a*.

5 Remembering that the isosceles triangle ACD consists of two congruent right-angled triangles, copy and complete the following:

a AB = ... *b* AC = ... *c* $\angle$ABC = ...
d $\angle$BAC = ... *e* $\angle$ACB = ...
f Area of $\triangle$ABC = area of $\triangle$ ...

6 The triangles PRS and XYZ in Figure 12 are isosceles.

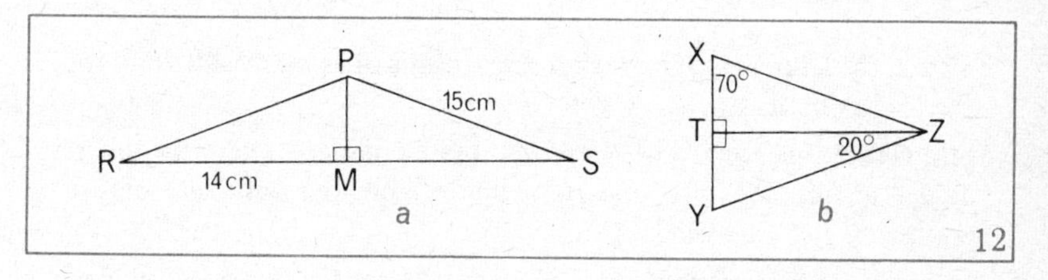

a For Figure 12*a*, (*1*) name the triangles from which $\triangle$PRS was formed;
(*2*) write down the lengths of RP, MS and RS.

b For Figure 12*b*, (*1*) name the triangles from which $\triangle$XYZ was formed;
(*2*) write down the sizes of angles XYZ, XZT and XZY.

7 Figure 13 shows the two ways in which an isosceles triangle fits its outline.

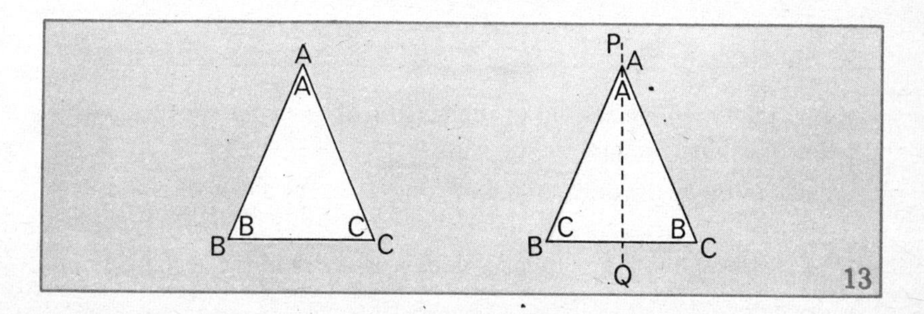

When $\triangle ABC$ is turned over about its axis of symmetry PQ, $A \rightarrow A$, $B \rightarrow C$, $C \rightarrow B$. Copy and complete:
$AB \rightarrow ...$, $\angle ABC \rightarrow ...$, $\triangle ABC \rightarrow ....$

8 In Figure 14, triangles PRS, XYZ and LMN are isosceles.

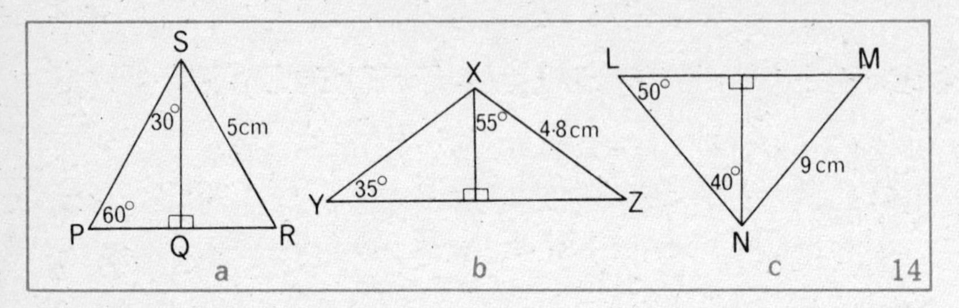

Sketch the triangles, and mark in the lengths of as many sides and the sizes of as many angles as you can.

9 Draw three isosceles triangles on squared paper—one with all its angles acute, one with one right angle, and one with one obtuse angle.

10a In an isosceles triangle DEF, DE = EF. Name the two equal angles.
b In an isosceles triangle STV, $\angle STV = \angle SVT$. Name the two equal sides.

11 Each diagram in Figure 15 contains a pair of congruent isosceles triangles. The units are centimetres.

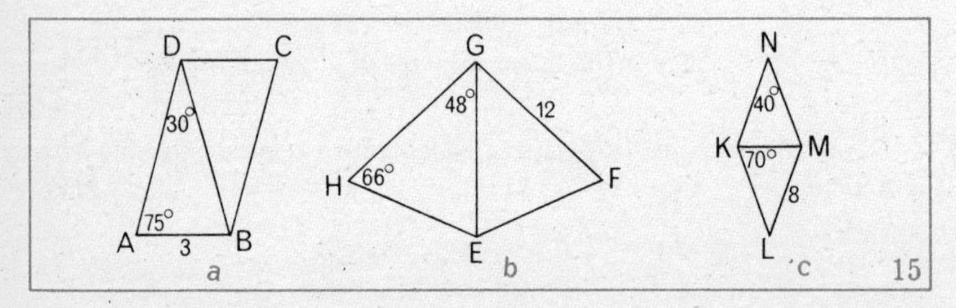

Copy the diagrams and fill in the lengths of as many lines and the sizes of as many angles as you can.

12 Where in the real world do you see objects in the shape of isosceles triangles?

13 Name two pairs of congruent isosceles triangles in the rectangle in Figure 16*a*.

14a Name four congruent isosceles triangles in the square in Figure 16*b*.
b Name another four congruent isosceles triangles in the square.

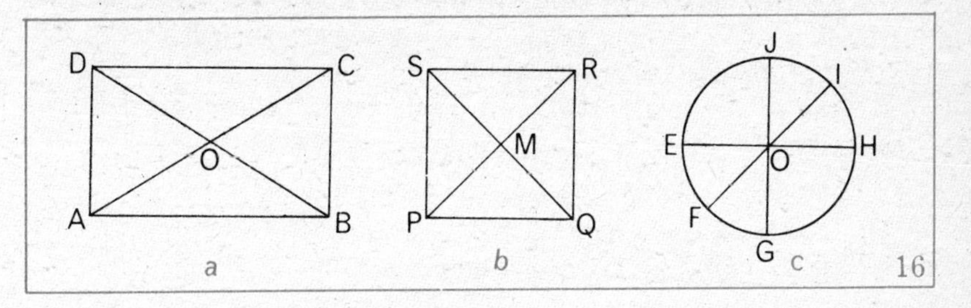

16

15 Copy the diagram of the wheel with some spokes in Figure 16*c*. By joining pairs of points, name pairs of congruent isosceles triangles. Can you find six pairs?

Summary—An isosceles triangle:

can be formed from two congruent right-angled triangles;
has two equal sides, and two equal angles opposite these sides;
has one axis of symmetry;
fits its outline in the two ways shown in Figure 13.

4 Drawing isosceles triangles

Suppose we are told that in triangle ABC, AB = AC. Is the triangle isosceles?

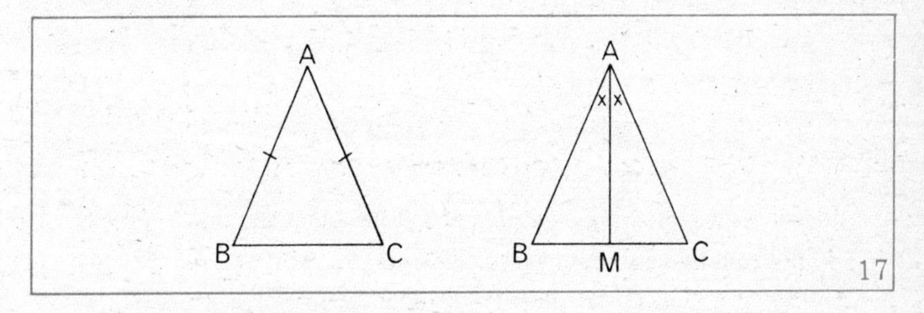

17

Draw AM to cut angle A into two equal angles as shown in Figure 17.

If we turn the triangle over about AM, AB → AC and △ABM → △ACM.

It follows that triangle ABC consists of two congruent right-angled triangles, which was our definition of an isosceles triangle.

So if a triangle has two equal sides, it is isosceles.

It can be shown also that if a triangle has two equal angles, it is isosceles.

Example.—Construct an isosceles triangle with:

a AB = AC = 8 cm; ∠BAC = 30°

b ∠PQR = ∠QPR = 70°; PQ = 4 cm

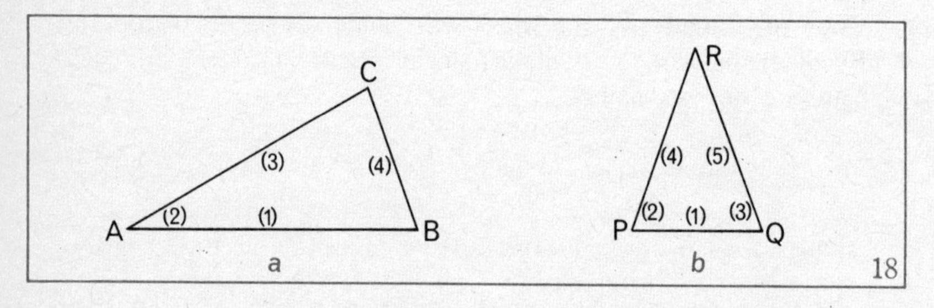

The numbers in Figure 18 indicate the order of construction.

Exercise 3

Construct the following triangles:

1 △ABC in Example ***a*** above.

2 △PQR in Example ***b*** above.

3 △ABC, AB = AC = 8 cm; ∠BAC = 50°.

4 △DEF, DE = DF = 10 cm; ∠EDF = 25°.

5 △GHK, GH = 6 cm; ∠GHK = ∠HGK = 65°.

6 △XYZ, XY = 9 cm; ∠XYZ = ∠YXZ = 40°.

Construct the following *quadrilaterals* (four-sided shapes):

7 ABCD, from two congruent isosceles triangles ACB and ACD with a common side AC. AC = 6 cm, ∠ACB = ∠CAB = 70°.

8 ABDC, from the same triangles as in question *7*, but sharing a common side BC.

5 Coordinates

Exercise 4A

1 On squared paper draw the isosceles triangle A(4, 4), B(10, 4), C(7, 11). Give:

a the names of two equal sides *b* the names of two equal angles
c the coordinates of the midpoint of AB.

2 Repeat question ***1*** for the triangle A(2, 12), C(15, 16), B(2, 20).

3 P is the point (3, 2), and Q is (15, 2). R is the point (r, 7) so that PQR is an isosceles triangle with RP = RQ.

a Draw the triangle on squared paper, and find r.
b Give the coordinates of the midpoint of PQ.

4 A is the point (4, 4), and B is (8, 4). C is the third vertex of an isosceles triangle ABC in which AC = BC.

a Draw four possible positions of triangle ABC, and write down the coordinates of C in each case.
b If C is the point (p, q), what can you say about p or q?

5 D is (5, 0), and E is (0, 5). F is the third vertex of an isosceles triangle DEF in which FE = FD.

a Give the coordinates of four possible positions of F.
b If F is the point (p, q), what can you say about p and q?

6 a Sketch an isosceles triangle AOB, where O is the origin, A is (6, 5), B lies on OX, and AO = AB.
b Give the coordinates of B, and name two equal angles in the figure.

7 Repeat question ***6*** for B on OY instead of OX.

Exercise 4B

1 A is the point (5, 5), and B is (2, 1). AB and AC are the equal sides of an isosceles triangle ABC. Find the coordinates of C if:

a BC is parallel to OX *b* BC is parallel to OY.

2 Repeat question ***1*** for A(5, 6) and B(2, 10).

3 Triangle ABC is isosceles, with AB = AC and BC parallel to the x-axis. D is the midpoint of BC.

a If A is the point (3, 5) and B is (1, 1), find the coordinates of D and C.

b If A is (15, 6) and C is (7, 10), find B and the lengths of AD and BC.

c If A is (9, 9), D is (9, 3), and BC = 4 units, find B and C. Calculate the area of triangle ABC.

4 A is the point (5, 7), and B is (11, 7). C can take any position such that triangle ABC is isosceles, with AC = BC.

a Write down the coordinates of six members of the set of possible positions of C.

b Which of these points are members of the set in ***a***? (8, 8), (7, 7), (8, 0), (8, 199).

c If (p, q) is a member of this set, write a true sentence about p or q.

d Can you find a point for which your sentence in ***c*** is true, but which is not a member of the set?

e For what positions of C is angle ACB a right angle?

f If C(x, y) makes angle ACB obtuse, what can you say about x and y?

5 Repeat question ***4*** for the points A(10, 0) and B(0, 10).

---

6 The sum of the angles of a triangle

Exercise 5

1 In Figure 19, triangle ***a*** has one right angle, ***b*** has two equal angles, ***c*** has three equal angles, ***d*** has no equal angles.

Draw triangles like these on squared paper.

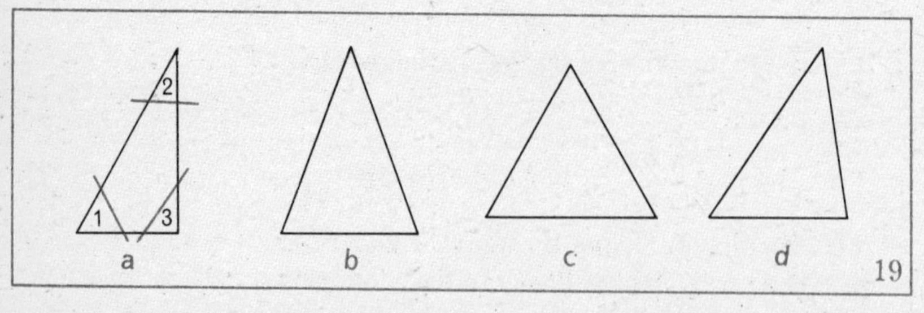

19

2 Cut out the first triangle, and cut off its angles as shown by the red lines.

Place the three angles together as shown in Figure 20. Does the sum of the three angles make a straight angle? If so, what is the sum of the angles of the right-angled triangle, in degrees?

3 Repeat question *2* for triangles ***b***, ***c*** and ***d***.

Questions *2* and *3* suggest that the sum of the angles of a triangle is 180°.

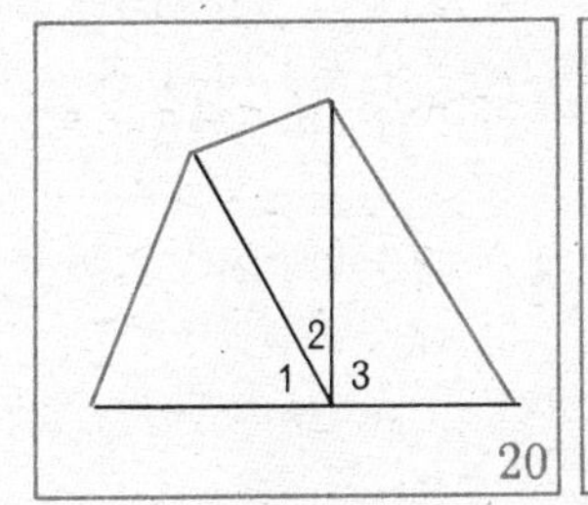

20

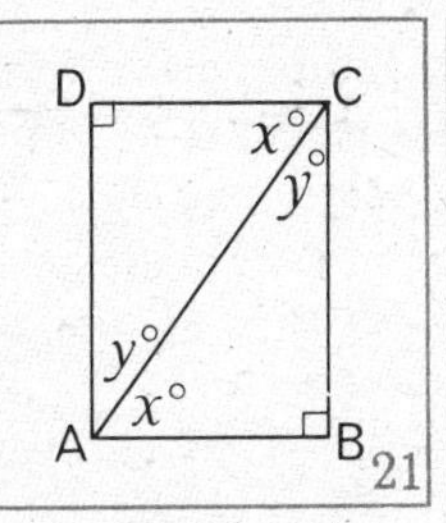

21

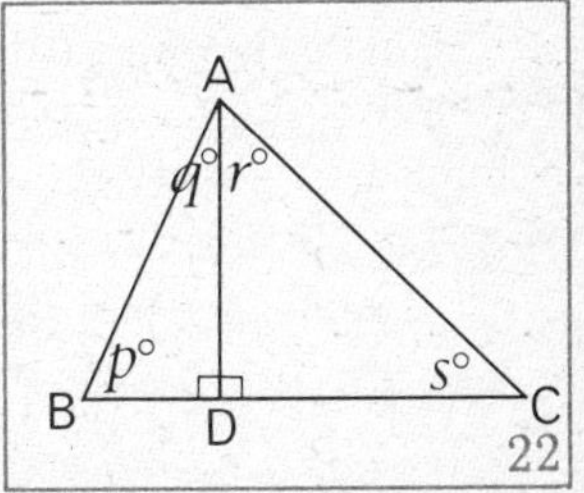

22

4 Figure 21 shows right-angled triangles ABC and ACD forming the rectangle ABCD. Copy and complete the following:

At corner A of the rectangle, $x° + y° = ...°$
So in triangle ABC, $\angle BAC + \angle ACB = ...°$
So $\angle BAC + \angle ACB + \angle ABC = ...°$.

You have shown that the sum of the angles of a right-angled triangle is 180°.

5 In Figure 22, triangle ABC is divided into two right-angled triangles. Copy and complete the following:

$$p° + q° = ...°$$
$$r° + s° = ...°$$
$$\text{so } p° + q° + r° + s° = ...°$$

That is, $\angle ABC + \angle BAC + \angle ACB = ...°$.

The sum of the angles of a triangle is 180°.

Exercise 6

1 Calculate the size of the third angle in each triangle in Figure 23.

2 In Figure 23, which triangles are:

a right-angled *b* acute-angled (with all its angles acute)
c obtuse-angled (with one obtuse angle)?

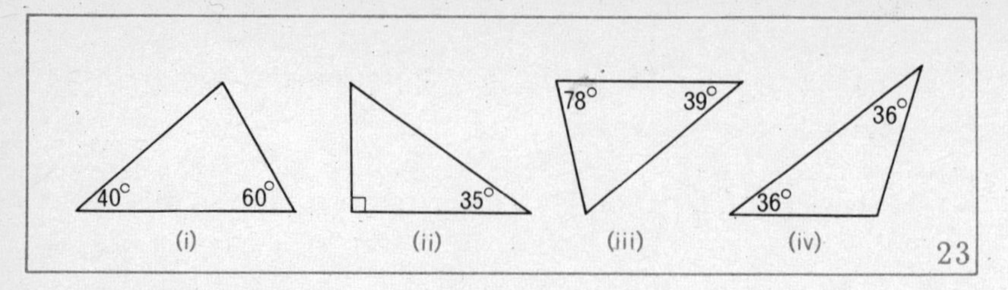

3 Calculate the size of the third angle in the triangles containing the following pairs of angles, and say whether each triangle is right-angled, acute-angled, or obtuse-angled.

a 40°, 50° *b* 85°, 85° *c* 39°, 49° *d* 60°, 60°
e 110°, 34° *f* 90°, 19° *g* 43°, 78° *h* 48°, 84°

4 Which of the triangles in question *3* were isosceles triangles?

5 Calculate x for each triangle in Figure 24.

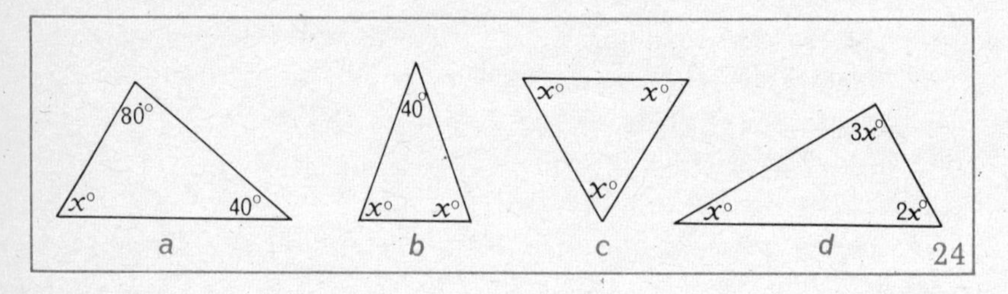

6 Draw a square with its two diagonals, and mark in the sizes of all the angles.

7 Sketch a rectangle, with its diagonals making an acute angle of 50° with each other. Mark the sizes of all the angles.

8 a Draw an isosceles triangle with one of its angles 120°.
b Draw in the axis of symmetry, and fill in the sizes of all the angles.

9 Say whether each of the following is true or false for angles in a triangle:

a The sum of the angles is equal to two right angles.
b If two angles are 88° and 22°, the third angle is 80°.
c It is possible for two of the angles to be right angles.
d If the three angles are equal, then each must be 60°.
e If one angle is obtuse, the others must be acute.
f The sum of two of the angles is always greater than the third angle.

7 Equilateral triangles

Figure 25 shows two 'Meccano' strips of equal length bolted together at A, and joined by elastic between B and C. As B and C are pulled apart the triangle ABC changes shape, but is always an isosceles triangle since AB = AC.

In one position, the middle triangle shown, AB = AC = BC. This special kind of isosceles triangle, with all its sides equal, is called an *equilateral* triangle.

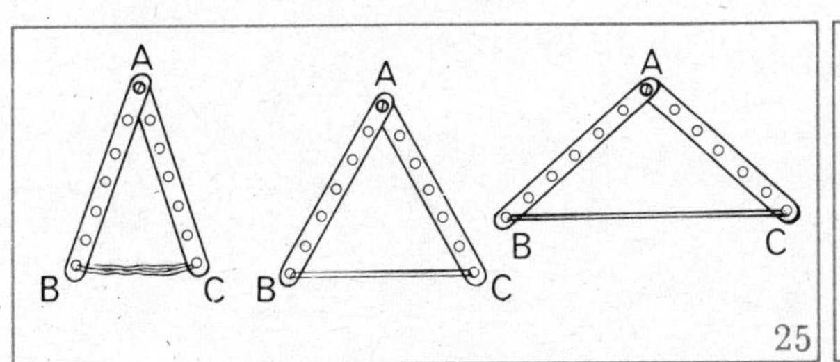

25

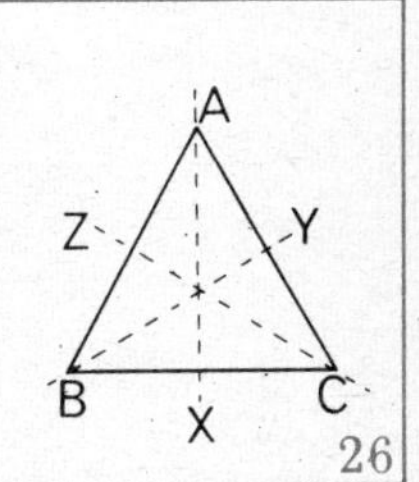

26

An equilateral triangle fits its outline in six ways, and has the three axes of symmetry shown in Figure 26.

Exercise 7

1 Copy Figure 26, and use your diagram to complete the following:

a Turning $\triangle ABC$ over about axis AX,

$AB \rightarrow \ldots$; and $\angle ABC \rightarrow \ldots$

So $AB = \ldots$; and $\angle ABC = \ldots$

b Turning $\triangle ABC$ over about axis BY,

$BC \rightarrow \ldots$; and $\angle ACB \rightarrow \ldots$

So $BC = \ldots$; and $\angle ACB = \ldots$

2 From question *1*, what can you deduce about the lengths of the sides and the size of the angles of an equilateral triangle?

3 Copy Figure 26 again, and let O be the point where AX, BY and CZ cross. For one fitting, $A \rightarrow B$, $B \rightarrow C$, $C \rightarrow A$. To obtain this we can think of the triangle turning about O. Through what angle

does OA turn? So what is the size of $\angle AOB$? And $\angle BOC$? And $\angle COA$? What is the sum of these three angles?
(Use a tracing of triangle ABC if you find this difficult to see.)

4 Figure 27 shows six congruent equilateral triangles fitting together to form a *hexagon.*

a Why do all the angles of the triangles fit together exactly at O?
b What size is $\angle AOD$? Name two lines equal in length to AD.
c How many lines are there equal in length to AB?
d What size is each angle of the hexagon, e.g. $\angle ABC$?

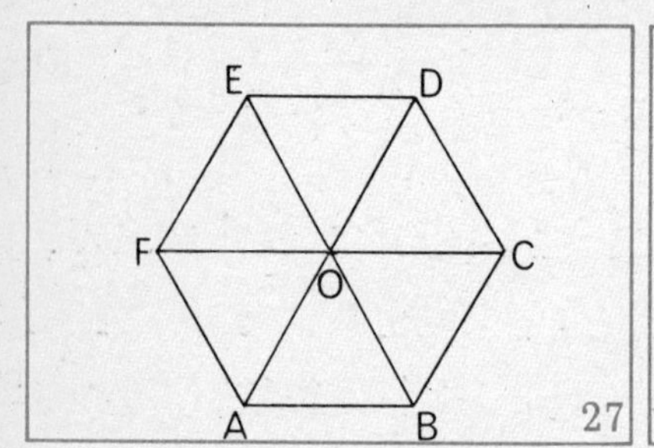

27

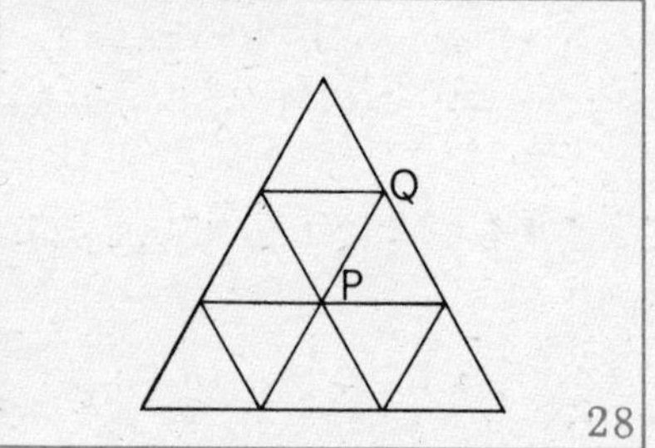

28

5 Figure 28 shows a tiling of congruent equilateral triangles.

a Why do the triangles fit together at P?
b Why do the sides of the triangles make a straight line at Q?
c If PQ = 1 cm, how many triangles are there in the figure with sides 1 cm long. How many with sides 2 cm long? How many with sides 3 cm long?

6 If $E = \{\text{triangles}\}$, $I = \{\text{isosceles triangles}\}$, $Q = \{\text{equilateral triangles}\}$ which of the following are true?

a $I \subset E$ *b* $I \subset Q$ *c* $Q \subset E$ *d* $Q \subset I$

Illustrate your answer with Venn diagrams.

7 *a* Draw a circle with centre P and radius 3 cm (see Figure 29). Mark a point Q on the circumference. With centre Q draw a second circle, the same size as the first, cutting the first one at R. With centre R draw a third circle, the same size as the first. Join PQ, QR, RP.

b Why is triangle PQR equilateral? Draw three more equilateral triangles in your diagram; call them QRA, RPB, PQC.
Why must ARB be a straight line? Describe triangle ABC.

c Draw three more circles of the same size in the diagram, with centres A, B and C. Hence draw nine more equilateral triangles in the diagram. Now add one more triangle at each corner to make a large equilateral triangle XYZ with sides four times those of triangle PQR.

d Cut out triangle XYZ, and fold it along AB, BC and AC to make a *triangular pyramid*, or *tetrahedron*, as shown in Figure 30.

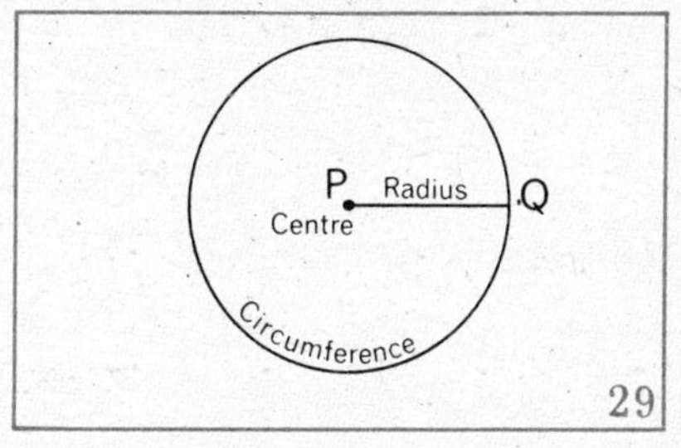

29

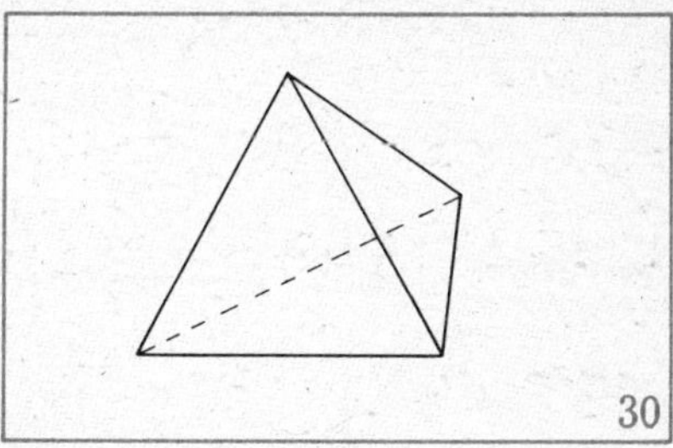
30

8 How many faces, corners and edges has the tetrahedron? If each edge is 12 cm long, what is the total length of the edges?

8 *The area of a triangle*

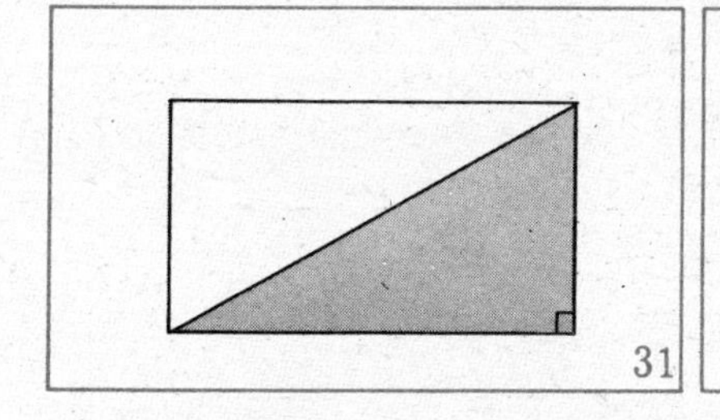
31

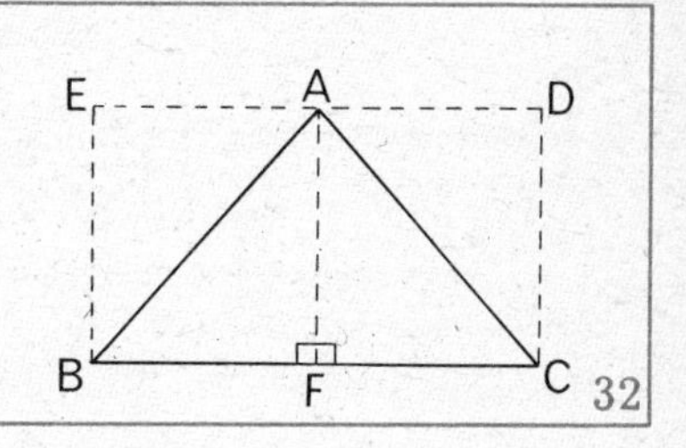

32

In Section 2 we found the area of a right-angled triangle by taking half the area of the surrounding rectangle; see Figure 31.

To find the area of triangle ABC in Figure 32 we divide the triangle into two right-angled triangles by drawing AF perpendicular to BC. We then complete the rectangles BFAE and FCDA.

$$\begin{aligned}\text{Area of } \triangle ABC &= \text{area of } \triangle ABF + \text{area of } \triangle ACF \\ &= \tfrac{1}{2} \text{ rectangle BFAE} + \tfrac{1}{2} \text{ rectangle FCDA} \\ &= \tfrac{1}{2} \text{ rectangle BCDE} \\ &= \tfrac{1}{2}\, BC \times CD \\ &= \tfrac{1}{2}\, BC \times FA\end{aligned}$$

So we can calculate the area of a triangle by using:

a Area of triangle = $\frac{1}{2}$ base × height
b The *formula* $\triangle = \frac{1}{2}bh$

FA is sometimes called an *altitude* of the triangle, and we say

Area of triangle = $\frac{1}{2}$ *base* × *altitude*

Exercise 8A

1 Calculate the area of each triangle in Figure 33 in two ways:

a by first calculating the area of the surrounding rectangle
b by using 'Area of triangle = $\frac{1}{2}$ base × height'.

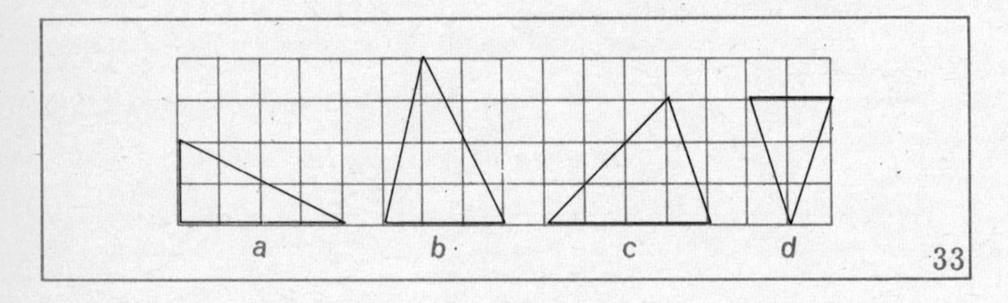

2 Calculate the area of each of the following triangles; the units are centimetres.

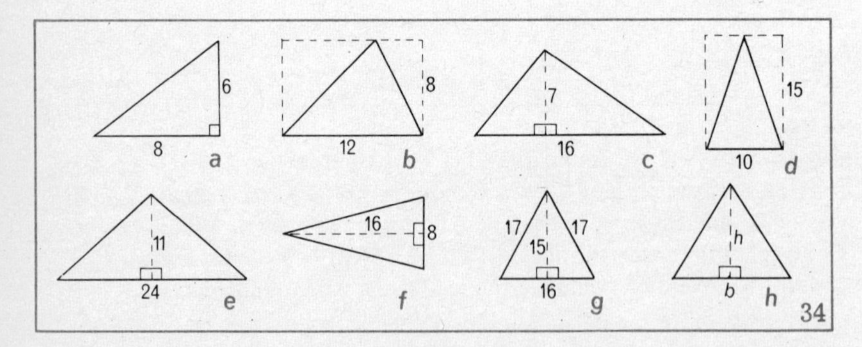

3 a Sketch an isosceles triangle with base 10 cm long, and height 6 cm. Calculate its area.

b Repeat ***a*** if the base is 6 cm long, and the height is 10 cm.

4 Calculate the areas of the triangles with these dimensions:

Length of base	18 cm	11 m	9 mm	85 cm	4·6 m
Height	8 cm	10 m	12 mm	30 cm	1·2 m

5 Sketch a triangle ABC, and draw in its three altitudes AD, BE and CF. Name the 'base' corresponding to each altitude.

6 Calculate the areas of the following shapes; the units are metres.

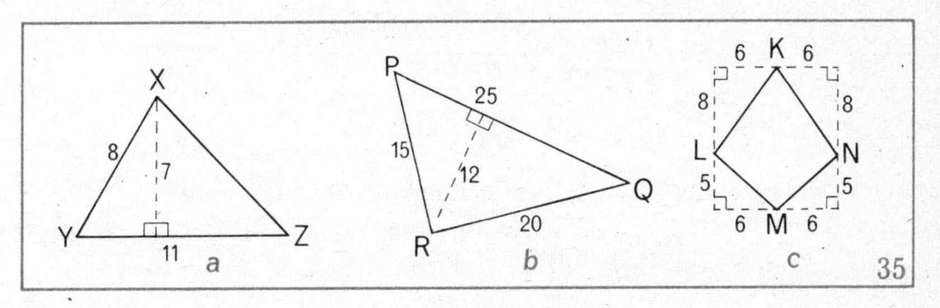

Exercise 8B

1 Calculate the areas of the triangles with the following dimensions:

Length of base	32 cm	11 m	27 cm	6·5 mm
Height	12 cm	9 m	16 cm	4·4 mm

2 Calculate the areas of the shapes in Figure 36; the dimensions are in metres. In *c* the two triangles are congruent.

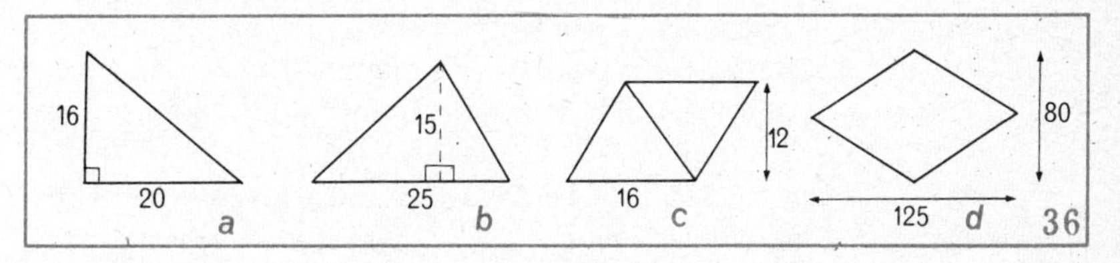

3 The diagonals of a square ABCD are 24 cm long. Calculate:

a the area of the isosceles triangle ABC *b* the area of the square.

4 A rectangle has length 18 cm and breadth 8 cm. Calculate the areas of the triangles which have one vertex at the centre of the rectangle when the diagonals are drawn.

5 In Figure 37*a*, triangle ABC is obtuse-angled at C, so the altitude from A lies outside the triangle.

a Name the altitude and the corresponding base.

b Use "Area of triangle = $\frac{1}{2}$ base × altitude" to calculate the area of $\triangle$ABC.

c Repeat ***a*** and ***b*** for triangles PQR and EFG in Figure 37.

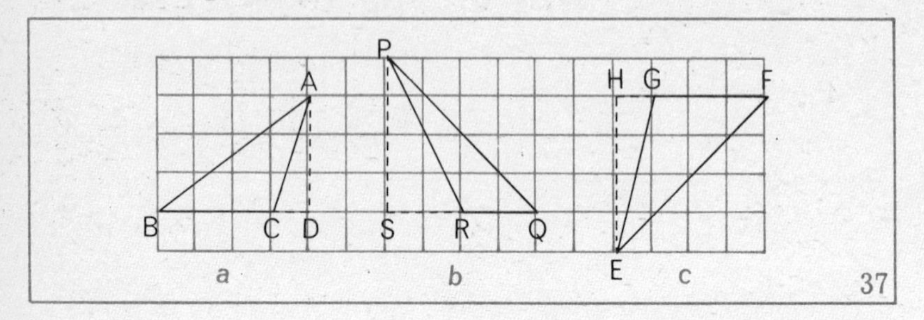

6 By drawing the surrounding rectangle for triangle ABC in Figure 37*a* can you explain why the formula '$\triangle = \frac{1}{2}bh$' is true for this obtuse-angled triangle?

9 The construction of triangles

In this section we are going to examine just what information must be given in order to be able to construct a required triangle.

Suppose the teacher has a cardboard triangle ABC. You are to consider how to make an accurate drawing of it. You can ask the teacher for the length of any side you like, and the size of any angle you like; but you can ask for only one piece of information at a time, and you are to stop asking when you have enough information to finish the drawing satisfactorily.

Sometimes the teacher may be awkward and refuse to give you what you want, perhaps offering you something else. Take it if it will do, refuse it if it will not.

The object is to decide *how much* information is necessary; and whether, for example, if three items suffice, any three items will do.

This question is often important in practice: for example a map-maker wants to get an accurate map without the expense of making unnecessary measurements; an engineer building a structure involving triangles must specify clearly the shape and size of the triangles; a draughtsman in his drawings must include enough information about lengths and angles to enable the fitter to make the component the drawing depicts, and since every straight-sided shape can be built up from triangles, if we master the art of 'specifying' the data required for a triangle, we shall be able to cope with all other figures.

Exercise 9

1 Draw a line AB 7 cm long. Using your compasses, take centre A and radius 5 cm, and draw an arc above AB. With centre B and radius 4 cm, draw an arc above AB to cut the first one at C. Join AC and BC to obtain triangle ABC.

2 Starting with AB 7 cm long in each case, try to draw a triangle for each of the following:

First radius (centre A)	5 cm	5 cm	5 cm
Second radius (centre B)	3 cm	2 cm	1 cm

3 How can you tell by looking at the lengths of the three lines whether or not a triangle can be drawn?

* * * * *

4 Draw a line AB 6 cm long. At A, draw $\angle BAX = 60°$, with AX as long as possible. On AX mark C so that AC = 3 cm, D so that AD = 4 cm, E so that AE = 5 cm, F so that AF = 6 cm. Join BC, BD, BE, BF.

5 Given '2 sides and the included angle' as in question ***4***, can you draw a triangle in every case?

* * * * *

6 Draw a line AB 6 cm long. At A, draw $\angle BAX = 30°$, with AX as long as possible. At B, draw $\angle ABC = 70°$, $\angle ABD = 90°$, $\angle ABE = 110°$, $\angle ABF = 130°$, $\angle ABG = 150°$, $\angle ABH = 170°$, with C, D, E, ... on AX where possible.

7 Given '2 angles and a side', when is it possible to draw a triangle?

* * * * *

8 Draw a triangle with angles of 50°, 60° and 70°. Now draw another triangle inside the first with angles of 50°, 60° and 70°. How many such triangles do you think you could draw?

Being given the sizes of three angles does *not* specify one unique shape *and* size of triangle.

* * * * *

A set of *congruent triangles* (i.e. with the same shape and size) can be drawn if three facts are given, as follows:

a three sides (where the sum of two of their lengths is greater than the third), or

b two sides and the included angle, or

c two angles and a corresponding side.

Exercise 10

Construct the following triangles as accurately as you can. In each case measure the other sides and angles.

	Base AB	BC	CA	∠ABC	∠BCA	∠CAB
1	10 cm	8 cm	6 cm	.	.	.
2	7·5 cm	7·5 cm	7·5 cm	.	.	.
3	9 cm	7 cm	.	50°	.	.
4	6 cm	.	8 cm	.	.	80°
5	12 cm	.	.	45°	.	45°
6	8 cm	.	.	50°	70°	.

7 Construct triangles congruent to the triangles in questions *1*, *3* and *5*.

Make scale drawings for the following; in each case say which three facts about the triangle are given.

8 The lengths of the sides of a triangular plot of ground are 50 m, 120 m and 130 m. Find the sizes of the angles at the corners.

9 An aircraft flies 400 km on a bearing of 010°, then 300 km on a bearing of 100°. Find the distance and bearing of its starting point from its finishing point.

10 A weather balloon is observed from two places on the same level, 40 km apart. The angles of elevation of the balloon from these places are 42° and 33°. Find the height of the balloon.

Summary

1 A rectangle can be divided into two congruent right-angled triangles.

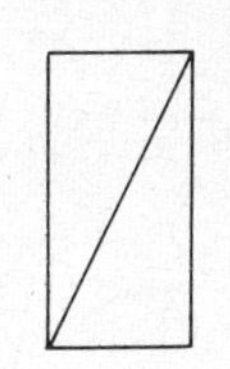

2 These two congruent right-angled triangles can form an isosceles triangle.

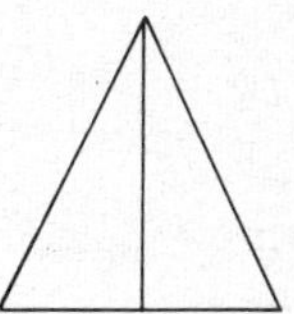

3 An isosceles triangle:
has two equal sides and two equal angles:
has one axis of symmetry;
fits its outline in two ways.

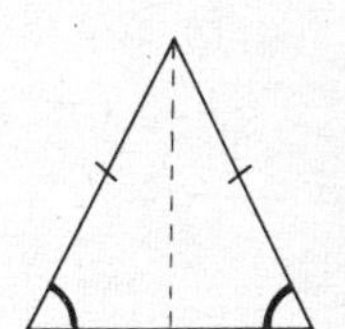

4 The sum of the angles of a triangle is 180°,
i.e. $\angle A + \angle B + \angle C = 180°$.

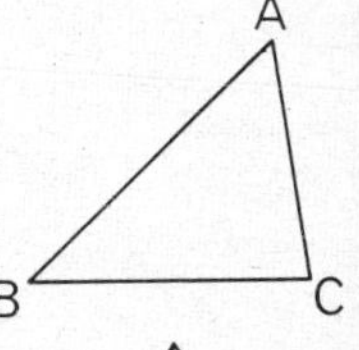

5 An equilateral triangle fits its outline in six ways. It has three equal sides, three equal angles (each 60°), three axes of symmetry.

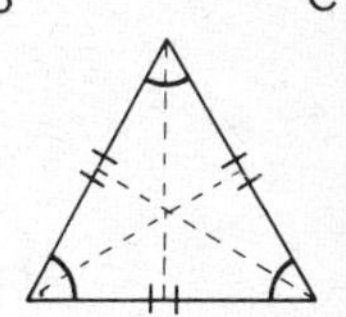

6 The area of a triangle is given by:
$\frac{1}{2}$ base × height, i.e. $\triangle = \frac{1}{2}bh$.

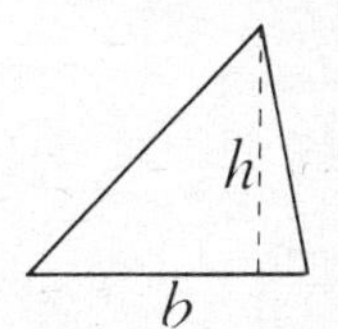

7 A set of congruent triangles can be specified by:
3 sides, 2 sides and the included angle, 2 angles and a corresponding side.

Topic to explore

The sum of the angles of a polygon

The sum of the angles of a *triangle* is 2 right angles. From any vertex of a *quadrilateral* it is possible to draw *one* diagonal, making *two* triangles; and the angle sum of the quadrilateral is 2×2 right angles.

Copy and complete the table below for 'polygons' with different numbers of sides.

Number of sides	Number of diagonals drawn through one vertex	Number of triangles formed	Sum of angles of polygon (in right angles)
3	0	1	2
4	1	2	4
5	2		
6			
7			
8			
20			
100			
	7		
		12	
			22
n			

Revision Exercises

Revision Exercises on Chapter 1
Rectangle and Square

Revision Exercise 1A

1 On squared paper draw diagrams to show that congruent tiles of the shapes shown in Figure 1 can be used to cover a plane completely without overlapping or leaving gaps.

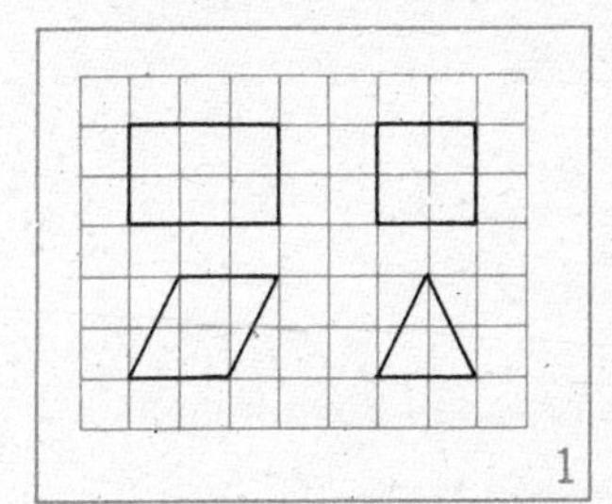

1

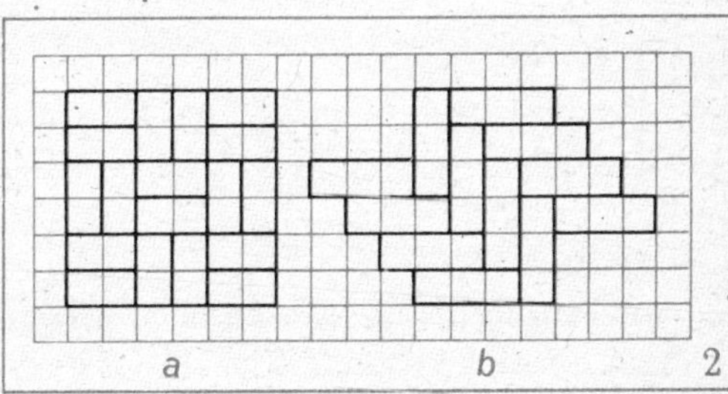

a b 2

2 Copy the tiling patterns in Figure 2 on to 5-mm squared paper and continue the patterns in all directions.

3 a Colour the tiles in Figure 2*a* red, green and blue so that a tile does not touch another of the same colour.

b Experiment with two- and three-colour patterns in Figure 2*b*.

4 A is the point (1, 1), B is (5, 1), C(1, 3), D(5, 3), and AB and CD are drawn. About what point will a half turn map A on to D, and B on to C. Under this half turn, what happens to CD, AC and DB?

5 Write down the coordinates of the point of intersection of the diagonals AD and BC in the figure ABCD in question *4*. What property of the diagonals does this show?

6 Draw a Venn diagram to illustrate the statement 'All squares are rectangles'.

7 If S is the set of all squares and R is the set of rectangles, which of the following is true? *a* $R \subset S$ *b* $S \subset R$ *c* $S = R$

8 A is the point (2, 1), B(2, 5), C(4, 5). If ABCD is a rectangle write down the coordinates of D.

9 A is (5, 3), B is (5, 5). If ABCD is a square write down the coordinates of C and D. (There are two possible answers for each.)

10 Sketch a rectangle with its diagonals, and mark in as many properties as you can. List the properties in words.

11 What properties, in addition to those in your answer for question ***10***, are true for a square?

12 Copy the rectangle and square in Figure 3, and fill in the lengths (in cm) of as many lines and the sizes of as many angles as you can.

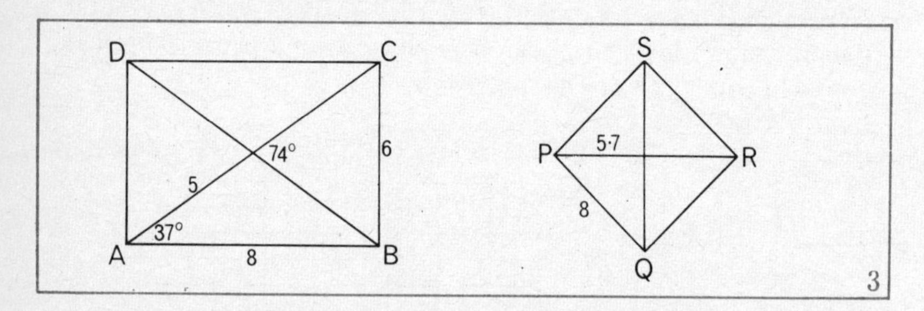

13 Use instruments to draw on plain paper a rectangle 8 cm long and 5 cm broad. Measure the lengths of its diagonals.

14 Construct on plain paper a square of side 6 cm. Measure the lengths of its diagonals.

15 On 5-mm squared paper draw a square with diagonals 10 units (5 cm long). Measure the lengths of the sides of the square.

16 State whether each of the following is true or false.

a Two lines perpendicular to the same plane must be parallel.
b Two lines which are parallel to the same line must be parallel to each other.
c A square fits its outline in eight ways.
d All the sides of a rectangle are equal.
e The sum of the angles of a rectangle is 360°.

17 Sketch a cuboid, and state the number of faces, corners, edges and space diagonals that it has.

18 A cube has a side 8 cm long. Calculate its

a volume ***b*** total surface area ***c*** total length of edges.

19 The length, breadth and height of a cuboid are all different.

a How many different lengths of edge has it?
b How many different shapes of face has it?
c How many different shapes of corner has it?

Revision Exercise 1B

1 Show by diagrams on squared paper that tiles of the shapes shown in Figure 4 can be used to cover the plane.

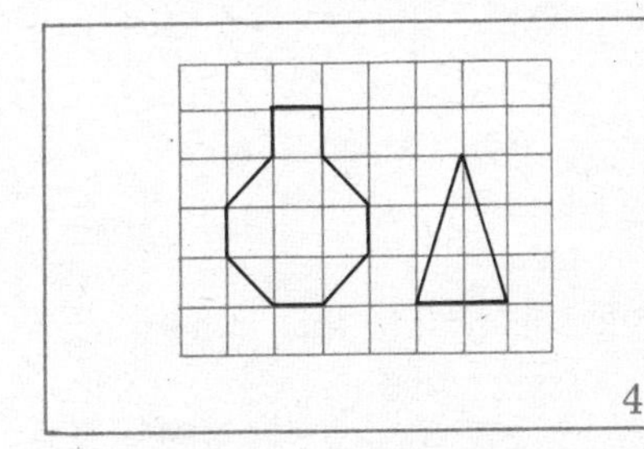
4

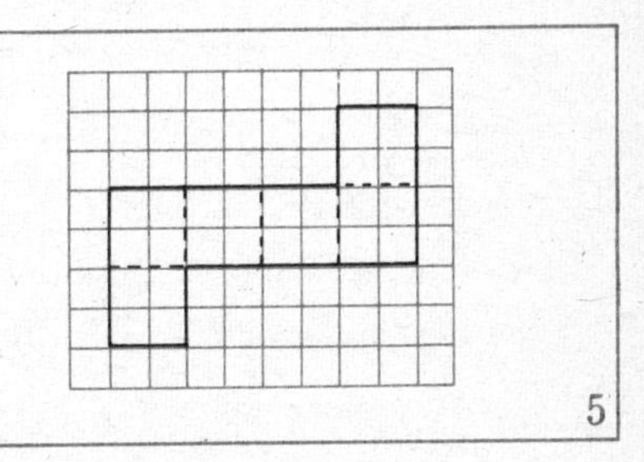
5

2 a Repeat question *1* for the shape in Figure 5.
b For what solid is the shape a net?

3 A is the point (2, 0), B is (3, 2), C(2, 4), D(1, 2). Show that the plane can be covered with shapes congruent to ABCD.
Write down the coordinates of the corners of the shape immediately above ABCD, and of the shape two places to the right of ABCD.

4 A is the point (3, 10), B is (0, 6) and C is (8, 0). Find the coordinates of D if ABCD is a rectangle.

5 On squared paper copy the line AB shown in Figure 6. Draw several lines the same length as AB and parallel to AB.
To get from A to B, how many units to the right do you have to go, and how many units up? Answer the same questions for the other parallels you drew.

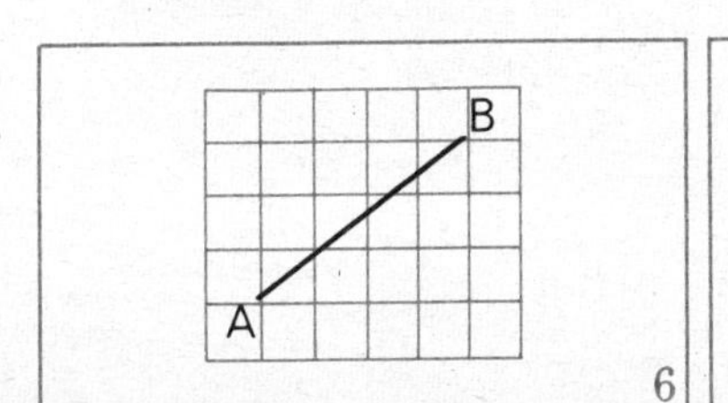

6

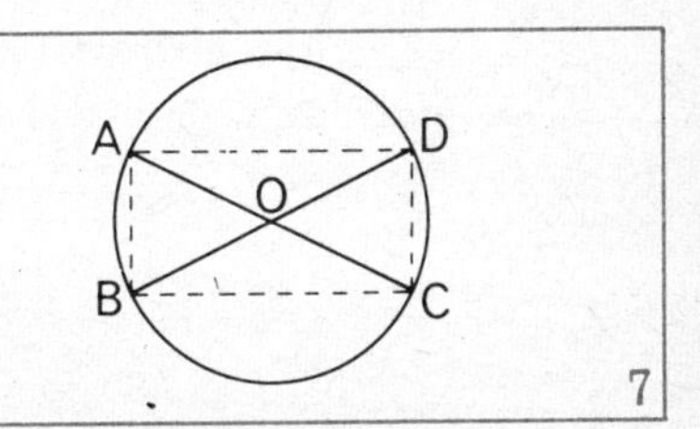

7

6 In Figure 7 AC and BD are diameters of the circle with centre O. What is the shape of ABCD?

If $\angle AOB = 46°$ sketch the figure and fill in the sizes of all the angles.

7 A is (1, 1), B(3, 2), C(4, 5), D(2, 4). Make any statements you can about AB and DC, BC and AD, AC and BD.

If AC and BD intersect at E, give the coordinates of E.

8 (4, 1) and (4, 5) are opposite vertices of a square. Write down the coordinates of the other two vertices.

9 List properties of a square that are not necessarily true for every rectangle.

10 Draw a straight line PR 8 cm long on plain paper. Construct a square with PR as one of its diagonals.

11 Say whether each of the following is true or false.

a The diagonals of a square divide the square into four congruent triangles.

b The area of a square of side 10 cm must be greater than the area of a rectangle one of whose sides is 10 cm long.

c A rectangle with all its sides equal can fit its outline in eight ways.

d A plumb line in Aberdeen is parallel to a plumb line in Paris.

12 The midpoints of the sides of a square ABCD are joined to form another square. If AB = 12 cm, calculate the areas of both squares.

13 A cuboid is 10 cm long, 8 cm broad and 6 cm high. On plain paper construct a rectangular face of this cuboid which is 10 cm by 6 cm, and measure the length of its diagonal.

Use this diagonal to construct another rectangle from which you can obtain the length of a space diagonal of the cuboid.

14 OABC is a rectangle with OA on the x-axis, and B is the point (3, 1). $AA_1B_1C_1$ is a rectangle with AA_1 on the x-axis, $AA_1 = 2OA$, $A_1B_1 = 2AB$; find the coordinates of B_1.

$A_1A_2B_2C_2$ is a rectangle with A_1A_2 on the x-axis, $A_1A_2 = 2AA_1$, $A_2B_2 = 2A_1B_1$; find the coordinates of B_2, and of the next two points B_3, B_4 in this sequence of rectangles.

Revision Exercises on Chapter 2
Triangles

Revision Exercise 2A

1 Sketch a rectangle ABCD with its diagonals intersecting at O. If the figure is given a half turn about O, where will the following go?

a OB *b* OC *c* AB *d* D *e* $\triangle$OBC *f* $\triangle$ABC

2 If in the rectangle in question *1*, AB = 12 cm and BC = 8 cm, calculate the areas of:

a rectangle ABCD *b* right-angled triangle BCD.

3 Say whether each of the following is true or false.

a In a right-angled triangle, one angle is a right angle.
b It is possible for a triangle to have two right angles.
c All right-angled triangles have the same shape.
d Every rectangle can be divided into two congruent right-angled triangles.

4 Plot the points O(0, 0), A(10, 0), B(10, 6) on squared paper.

a Calculate the area of triangle OAB.
b If D is the point (14, 0), calculate the area of triangle OBD.

5 On squared paper draw a square with area 36 square units. Draw a rectangle with the same area, but which is not a square. Draw a right-angled triangle with the same area. Can you find more than one rectangle and right-angled triangle?

6 The dotted lines in Figure 8 are axes of symmetry of the rectangle ABCD. Name four isosceles triangles in the figure, in each case writing down the names of their equal sides and angles.

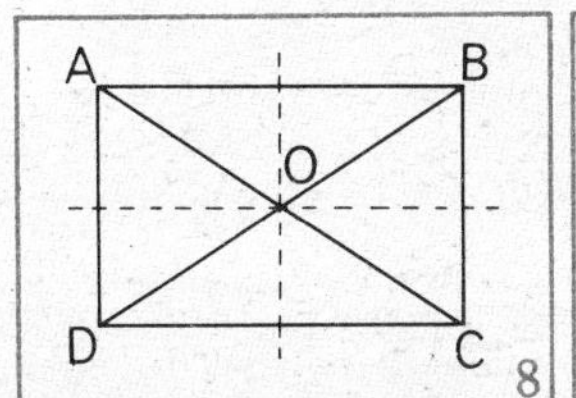

8

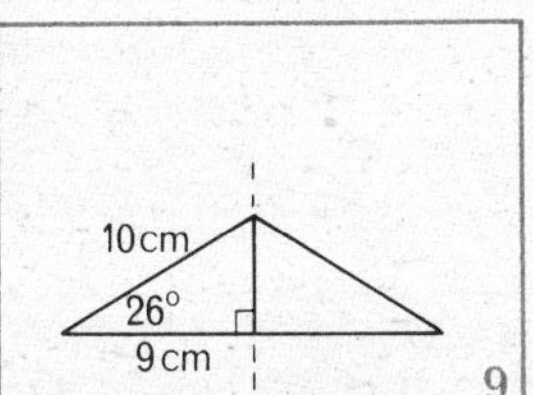

9

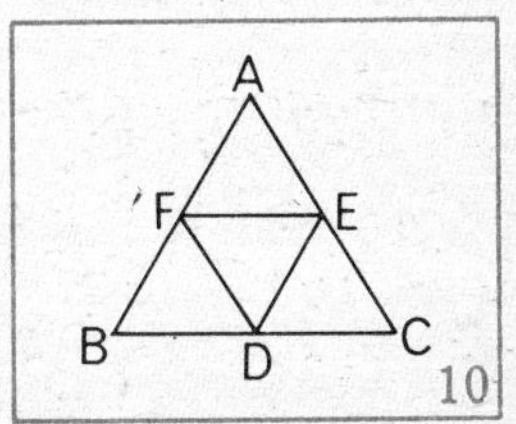

10

7 Copy the isosceles triangle in Figure 9, and fill in the lengths of as many sides and the sizes of as many angles as you can.

8 A rectangle of area 24 cm^2 is cut into two right-angled triangles which are placed side by side to form isosceles triangles in two different ways. Draw sketches, and write down the area of each isosceles triangle.

9 Plot the points A(4, 5), B(2, 0), C(6, 0). Explain why you think that $\triangle$ABC is isosceles. Name the equal sides and angles, and calculate the area of the triangle.

10 Figure 10 shows four congruent equilateral triangles.

a How many angles of 60° are there in the figure?
b Why is BDC a straight line?
c Name another equilateral triangle in the figure.

11 A regular tetrahedron (a pyramid all of whose faces are equilateral triangles) has edges 16 cm long. Sketch the tetrahedron, and calculate the total length of its edges. If the area of each face is 111 cm^2, calculate the total surface area.

12 Which of the following angle sizes are possible for a triangle ABC?

a $\angle A = 50°$, $\angle B = 60°$, $\angle C = 70°$
b $\angle A = 61°$, $\angle B = 60{\cdot}5°$, $\angle C = 59°$
c $\angle A = 60{\cdot}5°$, $\angle B = 60{\cdot}5°$, $\angle C = 59°$

13 In an isosceles triangle ABC, AB = AC.

a If $\angle B = 70°$, calculate angles A and C.
b If $\angle A = 70°$, calculate angles B and C.

14 An equilateral triangle is divided into two congruent right-angled triangles by one of its axes of symmetry. Draw a diagram, and fill in the sizes of all the angles.

15 Calculate the third angle of $\triangle$ABC in each of the following:

a $\angle A = 30°$, $\angle B = 70°$ *b* $\angle C = 21°$, $\angle B = 43°$
c $\angle A = p°$, $\angle B = q°$

16 Sketch Figure 8, with $\angle ODC = 29°$, and fill in the sizes of all the other angles in the diagram.

17 Calculate the areas of the following triangles:

Base	6 cm	8·8 m	8 mm	y cm	p cm
Height	4 cm	10 m	x mm	12 cm	q cm

18 Calculate the height of a triangle with area 84 cm^2, and base 6 cm long.

19 In rectangle ABCD, AC and BD intersect at O, AB = 18 cm and BC = 12 cm. Calculate the areas of the triangles which have a vertex at O.

20 Construct the following triangles, and measure their remaining sides and angles.

a AB = 5 cm, BC = 6 cm, AC = 7 cm
b AB = 6 cm, BC = 8 cm, $\angle$ABC = 110°
c BC = 10 cm, $\angle$ABC = 70°, $\angle$ACB = 45°.

Revision Exercise 2B

1 Figure 11 shows a rectangle, cut along one diagonal. Sketch the shapes you would obtain by fitting the two triangles together with the following pairs of angles alongside each other (not overlapping):

a A_1, C_2 and C_1, A_2 *b* A_1, D and B, C_2 *c* B, D and A_1, C_2

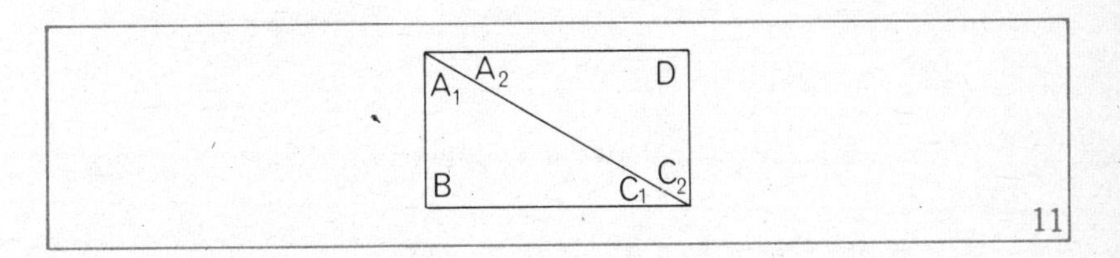

11

2 Sketch a right-angled triangle which has one side next to the right angle 9 cm long, and which has the same area as a square of side 6 cm. What length is the other side next to the right angle?

3 Plot the points A(1, 0), B(1, 10), C(6, 10), and calculate the area of $\triangle$ABC. Write down the coordinates of M, the midpoint of AC, and the coordinates of the image of B under a half turn about M.

4 A rectangle 8 cm long and 6 cm broad is cut along one diagonal. What are the heights of the two possible isosceles triangles that can be made? What is the length of the base of each?

5 Plot the points P(6, 2) and Q(2, 6). Mark points R, R_1, R_2 ... which make isosceles triangles PQR, etc., with RP = RQ.

Draw the line on which R, R_1, R_2, ... all lie. How many points on this line will give equilateral triangles PQR?

6 Construct isosceles triangles as follows:

a AB = AC = 8 cm, $\angle ABC = 70°$

b AB = AC, and the height of the triangle from A to BC is 6 cm

c AB = AC, the height from A to BC is 8 cm, and $\angle ABC = 70°$.

7 A is the point (2, 0), B(8, 0). By drawing $\angle BAC = 60°$, find C such that $\triangle ABC$ is equilateral. Estimate the coordinates of C.

What would be the dimensions of a rectangle formed from the two halves of $\triangle ABC$?

8 Figure 12 shows the net of a triangular pyramid from four congruent equilateral triangles. Sketch the pyramid. Which single lines in Figure 12 formed edges of the pyramid, and which pairs fitted together to form other edges?

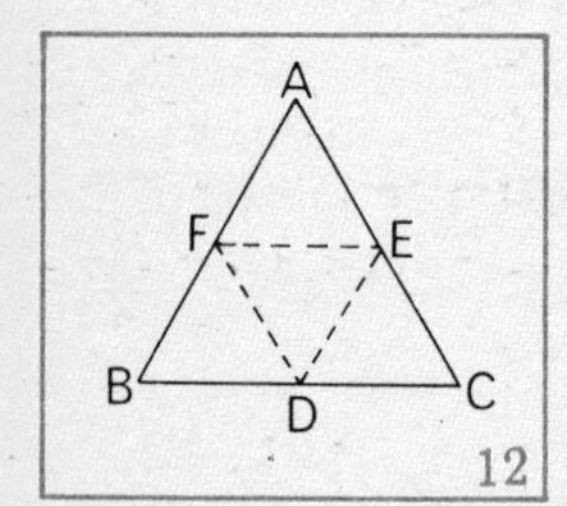

12

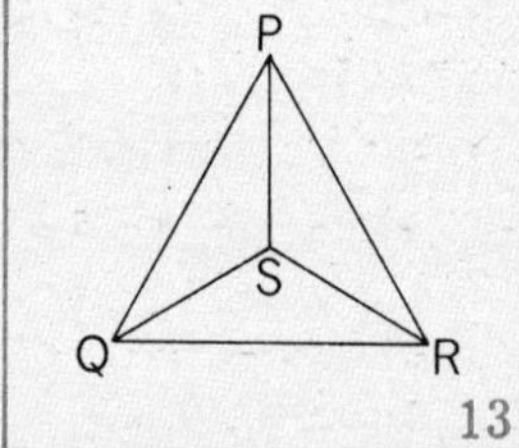

13

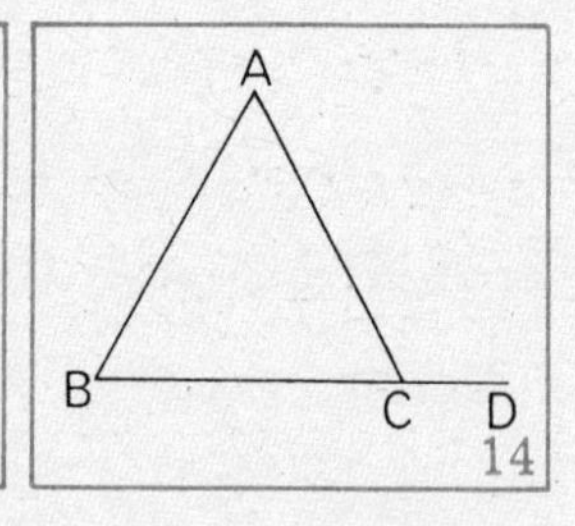

14

9 Why are no dotted lines shown in the sketch of the tetrahedron in Figure 13?

10 In Figure 14, calculate $\angle ACD$, given that:

a $\angle A = 70°$, $\angle B = 40°$ *b* $\angle A = 75°$, $\angle B = 45°$

c $\angle A = x°$, $\angle B = y°$

Can you draw a conclusion from your answers about the *exterior angle* ACD?

11 Calculate the sum of the angles of a quadrilateral ABCD. (Hint. Draw in one diagonal.)

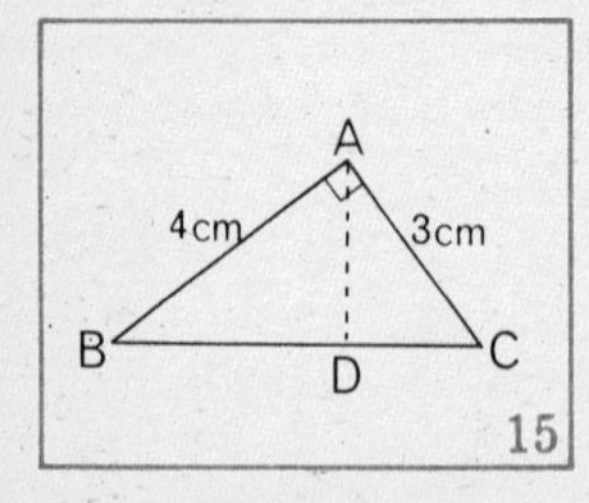

15

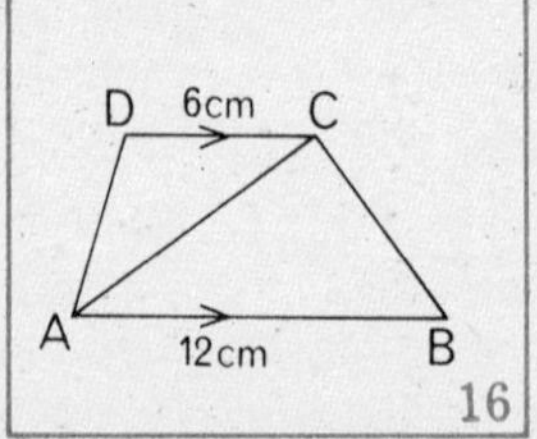

16

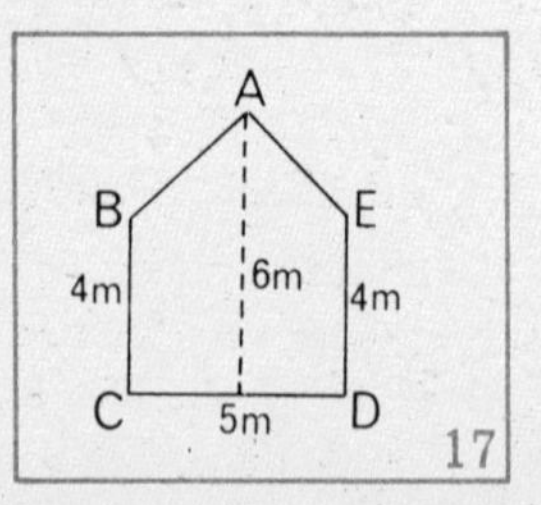

17

12 Figure 15 shows a triangle ABC with altitude AD. $\angle BAC = 90°$, AB = 4 cm, and AC = 3 cm. Calculate the area of $\triangle ABC$. Given that BC = 5 cm, calculate AD.

13 In Figure 16, AB is parallel to DC. If the distance between the parallel lines is 5 cm, calculate the area of ABCD.

14 In Figure 17, ABCDE represents the gable-end of a shed. Calculate its area.

15 Construct a triangle ABC satisfying the following conditions:

a BC = 7 cm, $\angle ABC = 43°$, $\angle BAC = 62°$.
b AB = 6 cm, $\angle ABC = 40°$, BC = 8 cm.
c AB = 8 cm, $\angle ABC = 40°$, AC = 6 cm; can you find two triangles?

16 A ship sails 5 km on a bearing of 070°, then 4 km on a bearing of 115°. By means of a scale drawing find the bearing and distance of the last point from the starting point.

17 Two boys on the same level as the base of a tower are due south of it, and are 60 m apart. They measure the angles of elevation of the top of the tower to be 40° and 23°. Find the height of the tower by means of a scale drawing.

18 A is the point (0, 4), B(2, 0), C(6, 2). Is $\triangle ABC$

a right-angled   *b* isosceles?

Explain why. Complete rectangle AODE, where O is the origin, D is (6, 0) and E is (6, 4). Hence or otherwise calculate the area of $\triangle ABC$. Try to deduce the approximate length of AB.

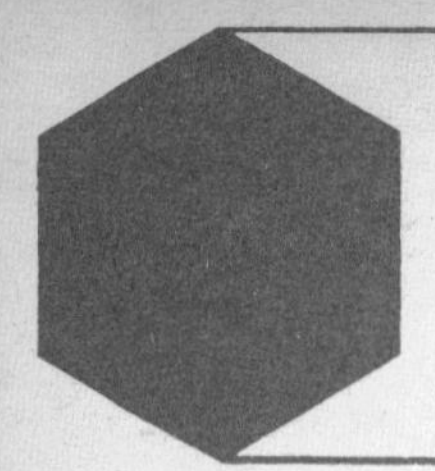

Cumulative Revision Section (Books 1 and 2)

Book 1 Chapter Summaries

Chapter 1 Cube and Cuboid

1 Each face of a *cuboid* is a *rectangle*.

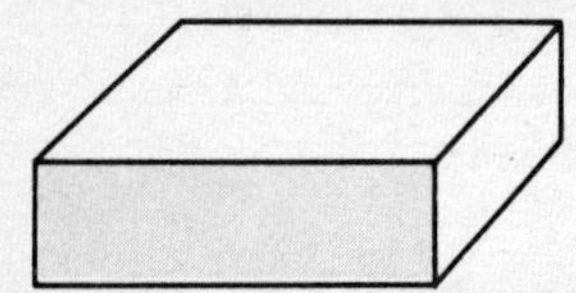

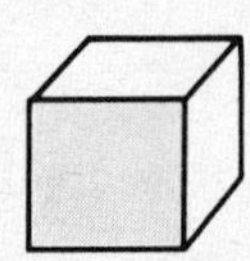

2 Each face of a *cube* is a *square*.

3 Cubes and cuboids each have 3 sets of *parallel* edges. They can be constructed from suitable *nets*.

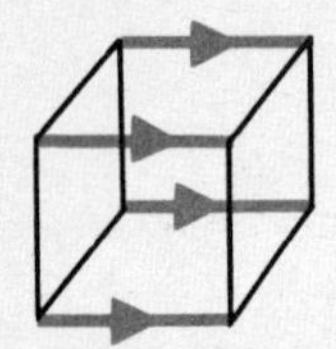

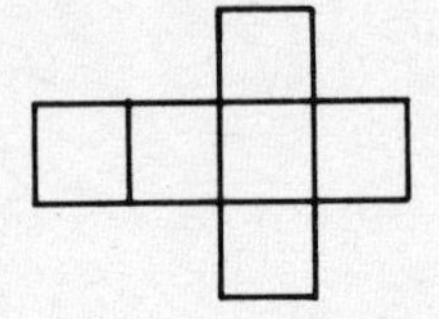

4 Tiles of the same shape and size are *congruent* to each other.

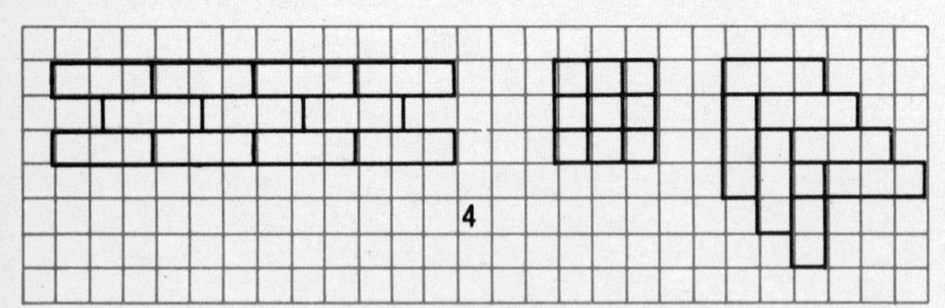

5 Some shapes have *half turn symmetry*.

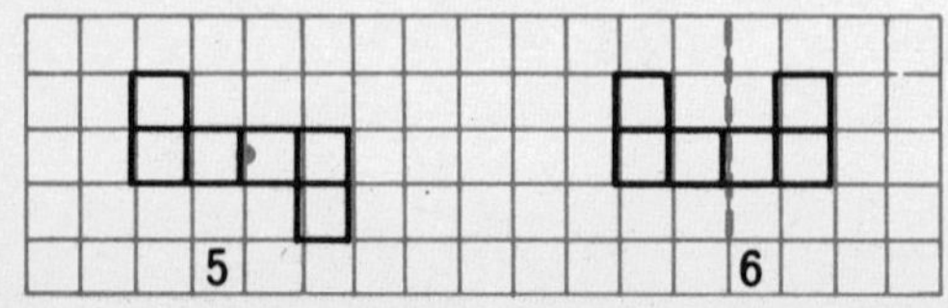

6 Some shapes have *line symmetry*.

Chapter 2 Angles

1 *Definitions*

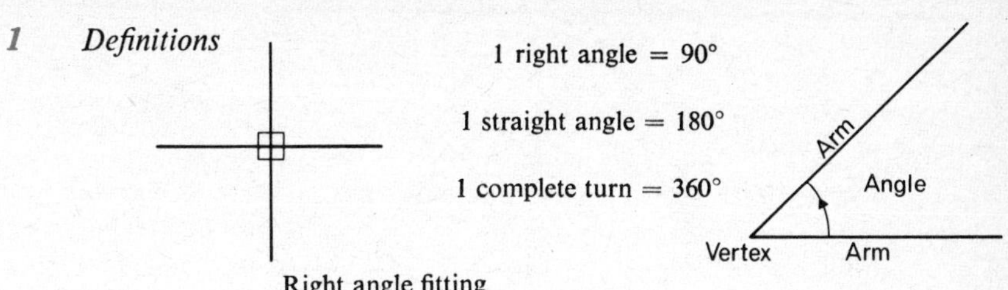

1 right angle = 90°

1 straight angle = 180°

1 complete turn = 360°

Right angle fitting property

2 *Kinds of angles*

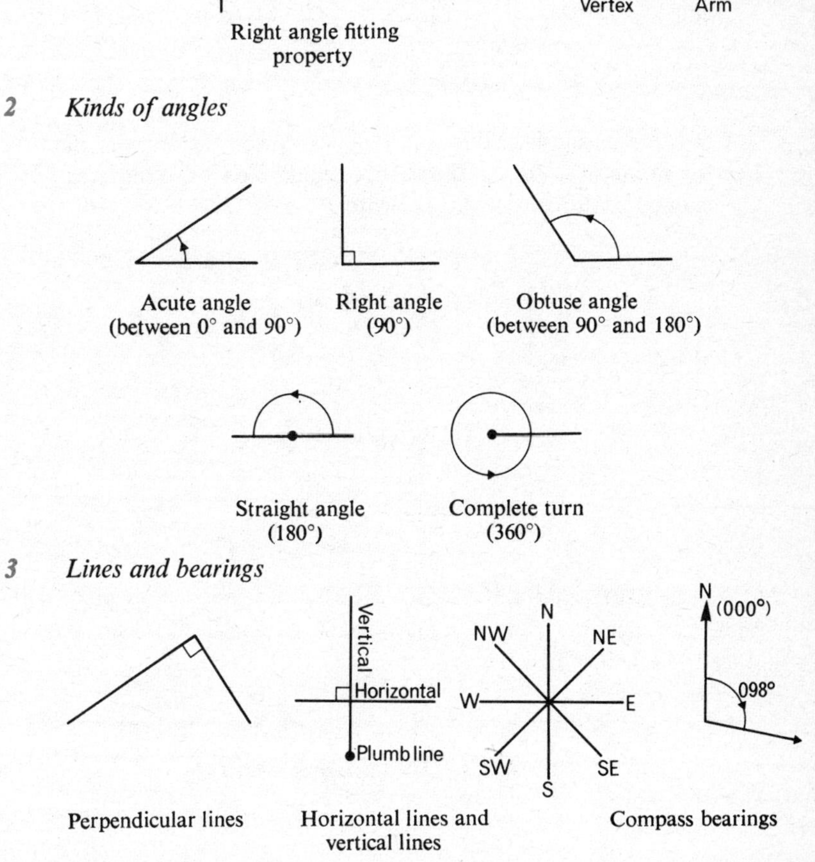

Acute angle (between 0° and 90°)

Right angle (90°)

Obtuse angle (between 90° and 180°)

Straight angle (180°)

Complete turn (360°)

3 *Lines and bearings*

Perpendicular lines

Horizontal lines and vertical lines

Compass bearings

4 *Related angles*

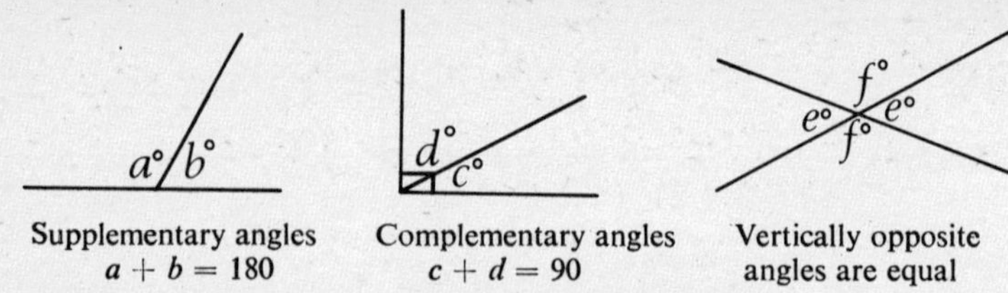

Supplementary angles $a + b = 180$ | Complementary angles $c + d = 90$ | Vertically opposite angles are equal

5 Problems on heights and distances may be solved by *scale drawing*.

Chapter 3 Coordinates

The position of a point is given in the XOY plane by its coordinates. For the point A(3, 2), 3 is the first coordinate or x-coordinate, 2 is the second coordinate or y-coordinate.

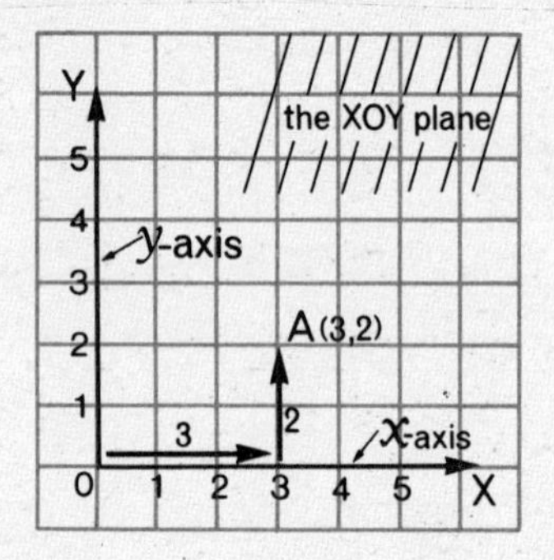

The axes OX and OY are often called Cartesian axes after the seventeenth-century French mathematician Descartes who introduced the idea of coordinates.

Cumulative Revision Exercises

Exercise A

1 Sketch a cube, a cone, and a pyramid.

2 A skeleton model of a cuboid is made of thin rods of lengths 4 cm, 5 cm and 6 cm. What is the total length of rod required?

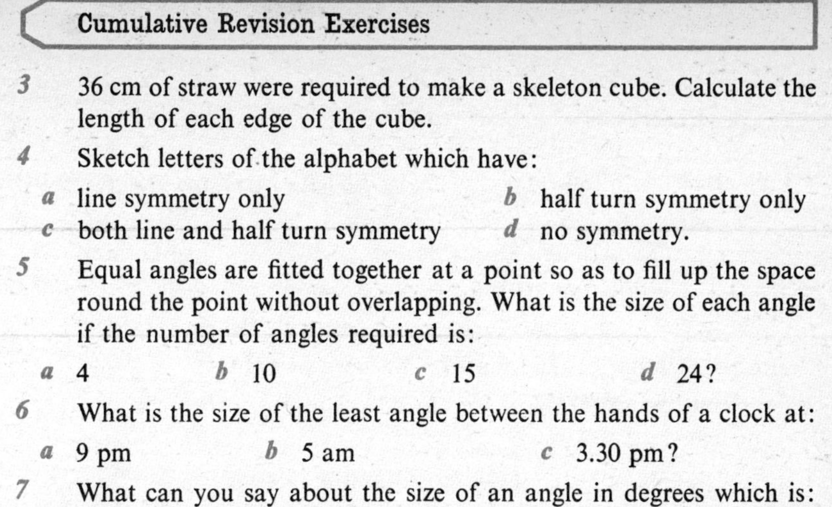

3 36 cm of straw were required to make a skeleton cube. Calculate the length of each edge of the cube.

4 Sketch letters of the alphabet which have:

a line symmetry only *b* half turn symmetry only
c both line and half turn symmetry *d* no symmetry.

5 Equal angles are fitted together at a point so as to fill up the space round the point without overlapping. What is the size of each angle if the number of angles required is:

a 4 *b* 10 *c* 15 *d* 24?

6 What is the size of the least angle between the hands of a clock at:

a 9 pm *b* 5 am *c* 3.30 pm?

7 What can you say about the size of an angle in degrees which is:

a acute *b* right *c* obtuse *d* straight *e* a complete turn?

8 What geographical directions correspond to bearings of:

a 090° *b* 225° *c* 270° *d* 000°?

9 Say whether the angles in the following pairs are complementary, supplementary, or neither.

a 40°, 140° *b* 60°, 130° *c* 60°, 30° *d* 133°, 47° *e* 72°, 18°

10 $\angle ABC = 32°$. Give the size of an angle which:

a is vertically opposite $\angle ABC$
b is complementary to $\angle ABC$
c is supplementary to $\angle ABC$
d makes up a complete turn with $\angle ABC$.

11 The angle of elevation of a church spire from a point 85 metres away horizontally from the base of the spire is 36°. Make a rough sketch to show these facts, and find the height of the spire by means of a scale drawing.

12a Show on a diagram the set of points
$\{(0, 3), (1, 4), (2, 2), (3, 7), (4, 4), (5, 2), (8, 8), (9, 2)\}$.

b List the members in this set which are equidistant from OX and OY.

c If (p, q) is equidistant from OX and OY, what can you say about p and q?

13 Plot the points A(2, 3), C(6, 5). A and C are opposite corners of a rectangle whose sides are parallel to the axes. Find the coordinates of the other two vertices.

14a Indicate on a diagram the line given by the set of points whose x-coordinate is 7.

b Is (7, 5) a member of this set? Is (5, 7)?

c Which of the following points lies to the left of the line: (4, 6), (8, 3), (9, 7), (5, 8)?

15 In a rectangle ABCD, AB is longer than BC. AC and BD intersect at O. Count the number of:

a right angles *b* acute angles *c* obtuse angles, in the figure.

16 In the diagram for question ***15***, name two pairs of:

a vertically opposite angles *b* complementary angles

c supplementary angles

17 If in the figure for question ***15***, AB = 16 cm, AO = 10 cm, BC = 12 cm, write down the lengths of all the other lines in the diagram.

18 If in the diagram for question ***15***, $\angle$AOD = 78°, fill in the sizes of all the other angles.

19 R is the point (0, 3), S is (6, 3), T is (6, 6). If RSTU is a rectangle give the coordinates of U.

20 A is the point (2, 3) and B is (6, 5), and AB is a diagonal of a rectangle with its sides parallel to the axes. Find the coordinates of the other two vertices.

21 P is (5, 0) and Q is (0, 5). TPQ is an isosceles triangle. Give the coordinates of four possible positions of T. Calculate the area of $\triangle$OPQ.

22 The base of $\triangle$ABC is 8 cm long, and its height is 6 cm. Sketch $\triangle$DEF which has the same area as $\triangle$ABC, but has a base twice as long. What is the height of $\triangle$DEF?

23 Construct isosceles triangles in which:

a AB = AC = 10 cm, and $\angle$BAC = 50°

b BC = 6 cm, $\angle$ABC = $\angle$ACB = 65°

c AB = AC = 12 cm, and BC = 8 cm.

24 A is (6, 2), and B is (3, 6). AB is given a half turn about (5, 5) into the position A′B′. Write down the coordinates of A′ and B′. What can you say about AB and B′A′? And about AB′ and BA′?

Exercise B

1 Sketch and name a solid which has:

a no edges *b* only one edge *c* only two edges.

2 What is the total length of straw needed to make a skeleton cuboid with a square base 4 cm long, and with a height of 5 cm?

3 Given four rectangular cards each 16 cm by 8 cm, what would be the dimensions of two more cards required to make a cuboid? Is there more than one answer? Sketch the cuboid or cuboids.

4 Equal angles are fitted together at a point so as to fill up the space completely without overlapping. How many angles are required if the size of each is:

a 10° *b* 15° *c* 24° *d* 90° *e* 120°?

5 What are the sizes of the *two* angles between the hands of a clock at:

a 2100 hours *b* 2145 hours *c* 2120 hours?

6 a Give the three-figure bearings of north, south and south-west.

b Two aircraft are flying on the same course of 212°. One alters course to 301°, and the other to 179°. Calculate the angle between their courses now.

7 OA, OB, OC, OD, OE are lines from a point O, in that order, and AOE is a straight angle. $\angle AOB = w°$, $\angle BOC = x°$, $\angle COD = y°$, $\angle DOE = z°$.

a If $\angle AOB = \angle COD = 58°$, and $\angle BOC = 40°$, find the size of $\angle DOE$.

b If $\angle BOD = 100°$, and $\angle AOB = \angle DOE$, find the size of $\angle BOE$.

c If $\angle AOC = 70°$, $w = x$ and $y = z$, find the size of $\angle BOD$.

d If $w = x$ and $y = z$, explain why $\angle BOD$ must be a right angle.

8 From a point 120 metres from the base of a building the angles of elevation of two parts of the building are 42° and 58°. By means of a scale drawing find how far the second part is above the first.

9 a Show on a diagram the set of points:
$S = \{(1, 2), (6, 3), (4, 8), (5, 10), (7, 2), (3, 5), (8, 16)\}$.

b If (x, y) denotes a member of this set, list the points for which $y = 2x$.

c List the members in this set for which x is greater than y.

10a A is the point (3, 3), B is (7, 3), C is (7, 6), D is (3, 6). What kind of figure is ABCD?

b Another shape BEFC is drawn congruent to ABCD (i.e. the same shape and size as ABCD). Write down the coordinates of E and F.

11a Indicate on a diagram the line given by the set of all points whose first coordinate is 10.

b If P(x, y) lies to the right of this line, can you say anything about x or y or both?

c If R(a, b) is a point such that a is less than 10, where does R lie relative to this line?

12a Indicate the line given by the set of all points whose y-coordinate is 5.

b If T(r, s) lies below this line, what can you say about r or s or both?

c If P(l, m) is a point such that m is greater than 5, where does P lie relative to this line?

13 Draw a rectangle ABCD on squared paper, with its diagonals intersecting at M. Why is it possible to draw a circle, centre M, to pass through A, B, C and D? Draw the circle.

14 If in the rectangle in question ***13*** $\angle$BAC = 42°, fill in the sizes of all the other angles in the figure. How many pairs of vertically opposite, complementary and supplementary angles are there?

15 A(4, 3) is one vertex of a square whose sides are 2 units long and are parallel to the axes. How many such squares can you draw? Write down the coordinates of all their vertices.

16 A is (3, 3), B(7, 3), C(7, 6), D(3, 6). What kind of shape is ABCD?
Calculate its area, and give the coordinates of the point where its axes of symmetry intersect.

17 In $\triangle$ABC, AB = 8 cm, BC = 6 cm, AC = 10 cm, and $\angle$B = 90°. Calculate the area of $\triangle$ABC. What would the area of the triangle be if all the sides were doubled?

18 In $\triangle$ABC, $\angle$C = 90°. Calculate the sizes of the other angles, a° and b°, if:

a $a = b$ *b* $a = 2b$ *c* $a = \frac{1}{4}b$ *d* $a-b = 45$

19 Calculate the sizes of the angles of an isosceles triangle if each of the equal angles is four times the size of the third angle.

20 A tetrahedron has an isosceles triangular base whose sides are 4 cm, 4 cm and 7 cm long. The other edges of the pyramid are all an equal whole number of centimetres long. Could the total length of all the edges be:

a 24 cm   *b* 27 cm   *c* 32 cm?

21 Explain why each angle of an equilateral triangle must be 60°.

22 Construct $\triangle PQR$ with PQ = 12 cm, PR = 13 cm, $\angle PQR = 90°$. Measure QR and calculate the area of the triangle.

23 AB and CD are equal and parallel lines. Find a point about which a half turn maps AB on to DC. Can you find more than one such point? What can you say about AC and BD?

24 Draw the net of a pyramid with a square base of edge 4 cm, and with sides in the form of isosceles triangles of height 4 cm. Calculate the total surface area of the pyramid.

By measuring *one* line in your diagram, calculate the total length of all the edges of the pyramid.

Arithmetic

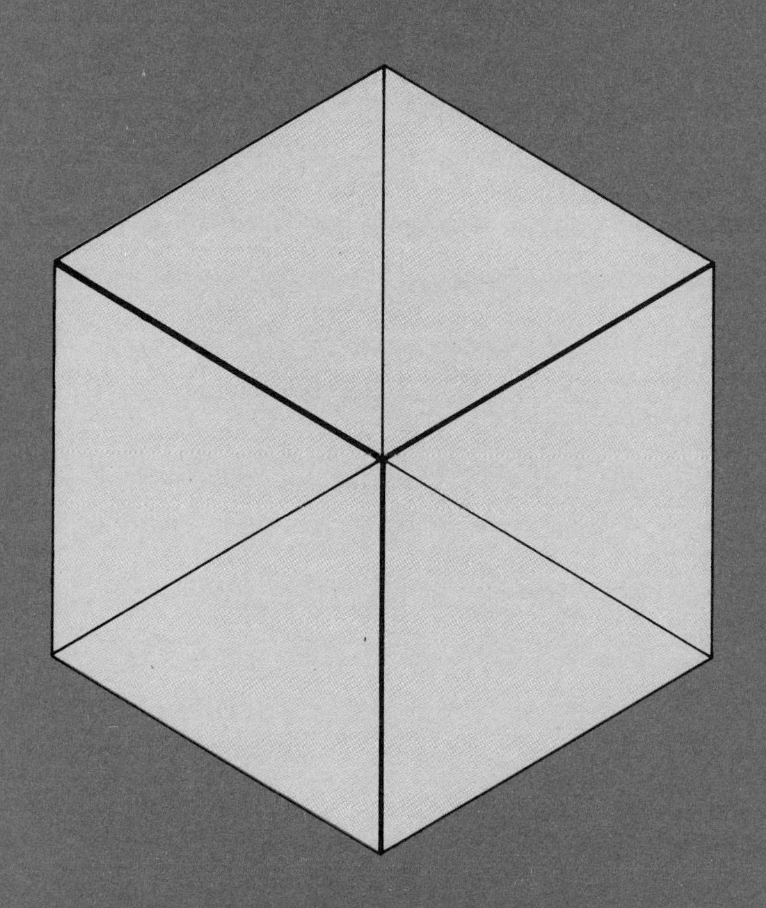

Arithmetic

Decimals and the Metric System

1 Notation

Can you see how this sequence of numbers has been formed?

1000, 100, 10, 1

If we arrange these numbers in the form of a series, as follows,

$$1000+100+10+1$$

we obtain their sum 1111, i.e. one thousand, one hundred and eleven.

This shows the meaning of the notation used in our decimal, or base 10, number system in which a figure has ten times the value of the same figure immediately to the right of it.

Exercise 1A

What numbers are given by the following?

1 80+3 *2* 100+30+5

3 400+20+7 *4* 200+2

5 2000+300+40+6 *6* 40000+3000+700+1

Explain the meaning of each of these numbers:

7 68 *8* 234 *9* 1324 *10* 24124

11a In the sequence ..., 1000, 100, 10, 1, ... how can we obtain the number immediately to the left of any given number?

b Write down the next three numbers to the left of 1000 in the sequence.

12 Give the next three terms to the left in the sequences:

a ..., 400, 40, 4, ... *b* ..., 50, 5, $\frac{5}{10}$, ...

c ..., 6000, 600, 60, ... *d* ..., 7, $\frac{7}{10}$, $\frac{7}{100}$, ...

13a In the sequence..., 1000, 100, 10, 1, ... how can we obtain the number immediately to the right of any given number?

b Write down the next 3 numbers to the right of 1.

14 Give the next three terms to the right in the sequences given in question *12*.

15 100 can be written as 10×10, or 10^2, a power of 10.
Write down the whole sequence of numbers from 1000 to $\frac{1}{1000}$, i.e. from 10^3 to $\frac{1}{10^3}$, expressing the numbers as powers of 10.

Exercise 1B

What numbers are given by the following?

1 $300+20+3$ *2* $700+7$

3 $4000+300+1$ *4* $20000+500+6$

Explain the meaning of each of these numbers:

5 542   *6* 1506   *7* 2020   *8* 98765

9 Give the next three terms to the left, and the next three terms to the right, in the sequences:

a ..., 800, 80, 8,...   *b* ..., 2000, 200, 20, ...   *c* ..., 9, $\frac{9}{10}$, $\frac{9}{100}$, ...

10 254 can be written as $(2 \times 10^2)+(5 \times 10)+4$. Express the following numbers in this way:

a 323   *b* 5000   *c* 6079   *d* 25846

The decimal point

If we denote hundreds by H, tens by T, units by U, tenths by t and hundredths by h we can show numbers under these headings as follows:

H	T	U	t	h
2	3	4		
	2	3	4	
		2	3	4

Each number contains 2, 3, 4 in that order, but each is a different number. To distinguish between whole numbers and fractions we insert a *decimal point* after the units figure, so that the above numbers can be written 234· (or just 234), 23·4, and 2·34 (read 'two point three four').

We can now give a number in three ways:

(i) As shown in the table (under the headings H, T, U, etc.).
(ii) As a decimal, 2·34.
(iii) As a mixed number, $2{\cdot}34 = 2+\frac{3}{10}+\frac{4}{100} = 2\frac{34}{100}$, or $2\frac{17}{50}$.

In the case of a decimal fraction like ·56 we will always insert a zero in front of the point, giving 0·56.

Example 1. $0{\cdot}8 = \frac{8}{10} = \frac{4}{5}$
Example 2. $1{\cdot}45 = 1\frac{45}{100} = 1\frac{9}{20}$

Exercise 2A

Read out the following numbers:

1	3·4	*2*	62·3	*3*	5·83	*4*	2·79
5	521·7	*6*	324·3	*7*	0·62	*8*	18·01

Express each of the following numbers as a decimal, and also as a mixed number:

	H	T	U	t	h
9			5	9	
10			8	5	
11		1	2	3	
12		2	3	6	
13			5	8	7
14	1	0	4	7	
15		2	3	1	2

Express each of the following numbers in decimal form:

16	$\frac{1}{10}$	*17*	$\frac{3}{10}$	*18*	$\frac{5}{10}$	*19*	$\frac{7}{10}$
20	$7\frac{9}{10}$	*21*	$3\frac{9}{100}$	*22*	$1\frac{9}{1000}$	*23*	$\frac{1}{5}$

Change the following into mixed numbers or fractions in their simplest form:

24	0·5	*25*	0·6	*26*	1·4	*27*	2·7
28	8·8	*29*	0·15	*30*	0·35	*31*	0·28
32	0·75	*33*	0·05	*34*	2·34	*35*	1·08

Exercise 2B

Read out the following numbers:

1	28·4	*2*	3·1416	*3*	9·42	*4*	304·6
5	0·75	*6*	0·09	*7*	800·21	*8*	1234·5

Express each of the following numbers as a decimal, and also as a mixed number:

	H	T	U	t	h
9		8	3	2	
10	2	4	9	1	
11		3	0	5	
12	1	0	2	9	
13		2	8	0	9
14	5	4	3	2	1

Express each of the following numbers in decimal form:

15 $\frac{9}{10}$ *16* $\frac{3}{100}$ *17* $\frac{5}{1000}$ *18* $1\frac{3}{10}$

19 $2\frac{7}{100}$ *20* $30\frac{15}{100}$ *21* $17\frac{23}{1000}$ *22* $10\frac{1}{50}$

Change the following into mixed numbers or fractions in their simplest form:

23 0·7 *24* 1·9 *25* 3·6 *26* 0·23

27 0·79 *28* 0·06 *29* 1·02 *30* 0·125

31 0·0357 *32* 12·375 *33* 11·806 *34* 20·005

2 *Measurement*

All measurements are approximate. Suppose we try to find the length of the line AB in Figure 1.

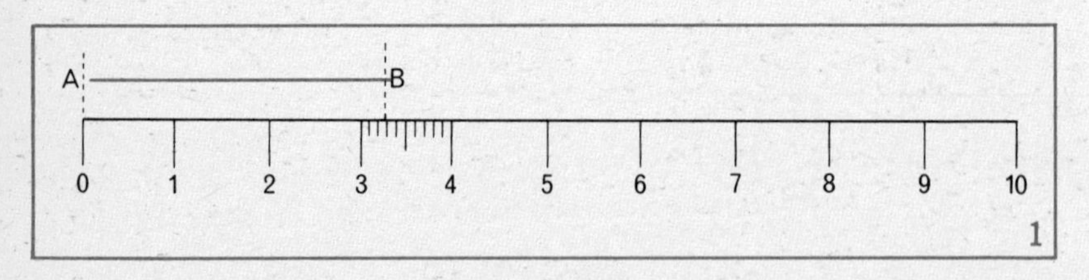

The ruler shows that the length is between 3·2 and 3·3 centimetres, and nearer 3·3 than 3·2.

We say that '*the length of AB is 3·3 cm, to the nearest tenth of a centimetre*', or simply that '*the length of AB is 3·3 cm*'.

Imagine that there are 10 divisions on the ruler between each tenth of a centimetre. How long would each one be? Under a magnifying glass they might appear as in Figure 2.

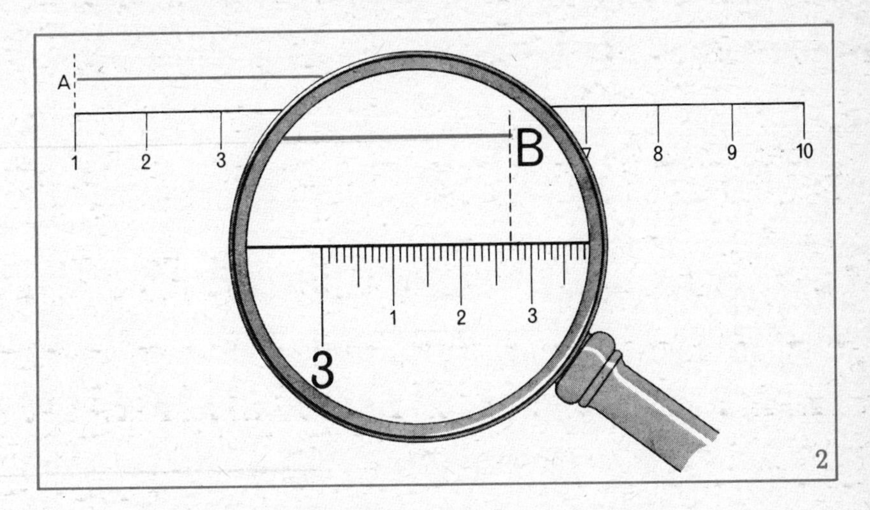

2

We can now see that the length of AB is 3·27 cm, *to the nearest hundredth of a centimetre.*

We can never find the exact length of AB, and always have to give the best possible approximation.

The breadth of this page might reasonably be measured to the nearest tenth of a centimetre, your height to the nearest centimetre, the distance between two towns to the nearest kilometre.

Exercise 3

1 Which unit of length (mm, cm, m or km) would you use to give:

a the length of a football pitch *b* the width of your desk
c the distance across an ocean *d* the thickness of a coin
e the distance to the moon *f* the height of a house?

2 *a* Using a ruler and pencil, draw lines of lengths 5 cm, 7 cm, 12·5 cm and 8·2 cm.

b Ask your neighbour to measure the lengths of the lines to the nearest tenth of a centimetre.

c What effect would a very blunt pencil have on the answers?

3 Measure the length and breadth of this page, to the nearest millimetre.

4 Measure the lengths of the lines in Figure 3, to the nearest tenth of a centimetre.

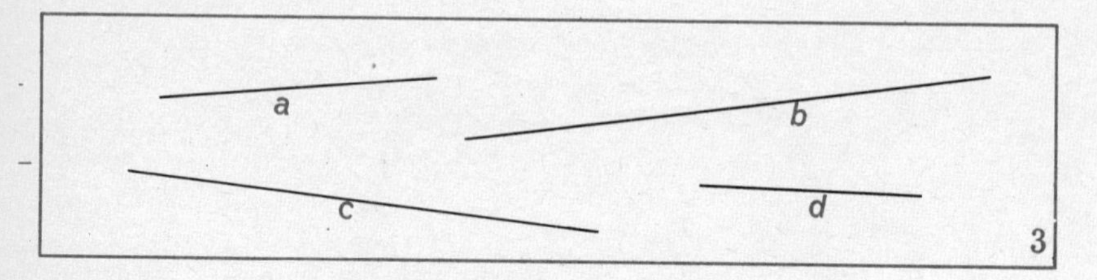

5 Measure the length of each side of the shapes in Figure 4.

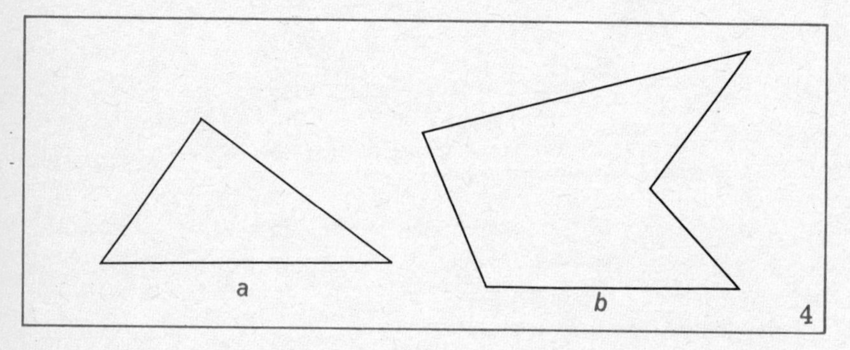

6 ***a*** Without measuring, draw lines whose lengths you estimate to be 6 cm, 9 cm and 10·5 cm.

b Measure the lengths of the lines to the nearest tenth of a centimetre.

---

3 Approximation

As we have seen, all measurement is approximate, and measures of length, mass, time, area, etc., should always be given to a reasonable degree of approximation. There are three main ways in which this is done; by *rounding off* to:

a the nearest appropriate unit
b an appropriate number of decimal places
c an appropriate number of significant figures.

a To the nearest unit

Suppose that a boy's height is measured as 164·2 cm. It would be reasonable to record this as 164 cm, rounded off to the nearest centimetre.

The rules for rounding off a number are as follows: If the following figure is greater than 5, increase the round-off figure by 1; if the following figure is 5, round off to the nearest even number; otherwise leave the round-off figure as it is.

Examples

a 14·7 kg = 15 kg, rounded off to the nearest kilogramme.
b 10·13 s = 10·1 s, rounded off to the nearest tenth of a second.
c 128·5 m = 128 m, rounded off to the nearest metre.
d 128·51 m = 129 m, rounded off to the nearest metre.

Exercise 4

Round off the following to the degree of approximation shown.

1 To the nearest cm: 15·8 cm, 8·4 cm, 6·5 cm, 88·8 cm

2 To the nearest kg: 18·3 kg, 64·9 kg, 13·5 kg, 19·1 kg

3 To the nearest tenth of a second: 11·17 s, 23·84 s, 56·35 s, 9·89 s

4 To the nearest km: 45·6 km, 1047·5 km, 234·5 km, 8·2 km

5 To the nearest ml: 125·6 ml, 80·3 ml, 18·8 ml, 7·1 ml

6 To the nearest mm: 11·47 cm, 14·739 cm, 0·864 cm, 10·98 cm

7 To the nearest cm^2: 51·1 cm^2, 100·5 cm^2, 7·7 cm^2, 6047·4 cm^2

8 To the nearest hundredth of a second: 10·147 s, 54·606 s, 28·015 s

b Decimal places

Approximations are not only applied to measurements; sometimes it is convenient to round off a decimal by expressing it to a given number of *decimal places*.

For example, 5·20735 = 5·2074, rounded off to 4 decimal places
= 5·207, rounded off to 3 decimal places
= 5·21, rounded off to 2 decimal places
= 5·2, rounded off to 1 decimal place.

Exercise 5

1 Round off each of the following to 1 decimal place:
4·71, 9·68, 27·82, 0·09, 8·92, 8·45, 6·75

2 Round off each of the following to 2 decimal places:
5·327, 0·048, 1·053, 0·0087, 3·149, 8·025

3 Round off each of the following to 3 decimal places:
6·5432, 12·0256, 0·07943, 0·5356, 1·0075

4 Round off 0·14285 to:
a 4 decimal places *b* 3 decimal places *c* 2 decimal places

5 $\frac{1}{3}$ = 0·3333.... Express $\frac{1}{3}$ as a decimal fraction rounded off to 2 decimal places.

6 $\frac{2}{3}$ = 0·6666.... Express $\frac{2}{3}$ as decimal fractions rounded off to 2 decimal places, and to 3 decimal places.

7 $\frac{7}{11}$ = 0·636363.... Express $\frac{7}{11}$ as decimal fractions rounded off to 2, 3, and 4 decimal places.

c Significant figures

Suppose that you measure the length of a window with a steel tape and find that it is 67·3 cm long. If the tape measure was made of linen you could not measure the length so accurately, and would probably have to say that the window was 67 cm long.

A convenient way to indicate the degree of approximation is by means of the number of figures used. We say that 67·3 cm has 3 *significant figures*, and that 67 has 2 *significant figures*.

Exercise 6

Give the number of significant figures in each of these:

1	2·2 cm	*2*	13 years	*3*	154 months
4	27·4 g	*5*	8 kg	*6*	6 cm^2
7	32·4 litres	*8*	12·7 m^2	*9*	9·9 s
10	56·2 g	*11*	1256 kg	*12*	4·58 m
13	4 mm	*14*	15·8 km/h	*15*	1258·2 cm

* * * * *

A zero is a significant figure except when it is used simply to indicate the position of the decimal point.

Examples

a 504 cm. 3 significant figures.
b 5·04 m. This is the same length, expressed in a different unit.
c 2·40 m. The 0 indicates that the length has been measured to the nearest hundredth of a metre, and is significant. 3 significant figures.
d 0·45 m. The 0 shows the position of the point, and is not significant. 2 significant figures.
e 0·0810 km. The first two 0s show the position of the point, and are not significant. The third 0 shows that the length has been measured to one tenth of a metre, and is significant. 3 significant figures.

Exercise 7

Give the number of significant figures in each of these:

1	23	*2*	2·3	*3*	2·03	*4*	2·003
5	0·003	*6*	0·40	*7*	0·303	*8*	1·555
9	0·0404	*10*	1047	*11*	0·0001	*12*	3·690
13	2·54	*14*	0·25	*15*	10·03	*16*	1234
17	205	*18*	2·05	*19*	2·050	*20*	8009

Exercise 8

1 Round off to 2 significant figures:
9·63, 1·87, 13·1, 25·9, 5·02, 2·97

2 Round off to 3 significant figures:
48·21, 2·465, 0·4910, 0·4991

3 Round off to 4 significant figures:
1·2865, 0·12964, 0·05378, 124·651, 0·005816

4 Round off 3·1415926 to:
a 3 significant figures *b* 4 significant figures *c* 5 significant figures

5 Repeat question ***4*** for the number 2·71828.

6 Round off the following to the number of figures shown in brackets:

a	6·135 (2)	*b*	5·007 (3)	*c*	18918 (2)	*d*	18918 (3)
e	0·00518 (2)	*f*	4·821 (1)	*g*	10·001 (4)	*h*	3·1416 (3)

4 *Addition and subtraction*

When adding numbers in decimal form we must be careful to keep the hundreds in one column, the tens in another, the units in another, the decimal points in another and so on, as you have already seen when adding money, etc.

Example 1. Add 23·46, 102·019, and 0·96.

```
 23·460
102·019
  0·960
-------
126·439
-------
```

Notice how easy it is to add decimal fractions. We use this fact to help us to add and subtract quantities expressed in the various units of measure in the metric system.

Example 2. Find the perimeter of a triangle with sides of lengths 3·5 cm, 5·7 cm, and 7·6 cm.

Perimeter = 3·5+5·7+7·6 = 16·8 cm.

Example 3. Two stations are 12·05 km and 15·82 km from the centre of a city. How much farther is one from the centre than the other?

Difference of distance = 15·82−12·05 = 3·77 km.

Exercise 9A

Add:

1 2·8 and 3·7

2 6·35 and 2·06

3 1·51 and 2·04

4 1·75, 2·32 and 7·18

5 0·065, 0·703, and 0·987

6 5·6, 3·24, and 0·732

Subtract:

7 12·5 from 17·2

8 4·78 from 9·64

9 0·943 from 1·072

10 Simplify:

a 6·75+8·61−2·75 *b* 12·84−9·96+8·16 *c* 0·965+0·068−0·473

11 Find the perimeter of a triangle with sides 3·8 cm, 4·7 cm, and 5·6 cm long.

12 1·25 litres of liquid are poured from a flask containing 3·15 litres. What volume of liquid is left?

13 Find the perimeter of a triangle with sides 6·3 cm, 7·9 cm, and 8·5 cm long.

14 A man paid accounts of £2·58, £5·64, £7·85½, and £0·23. What change did he receive from £20?

15 The perimeter of a triangle is 28·3 centimetres. If two of the sides are 9·6 centimetres and 12·3 centimetres long, find the length of the third side.

Exercise 9B

Add:

1 6·95 and 7·39

2 17·84 and 5·92

3 156·9 and 3·87

4 3·54, 13·27, and 0·08

5 164·3 and 16·08

6 29·38, 471·6, and 0·681

Subtract:

7 28·385 from 32·196

8 2·973 from 13·56

9 0·938 from 1·002

10 Simplify:

a 8·35 + 2·96 − 4·79

b 15·64 − 13·05 + 17·23

c 0·634 + 0·791 − 0·308

11 The tank of a car holds 20 litres of petrol. On a journey, 9·85 litres of petrol are used. What volume of petrol is left in the tank?

12 In a Diving Competition the gold medallist scored 49·68, 47·52, and 60·03 points, and the silver medallist scored 46·86, 51·84, and 55·38 points in the final. Find the total score of each. By how many points was the gold medallist the winner?

13 A motorist travels 236·5 km, 168·7 km, 97·6 km, and 302·8 km. How much farther does he have to travel to complete a total distance of 1000 km?

14 A man paid accounts of 12·75 dollars, 25·40 dollars, 18·87 dollars, and 39·17 dollars. What change did he receive from a 100-dollar bill?

15 Two radioactive particles follow zigzag paths before losing their activity. Atom C travels 48·6 cm, then 21·9 cm, then 8·7 cm. Atom D travels 53·4 cm, 18·7 cm, and 6·2 cm. Which atom travels farther, and by how much?

5 Multiplication

Example 1. Find the value of $3{\cdot}48 \times 10$.

When we multiply 3·48 by 10 the hundredths become tenths (since $10 \times \frac{1}{100} = \frac{1}{10}$), the tenths become units, and the units become tens as shown:

	H	T	U	t	h	
			3	4	8	× 10
=		3	4	8		

i.e. $3{\cdot}48 \times 10 = 34{\cdot}8$.

Notice that multiplication by 10 can be effected by moving the figures one place to the left, keeping the decimal point fixed (so that the number is 10 times greater). Similarly, to multiply by 100, move the figures two places to the left; and so on.

or Multiplication by 10 can be effected by moving the decimal point one place to the right of its original position (so that the number is 10 times greater). Similarly, to multiply by 100, move the point two places to the right; and so on.

Example 2. Evaluate $6{\cdot}72 \times 600$.

We can multiply by 100, then by 6, so that

$$\begin{aligned} &6{\cdot}72 \times 600 \\ =\ &672 \times 6 \\ =\ &4032 \end{aligned}$$

Example 3. Evaluate $1{\cdot}5 \times 0{\cdot}13$.

$$\begin{aligned} &1{\cdot}5 \times 0{\cdot}13 \\ =\ &1\tfrac{5}{10} \times \tfrac{13}{100} \\ =\ &\tfrac{15}{10} \times \tfrac{13}{100} \\ =\ &\tfrac{195}{1000} \\ =\ &0{\cdot}195 \end{aligned}$$

Notice that 1·5 is a whole number of tenths and 0·13 is a whole number of hundredths, so that the product is bound to be a whole number of thousandths. Also, in 1·5 there is 1 figure after the point, in 0·13 there are 2 figures after the point, and in the answer there are 3 figures after the point.

We can multiply 1·5 by 0·13 more briefly as shown.

```
   1·5    (1 figure after the point)
  0·13    (2 figures after the point)
  ----
    45
   150
  ----
 0·195    (So 3 figures after the point)
```

Exercise 10A

Multiply:

1. Each of these by 10: 3·4, 17·6, 0·84, 293·1, 0·0007
2. Each of these by 100: 1·94, 2·7, 0·06, 31·66, 0·972
3. Each of these by 10, by 100, by 1000: 24·7, 0·34, 1·9876, 0·0065
4. 5·5 by 2, by 20, by 200
5. 0·34 by 8, 80, by 800
6. 0·06 by 300, by 3000
7. 2·3 by 5, by 50, by 500
8. 2·3 by 2·1
9. 11·7 by 1·2
10. 2·5 by 2·5
11. 0·65 by 0·18
12. 0·34 by 0·28
13. 1·23 by 2·01
14. 1·07 by 0·023
15. 365 by 2·5
16. 6·87 by 1·8; then write down the value of 687×18
17. 0·47 by 6·5; then write down the value of $47 \times 0{\cdot}65$

Exercise 10B

Multiply:

1. Each of these by 10, by 100, by 1000:
   13·4, 6·85, 12·85, 0·04, 0·64, 0·0082
2. Each of these by 3, by 30, by 3000:
   2·8, 3·75, 12·34, 0·087, 0·0052
3. Each of these by 6, by 60, by 6000:
   1·5, 3·46, 9·06, 0·209, 0·0037
4. Each of these by 5, by 500:
   1·1, 7·3, 12·5, 0·78
5. 6·4 by 4·9
6. 15·23 by 14·6
7. 0·67 by 0·32
8. 0·378 by 0·092
9. 8·29 by 1·7; then write down the value of 82·9 multiplied by 17; of 829 multiplied by 0·0017

10 5·73 by 0·34; then write down the value of 57·3 multiplied by 340; of 573 multiplied by 0·00034

Rough estimates of answers (using one significant figure of each number)

We can estimate the value of $27{\cdot}5 \times 5{\cdot}4$ roughly by using one significant figure of each number.

Then we have 30×5, i.e. 150, as an estimate.

Rough estimates of answers ensure that we have put the decimal point in the correct place after a calculation, and are essential when using slide rules and calculating machines.

We can indicate that we are using approximate values by means of the symbol $\doteqdot$.

Example 1

$445 \times 0{\cdot}316$
$\doteqdot 400 \times 0{\cdot}3$
$= 120{\cdot}(0)$
$\doteqdot 100$

Example 2

$0{\cdot}73 \times 0{\cdot}089$
$\doteqdot 0{\cdot}7 \times 0{\cdot}09$
$= 0{\cdot}06(3)$

Exercise 11

1 Round off each of the following numbers to 1 significant figure: 3·7, 1·26, 2·52, 0·84, 45, 235.

In the following examples, first make a rough estimate of each answer (using 1 significant figure in each number), and then carry out the calculation.

2 Calculate the area of a rectangle 7·2 cm long and 2·8 cm broad.

3 Calculate the area of a rectangle 8·5 cm long and 2·6 cm broad.

4 A square has side 1·8 cm long. Find its perimeter and its area.

5 Find the total number of hours of sunshine in a week in which the average amount of sunshine was 2·75 hours per day.

6 Calculate the volume of a cuboid 1·2 cm long, 0·8 cm broad, and 2 cm high.

7 Calculate the volume of a cube of side 1·4 cm.

8 A rectangle is 4·5 cm long and 2·2 cm broad. Find its perimeter and its area.

9 Floor tiles are square, with their sides 225 mm long. Calculate:

a the length of a tile in cm *b* the area of a tile in cm².

10 A cuboid has length 6 cm, breadth 4·2 cm, height 1·8 cm. Find:

a the total length of all its edges *b* the total surface area

c its volume.

11 A jeweller packs valuables into a rectangular box 8·7 cm long, 6·9 cm broad, 4·3 cm high. What length of tape is required to bind all the edges?

12 Find the perimeter and area of each of the shapes in Figure 5, in which the units are centimetres and all the angles are right angles.

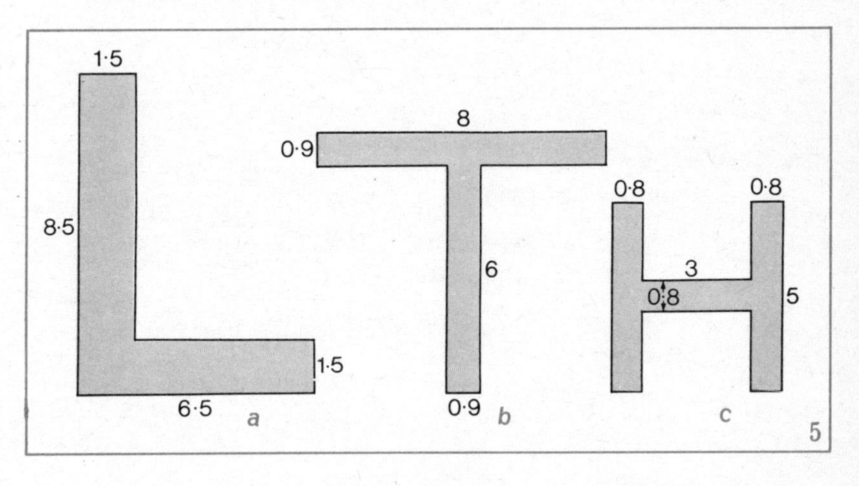

6 Division

Example 1. 56·8 ÷ 10.

When we divide 56·8 by ten the tens become units (since 10÷10 = 1), the units become tenths, the tenths become hundredths, as shown.

	H	T	U	t	h	
		5	6	8		÷ 10
=			5	6	8	

i.e. 56·8÷10 = 5·68.

Notice that division by 10 can be effected by moving the figures one place to the right, keeping the decimal point fixed (so that the number is 10 times smaller). Similarly, to divide by 100, move the figures two places to the right; and so on.

or Division by 10 can be effected by moving the decimal point one place to the left of its original position (so that the number is 10 times smaller). Similarly, to divide by 100, move the point two places to the left; and so on.

Example 2. $7{\cdot}854 \div 300$.

We can divide by 100, then by 3.

$$\frac{7{\cdot}854}{300} = \frac{0{\cdot}07854}{3} = 0{\cdot}02618$$

Example 3. $66{\cdot}957 \div 0{\cdot}11$.

The method is to make the divisor a whole number by multiplying the numerator and denominator by the appropriate power of 10.

$$\frac{66{\cdot}957}{0{\cdot}11} = \frac{66{\cdot}957 \times 100}{0{\cdot}11 \times 100} = \frac{6695{\cdot}7}{11} = 608{\cdot}7$$

Example 4. $11{\cdot}6 \div 0{\cdot}7$, giving the answer to 2 decimal places.

$$\begin{aligned} \frac{11{\cdot}6}{0{\cdot}7} &= \frac{116}{7} \\ &= 16{\cdot}571 \ldots \\ &= 16{\cdot}57 \text{ to 2 decimal places.} \end{aligned}$$

Example 5. $0{\cdot}2035 \div 0{\cdot}53$, giving the answer to 2 significant figures.

$$\begin{aligned} \frac{0{\cdot}2035}{0{\cdot}53} &= \frac{20{\cdot}35}{53} \\ &= 0{\cdot}383 \ldots \\ &= 0{\cdot}38 \text{ to 2 significant figures} \end{aligned}$$

```
     0·383
53)20·350
   15 9
   ----
    4 45
    4 24
    ----
      210
      159
      ---
```

Exercise 12A

Divide:

1 Each of these by 10: 57·3, 6·21, 365, 0·4, 0·009.

2 Each of these by 100: 2487, 63·81, 1·9, 243·1, 0·06.

3 Each of these by 10, by 100, by 1000: 2864, 357·6, 2·85, 0·04.

4 460 by 2, by 20, by 200 *5* 5·7 by 3, by 30, by 300

6 81 by 5, by 50, by 500 *7* 31·78 by 0·7

8 4·84 by 2·2 *9* 0·925 by 0·37 *10* 0·0782 by 0·23

11 23·8 by 1·7 *12* 57·6 by 1·5.

13 0·657 by 0·11, to 2 decimal places.

14 0·0584 by 0·37, to 3 significant figures.

15 10000 by 600, to 3 decimal places.

Exercise 12B

Divide:

1 Each of these by 10, by 100, by 1000: 1684, 34·65, 2·93, 0·74, 0·002.

2 Each of these by 3, by 30, by 3000: 261, 12·6, 3·42, 0·0072.

3 Each of these by 5, by 50, by 5000: 425, 12·2, 1·53, 0·04.

4 64·71 by 0·9 *5* 5·75 by 2·3 *6* 0·765 by 0·34

7 6·82 by 0·011; then write down the value of 682 divided by 0·00011.

8 73·85 by 0·35; then write down the value of 7·385 divided by 3·5.

9 0·387 by 0·28, to 2 decimal places.

10 1·607 by 2·6, to 2 decimal places.

11 0·053 by 0·87, to 2 significant figures.

12 0·00238 by 7·9, to 2 significant figures.

Rough estimates of answers (using one significant figure of each number)

Example 1. $\frac{28{\cdot}7}{5{\cdot}32} \div \frac{30}{5} = 6$

Example 2. $\frac{0\cdot0843}{0\cdot291} \div \frac{0\cdot08}{0\cdot3} = \frac{0\cdot8}{3} \doteqdot 0\cdot3$

Example 3. $\frac{1\cdot42}{3142} \div \frac{1}{3000} = \frac{0\cdot001}{3} \doteqdot 0\cdot0003$

Exercise 13

In the following examples, first make a rough estimate of each answer (using 1 significant figure in each number), and then carry out the calculation.

1 a Divide 11·76 by 0·7 *b* Divide 23·2 by 0·08

2 How many pieces of wire 3·6 cm long can be cut from 54 cm of the wire?

3 A school kitchen has a bin containing 6·4 kilogrammes of flour. A recipe uses 160 grammes of flour.
How many times can the recipe be made, using up all the flour?

4 If the area of a rectangle is 0·91 cm^2 and its length is 1·3 cm, find its breadth.

5 In a month of 30 days the total rainfall was 18 cm. What was the average daily rainfall?

6 The volume of a cuboid is 1·75 cm^3 and the area of its cross-section is 1·25 cm^2. Find its height.

7 After 400 turns a machine drill penetrates to a depth of 18 centimetres. Find its penetration in 1 turn.

8 How many whole pieces of wire 9·7 cm long can be cut from 150 metres of the wire?

9 In a certain holiday week the hours of sunshine each day at a seaside town were 9·8, 7·5, 1·9, 0·6, 11·3, 10·2, 6·3.

a Find the average number of hours of sunshine per day.
b How many more hours of sunshine did the sunniest day have than the average?

10 A square has a perimeter of 2·8 m. Find

a the length of its side *b* its area.

11 A rectangle has area 10·8 cm^2 and length 4·5 cm. Find its breadth.

12 The product of two numbers is 3·696. If one of the numbers is 2·31, find the other number.

Exercise 14

This exercise is intended for pupils familiar with calculating machines; various calculating devices will be considered later in the course.

Add the following:

1 56·8, 37·9, 88·6

2 9·37, 8·97, 4·86, 7·65

3 987·4, 869·3, 578·8

4 9·876, 4·777, 6·895, 0·909, 3·456

Subtract:

5 68 359 from 77 777

6 79·89 from 83·09

7 0·2936 from 0·3091

8 5·765 from 6·014

Evaluate:

9 84·63 + 97·28 − 79·95

10 125·9 + 764·8 − 841·9

Multiply:

11 636 by 35

12 6·36 by 3·5

13 94·87 by 9

14 68·57 by 8·5

15 23·56 by 85·2

16 9·999 by 9·999

Divide (giving answers to 1 decimal place):

17 57·60 by 5·90

18 234·90 by 88·70

19 0·948 by 0·497

20 98·76 by 532

---

7 Relations between decimal and common fractions

We are already familiar with the decimal form of some common fractions. For example, $\frac{1}{10} = 0{\cdot}1$, $\frac{7}{10} = 0{\cdot}7$, $\frac{3}{100} = 0{\cdot}03$. But we can express any fraction in decimal form by dividing the numerator by the denominator as follows:

$$\tfrac{5}{8} = 5 \div 8 = 0{\cdot}625 \qquad \begin{array}{r} 8\overline{)5{\cdot}000} \\ \hline 0{\cdot}625 \end{array}$$

Exercise 15

Express each of the following as a decimal fraction:

1	$\frac{1}{2}$	*2*	$\frac{1}{4}$	*3*	$\frac{3}{4}$	*4*	$\frac{3}{10}$	*5*	$\frac{9}{10}$	*6*	$\frac{7}{8}$
7	$\frac{1}{5}$	*8*	$\frac{3}{5}$	*9*	$\frac{31}{100}$	*10*	$\frac{4}{5}$	*11*	$\frac{3}{8}$	*12*	$\frac{7}{100}$
13	$\frac{1}{8}$	*14*	$\frac{3}{16}$	*15*	$\frac{1}{3}$	*16*	$\frac{1}{7}$	*17*	$\frac{2}{11}$	*18*	$\frac{1}{13}$

No doubt you discovered that the fractions in questions ***15–18*** cannot be expressed exactly as decimal fractions. Keep dividing in question ***16*** until you see the pattern in the answer. Can you suggest a rule which describes whether a fraction can be given as an exact decimal fraction or not?

The following fractions occur frequently; copy them and fill in their decimal equivalents:

$\frac{1}{10} =$	$\frac{1}{8} =$	$\frac{1}{5} =$		
$\frac{3}{10} =$	$\frac{3}{8} =$	$\frac{2}{5} =$	$\frac{1}{4} =$	
$\frac{5}{10} =$	$\frac{5}{8} =$	$\frac{3}{5} =$	$\frac{3}{4} =$	$\frac{1}{2} =$
$\frac{7}{10} =$	$\frac{7}{8} =$	$\frac{4}{5} =$		
$\frac{9}{10} =$				

8 Standard Form (Scientific Notation)

We have already seen that

10^2 stands for 10×10, i.e. 100,

10^3 stands for $10 \times 10 \times 10$, i.e. 1000, and so on.

The raised numeral is called an *index* (plural *indices*).

10^2 is read '10 squared' or '10 to the power 2'.

10^3 is read '10 cubed' or '10 to the power 3'.

In the same way, 10^5 stands for $10 \times 10 \times 10 \times 10 \times 10$ and is read '10 to the power 5'.

Note. 10^1 is defined as 10.

We say that 100 is the *second power* of 10 because $100 = 10^2$. In the same way, $32 = 2 \times 2 \times 2 \times 2 \times 2 = 2^5$, so we can say that 32 is the *fifth power* of 2.

Example. Write $5{\cdot}72 \times 10^4$ in the usual whole-number form.

$$5{\cdot}72 \times 10^4 = 5{\cdot}72 \times 10000 \quad . \quad . \quad . \quad . \quad . \ (1)$$

so $$5{\cdot}72 \times 10^4 = 57200.$$

Can you see an easy rule for writing down the answer without writing line (1)? (What has happened to the position of the decimal point?)

Exercise 16A

1 What is the *index* in each of the following?
5^2, 8^3, 3^7, 7^3, 10^2, 10^{100}, 6^1, 9^4, 16^2, 25^5

2 Copy and complete the following:

a 5^2 = ..., so 25 is the ... power of 5
b 3^3 = ..., so 27 is the ... power of ...
c 10^4 = ..., so ... is the ... power of ...

3 Which is greater, 3^2 or 2^3?

4 Calculate *a* 9^2 *b* 5^3 *c* 2^6 *d* 10^7

5 Look carefully at your answer to question ***4d*** (count the zeros!). Can you write down the value of each of the following without actually multiplying?

a 10^3 *b* 10^5 *c* 10^6 *d* 10^4 *e* 10^9

6 Express each of the following as a whole number or decimal (see Worked Example).

a 2×10^2 *b* 12×10^3 *c* $2{\cdot}5 \times 10^2$ *d* $12{\cdot}9 \times 10^3$
e $3{\cdot}7 \times 10^4$ *f* $5{\cdot}8 \times 10^5$ *g* $9{\cdot}1 \times 10^7$ *h* $8{\cdot}6 \times 10^4$
i $2{\cdot}85 \times 10^2$ *j* $6{\cdot}12 \times 10^4$ *k* $6{\cdot}71 \times 10^8$ *l* $1{\cdot}215 \times 10^{10}$
m 4×10^1 *n* $4{\cdot}6 \times 10^1$ *o* $4{\cdot}68 \times 10$ *p* $4{\cdot}685 \times 10$

Exercise 16B

1 Copy and complete the following:

a 2^6 = ..., so 64 is the ... power of 2
b 5^4 = ..., so ... is the ... power of ...
c 10^9 = ..., so ... is the ... power of ...

2 Calculate *a* 3^4 *b* 6^3 *c* 2^{10} *d* 10^8

3 Arrange in order of size, greatest first, 2^9, 3^6, 5^4.

4 Without actually multiplying, write down the value of:

a 10^2 *b* 10^6 *c* 10^{10} *d* 10^4 *e* 10^7

5 Express each of the following as a whole number or decimal (see Worked Example).

a 5×10^3	*b* $5{\cdot}7\times10^3$	*c* 83×10^4	*d* $83{\cdot}6\times10^4$
e $4{\cdot}9\times10^2$	*f* $1{\cdot}1\times10^5$	*g* $9{\cdot}6\times10^7$	*h* $3{\cdot}7\times10^4$
i $3{\cdot}75\times10^2$	*j* $8{\cdot}51\times10^6$	*k* $3{\cdot}142\times10^5$	*l* $3{\cdot}142\times10^2$
m 9×10^1	*n* $7{\cdot}9\times10^1$	*o* $5{\cdot}79\times10$	*p* $1{\cdot}579\times10$

Very large numbers

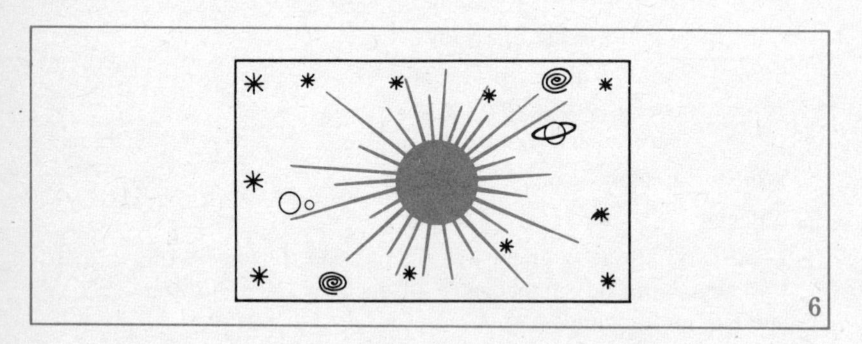

6

Scientists, engineers, accountants, and mathematicians often have to deal with very large numbers. For example

a The Earth weighs about 6600000000000000000000 tonnes
b Savings Bank deposits total £2185000000
c The speed of light is about 300000000 metres per second.
d The largest prime number known at present is $2^{11213}-1$, and has 3376 figures in it.

There are two difficulties in reading such large numbers; one is that it is not easy to compare quickly the sizes of two of them; the other is that it is often not certain how many significant figures there are. The index notation provides a way of overcoming these difficulties. We shall write such numbers in *standard form*. (This is sometimes called *scientific notation.*)

A large number in standard form is in the form:

(A number greater than or equal to 1, but less than 10)
$\times$ (a power of 10)

This can be written

$$a \times 10^n, \quad \text{where} \quad 1 \leqslant a < 10 \quad \text{and} \quad n \in N.$$

Example 1. Write $8{\cdot}3 \times 10^6$ as a whole number. As you saw in Exercises 16A and B (by counting the movements of the decimal point)

$$8{\cdot}3 \times 10^6 = 8\,300\,000$$

Example 2. Express each of the measurements ***a***, ***b***, ***c*** above in standard form.

a We want to write

$$6\,600\,000\,000\,000\,000\,000\,000 = 6{\cdot}6 \times 10^n$$

Reversing the method of Example 1, we ask, 'How many places must we move the point to turn 6·6 into the original number?' The answer is 21. So the earth's weight is about $6{\cdot}6 \times 10^{21}$ tonnes.

b £2 185 000 000 = £$2{\cdot}185 \times 10^9$

c 300 000 000 metres per second = 3×10^8 metres per second.

Summing up, the standard form of a number has 1 figure before the decimal point, and is expressed as follows:

$$136000 = 1{\cdot}36 \times 10^5$$

Example 3. Express 8837 in standard form, rounding off your answer ***a*** to 3 significant figures ***b*** to 2 significant figures

$$\begin{aligned} 8837 &= 8{\cdot}837 \times 10^3 \\ &= 8{\cdot}84 \times 10^3, \text{ to 3 significant figures} \\ &= 8{\cdot}8 \times 10^3, \text{ to 2 significant figures.} \end{aligned}$$

Exercise 17A

1 Write the following as whole numbers in their usual form:

a 3×10^2 *b* $2{\cdot}4 \times 10^3$ *c* $4{\cdot}25 \times 10^5$ *d* $5{\cdot}0 \times 10^6$

2 Write the following numbers in standard form:

a 50 *b* 500 *c* 5000 *d* 50 000
e 200 *f* 56 *g* 3579 *h* 23·4
i 50 700 *j* 6000 *k* 7 500 000 *l* 100 million

3 Re-write each of the following using scientific notation:

a The distance from the earth to the moon is 380 000 km.
b The area of Canada is 9 400 000 square kilometres.

c The population of Africa is about 216000000.
d A company's sales in 1970 were 324 million dollars.
e The nearest star in the night sky is about 40 million million kilometres from the earth.

4 Write the numbers in the following statements in standard form so as to show the number of significant figures stated.

a The radius of the earth is about 6340 km. (3)
b He wrote a cheque for £500. (3)
c The car was travelling at 100 km per hour. (2)
d She is 12 years old. (2)
e He is 80 years old. (2)

5 Express the following in standard form, rounding off each to 3 significant figures:

a 2468 *b* 303030 *c* 234 *d* 13579
e 10101 *f* 1728 *g* 33·92 *h* 33·98

6 A recent World Cup football match attracted $1{\cdot}08 \times 10^5$ spectators. How many watched the game?

7 Find some more statements containing large numbers, and express the numbers in standard form.

Exercise 17B

1 Write the following in whole-number or decimal form:

a 8×10^6 *b* $9{\cdot}97 \times 10^2$ *c* $7{\cdot}00 \times 10^4$ *d* $7{\cdot}00 \times 10^1$

2 Write the following numbers in standard form:
a 1000 *b* 23 *c* 48700 *d* 9999
e 48·6 *f* 48·63 *g* 810000 *h* 6 million million

3 Re-write each of the following statements using scientific notation:

a The population of Cairo in 1970 was 2300000.
b The distance of Venus from the sun is 107000000 km.
c The height of the Great St Bernard Pass is 2500 metres.
d The volume of the sun is 1330 million million km³.

4 Write the numbers in the following statements in the form $a \times 10^n$ where $1 \leqslant a < 10$, $n \in N$, rounding them off to the number of significant figures stated:

a The circumference of the earth is 38400 km. (3)
b The bank granted a loan of £6500. (4)

c The maximum speed of my car is 200 km/h. (2)
d This is page 181. (3)

5 Express the following numbers in standard form, rounded off to two significant figures:

a 37514 *b* 29·36 *c* 29·63 *d* 173 million

6 Give the answer to each of the following in standard form, rounded off to 2 significant figures:

a $374{\cdot}8+19{\cdot}87-190{\cdot}5$
b $(3{\cdot}6\times10^3)$ multiplied by $3{\cdot}14$
c $(5{\cdot}2\times10^4)$ multiplied by $(2{\cdot}5\times10^2)$
d $(3{\cdot}8\times10^7)$ divided by $(1{\cdot}9\times10^3)$
e The distance from the earth to the moon is $3{\cdot}8\times10^5$ km. How many hours will the journey take at an average speed of 3000 kilometres per hour?

---

9 *The metric system*

You have already met the units of length in Book 1 Chapter 2, and again in Section 2 of this Chapter. The metric table of length is:

10 millimetres = 1 centimetre
10 centimetres = 1 decimetre
10 decimetres = 1 metre
10 metres = 1 decametre
10 decametres = 1 hectometre
10 hectometres = 1 kilometre

Thus the prefix 'milli' means $\frac{1}{1000}$, 'centi' means $\frac{1}{100}$, 'deci' means $\frac{1}{10}$, and 'kilo' means 1000. *The same prefixes are used in all metric systems of measurement*. Other prefixes that you may come across are 'mega' meaning 1000000 (as in megatonne and megawatt); and 'micro' meaning $\frac{1}{1000000}$ (as in microsecond).

However, the following metric units are most frequently used:

For *length*, 1 cm = 10 mm
1 m = 100 cm = 1000 mm
1 km = 1000 m

For *volume*, 1 litre = 1000 ml = 1000 cm^3
For *mass*, 1 g = 1000 mg
1 kg = 1000 g
1 tonne (metric ton) = 1000 kg

Also, for *money*, in Britain £1 = 100 pence
in France 1 franc = 100 centimes
in U.S.A. 1 dollar = 100 cents
in Norway 1 krone = 100 öre
in Germany 1 mark = 100 pfennige

Since 10 is the base of all these systems, the various measures may be readily expressed in the decimal form.

For example, 6 cm 7 mm = $6\frac{7}{10}$ cm = 6·7 cm
5 m 28 cm = $5\frac{28}{100}$ m = 5·28 m
2 km 59 m = $2\frac{59}{1000}$ = 2·059 km

Exercise 18A

Express:

1 in cm: *a* 1 m 52 cm *b* 1 m 47 cm *c* 5 m 9 cm
d 6 cm 8 mm *e* 25 cm 2 mm *f* 19 cm 9 mm

2 in m: *a* 6 m 48 cm *b* 40 m 50 cm *c* 17 m 8 cm

3 in km: *a* 5 km 283 m *b* 10 km 35 m *c* 1 km 1 m

4 in francs: *a* 5 francs 25 centimes *b* 81 francs 9 centimes

5 in dollars: *a* 10 dollars 10 cents *b* 57 dollars 5 cents

6 in kg: *a* 2 kg 486 g *b* 5 kg 48g

7 in litres: *a* 3 litres 673 ml *b* 2 litres 2 ml

8 A brick weighs 2 kg 486 g. How many grammes is this?

9 A book costs 1 dollar 85 cents. How many cents is this?

10 A flask contains 2 litres 125 ml of liquid. How many millilitres is this?

11 A window is 1 m 38·5 cm wide. Express this in centimetres.

12 A rail ticket costs 5 francs 85 centimes. How many centimes is this? How many francs is it?

13 A railway truck can carry 15 500 kg.

a How many tonnes is this? (1 tonne = 1000 kg)
b How many tonnes can a train with 24 trucks take?

Exercise 18B

Express:

1 in cm: *a* 4 cm 7 mm *b* 2 m 27 cm 4 mm *c* 5 m 2 cm 9 mm

2 in m: *a* 3 m 47 cm *b* 27 m 7 cm *c* 31 m 2 cm

3 in km: *a* 5 km 297 m *b* 11 km 23 m *c* 7 km 7 m

4 in marks: *a* 3 marks 25 pfennige *b* 854 marks 5 pfennige

5 in francs: *a* 35 francs 64 centimes *b* 128 francs 3 centimes

6 in kg: *a* 24 kg 135 g *b* 82 kg 7 g

7 in litres: *a* 5 litres 278 ml *b* 876 cm^3

8 A book costs 3 dollars 35 cents. How many cents is this?

9 The distance a pupil has to walk to school each morning is 1 km 34·3 m. How many metres is this?

10 The mass of the hammer that is used in international athletic matches is 7 kg 257 g. Express this in kg.

11 Three flasks contain 2 litres 38 ml, 12 litres 107 ml, and 9 litres 8 ml of liquid. How many millilitres of the liquid are there altogether?

12 A box contains 72 tins of meat each weighing 225 grammes. Find the total mass in kilogrammes.

13 A van can carry 2500 kg. How many journeys must it make to transport 18 tonnes? (1 tonne = 1000 kg)

---

10 Calculation of percentages of money

Method 1. Some percentages are equivalent to simple fractions which can be used in calculations.

$2\frac{1}{2}\% = \frac{1}{40}$ $5\% = \frac{1}{20}$ $10\% = \frac{1}{10}$ $20\% = \frac{1}{5}$

$25\% = \frac{1}{4}$ $50\% = \frac{1}{2}$ $75\% = \frac{3}{4}$

$33\frac{1}{3}\% = \frac{1}{3}$ $66\frac{2}{3}\% = \frac{2}{3}$

$100\% = 1$

Example. 75% of £3·76 $= \frac{3}{4} \times$ £3·76
$= \frac{1}{4} \times$ £11·28
$=$ £2·82

Method 2. In all other cases we may proceed as follows:

Example 1. 7% of £4·83 $= \frac{7}{100} \times$ £4·83
$= 7 \times$ £0·0483
$=$ £0·3381
$=$ 34 p, to the nearest penny

Example 2. $3\frac{1}{2}\%$ of £497 $= \dfrac{3\frac{1}{2}}{100} \times$ £497

	$3 \times 4{\cdot}97 =$	14·91
$= 3\frac{1}{2} \times$ £4·97	$\frac{1}{2} \times 4{\cdot}97 =$	2·485
$=$ £17·39$\frac{1}{2}$	$3\frac{1}{2} \times 4{\cdot}97 =$	17·395

Exercise 19A

Calculate to the nearest penny:

1 25% of £4·72 *2* $33\frac{1}{3}\%$ of £59·10 *3* 20% of £86

4 6% of £374 *5* $3\frac{1}{2}\%$ of £85 *6* $1\frac{1}{4}\%$ of £3940

7 A hotel bill amounts to £13·80, and 10% is added as a service charge. How much must be paid?

8 A television set is priced at £262 but $2\frac{1}{2}\%$ is deducted for prompt payment. How much is actually paid?

Exercise 19B

Calculate to the nearest penny or to 3 significant figures:

1 25% of £5·84 *2* $66\frac{2}{3}\%$ of £57·33 *3* 80% of 16 metres

4 9% of 270 km *5* $5\frac{1}{2}\%$ of 52 g *6* $2\frac{3}{4}\%$ of £17·28

7 A hotel bill amounts to £17·60, and $12\frac{1}{2}\%$ is added as a service charge. How much must be paid?

8 A discount of $2\frac{1}{2}\%$ is allowed for prompt payment. How much is actually paid on a bill which amounts to £13·83?

Summary

1 *Notation*

H	T	U	t	h			
		4	6		=	4·6	= $4\frac{6}{10}$
1	0	5	7	9	=	105·79	= $105\frac{79}{100}$
				3	=	0·03	= $\frac{3}{100}$

2 *Measurement*

All measurements are approximate.

A length of 3·3 cm means that the length is 3·3 cm to the nearest tenth of a centimetre.

3 *Approximation*

Rounding off

If the next figure is greater than 5, increase the round-off figure by 1.

If the next figure is 5, round off to the nearest even number; e.g. 36·75 = 36·8 to 1 *decimal place*, or to 3 *significant figures*.

4 *Addition and subtraction*

Examples:

```
  53·40        53·40
 +31·82       -31·82
 ------       ------
  85·22        21·58
```

5 *Multiplication*

a To multiply by 10, move the figures one place left (*or* move the decimal point one place right), making the number 10 times greater. Similarly, two places for 100; and so on.

b *Example*. 47·6 × 4·3 = 204·68

Estimate: 50 × 4 = 200

```
   47·6  (1 figure after point)
    4·3  (1 figure after point)
  ------
   1428
  19040
  ------
 204·68  (2 figures after point)
```

6 Division

a To divide by 10, move the figures one place right (*or* move the decimal point one place left), making the number 10 times smaller. Similarly, two places for 100; and so on.

b *Example.* $0{\cdot}276 \div 4{\cdot}3$

$$= \frac{0{\cdot}276}{4{\cdot}3} = \frac{2{\cdot}76}{43} = 0{\cdot}064\ldots$$

Estimate

$$\frac{0{\cdot}3}{4} \doteqdot 0{\cdot}07$$

```
     0·064
43)2·76
   258
    180
    172
      8
```

7 Expressing a common fraction as a decimal

$\frac{5}{7} = 0{\cdot}714$, to 3 significant figures.

```
7)5·0000
  0·7142
```

8 Standard form

$a \times 10^n$ where $1 \leqslant a < 10$, and $n \in N$.

$2300 = 2{\cdot}3 \times 10^3$

$608\,000 = 6{\cdot}08 \times 10^5$

$12478 = 1{\cdot}25 \times 10^4$, to 3 significant figures.

9 The metric system

km —1000— m —1000— mm; m —100— cm —10— mm

Length: 1 cm = 10 mm
1 m = 100 cm = 1000 mm
1 km = 1000 m

Volume: 1 litre = 1000 ml = 1000 cm^3

Mass: 1 g = 1000 mg
1 kg = 1000 g
1 tonne (metric ton) = 1000 kg

10 Calculating a percentage of money

Example. $4\frac{1}{2}\%$ of £6·21

$$= \frac{4\frac{1}{2}}{100} \times £6{\cdot}21$$

$$= 4\tfrac{1}{2} \times £0{\cdot}0621$$

= 28 p, to the nearest penny

```
£0·0621
      4
 0·2484
 0·0310
 0·2794
```

Computers and Binary Arithmetic

1 The binary system—counting in twos

Throughout the centuries every civilization has found it necessary to develop systems for:

a recording written information, which required an alphabet and grammar, to provide a language of words;

b recording numbers and making calculations, which required a set of numerals and a counting system, to provide a language of number.

Long ago man based his counting system on *ten*, probably because he had *ten* fingers which helped him with simple calculations. Our present *Decimal System* has ten figures:

0, 1, 2, 3, 4, 5, 6, 7, 8, 9

and base 10, which enables us to represent every number in units, tens, and powers of ten (hundreds, thousands, ...).

For example, 345 means 3 hundreds, 4 tens and 5 units;
2307 means 2 thousands, 3 hundreds and 7 units.

Modern computers can act as very fast calculating machines. How long would you take to multiply 987 654 321 by 135 798 642? Computers can carry out thousands of such calculations in a second.

Computers work electrically (or electronically) and their operation depends on the fact that an electric current can give a signal in two ways; for example, an electric light bulb can be 'on' or 'off' depending on whether or not a current is flowing in the circuit. So computers need only two figures, 0 and 1, for their working instructions.

The *Binary System* of calculation has only two figures, 0 and 1, and is a base-2 system, so that every number can be represented in units, twos, and powers of two (fours, eights, sixteens, ...).

For example, in this system, 111 means 1 four, 1 two and 1 unit,
and 1010 means 1 eight and 1 two.

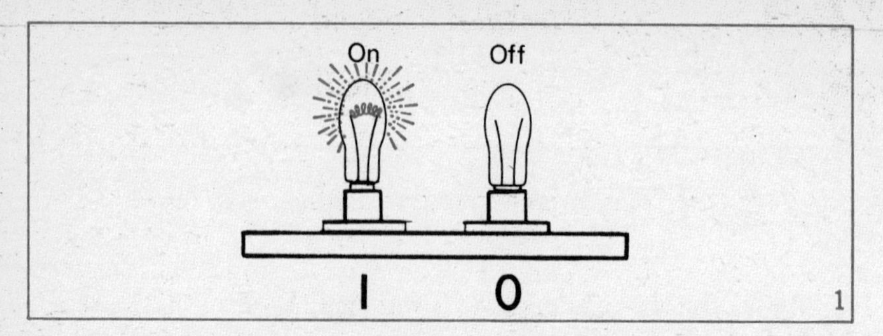

1

Just as we used the headings HTU for numbers in the decimal system, we can use SEFTU (Sixteens, Eights, Fours, Twos, Units) for numbers in the binary system.

The number of dots in each group below can be represented by *numerals* in both the decimal and binary systems as follows:

Number of objects			*Decimal form* TU		*Binary form* SEFTU
·	represented	by	1	or	1
··	,,	,,	2	,,	10
···	,,	,,	3	,,	11
::	,,	,,	4	,,	100
:·:	,,	,,	5	,,	101
:::	,,	,,	6	,,	110
:::·	,,	,,	7	,,	111
::::	,,	,,	8	,,	1000
::::·	,,	,,	9	,,	1001
:::::	,,	,,	10	,,	1010

Where necessary we indicate the base of the number system we are using by showing it like this: 10_{two}, 10_{ten}.

Example. Convert 10101 from binary to decimal form.

S E F T U
1 0 1 0 1 means $16+4+1 = 21$ in the decimal system.

$$\text{i.e. } 10101_{two} = 21_{ten}$$

(This reads 'one, zero, one, zero, one to base two is equal to twenty-one to base ten', or 'one, nought, one, nought, one to base two is equal to twenty-one to base ten'.)

Exercise 1A

1 The following numbers are in decimal form. Write down their meanings as hundreds, tens, units, etc.

a 27 *b* 234 *c* 702 *d* 2135

2 The following numbers are in binary form. Write down their meanings as eights, fours, twos, units, etc. Then give each in decimal form.

a 101 *b* 110 *c* 1100 *d* 1111

3 Convert each of the following from binary to decimal form.

a 1 *b* 11 *c* 100 *d* 111
e 1001 *f* 1011 *g* 1101 *h* 10001

4 Convert the following from binary to decimal form. What do you notice about the results?

a 1 *b* 10 *c* 100 *d* 1000 *e* 10000

5 Write out the first twelve natural numbers in binary form.

6 Which number in the binary system comes immediately after each of the following?

a 1100 *b* 1010 *c* 1001 *d* 10011

7 Which number in the binary system comes immediately before each of the following?

a 1001 *b* 1011 *c* 1010 *d* 1110

Exercise 1B

1 The following numbers are in decimal form. Write down their meaning as hundreds, tens, units, etc.

a 75 *b* 231 *c* 1234 *d* 50721

2 The following numbers are in binary form. Write down their meanings as eights, fours, twos, units, etc. Then give each in decimal form.

a 1000 *b* 1101 *c* 11010 *d* 10111

3 Convert each of the following from binary to decimal form.

a 1111 *b* 1110 *c* 100010 *d* 111111

4 Convert the following from binary to decimal form. What do you notice about the sequence of numbers you get?

a 1 *b* 11 *c* 111 *d* 1111 *e* 11111

5 Give the next three numbers in the sequence in question **4**, both in binary and in decimal form.

6 Write out the natural numbers from 12 to 20 using binary form.

7 Which number in the binary system comes immediately before each of the following, and which number comes immediately after each?

a 100000 *b* 101010 *c* 100101 *d* 111111

* * *

Example. Convert 26 from decimal to binary form.

We do this by arranging 26_{ten} as powers of 2, in order. Two methods are given below.

a We find the largest power of 2 less than 26, then the next largest, etc. (It is worth memorizing the powers: 1, 2, 4, 8, 16, 32, 64.)

$26 = 16+10$

$= 16+8+2$

so T U = S E F T U

2 6 = 1 1 0 1 0

i.e. $26_{ten} = 11010_{two}$

b

2	26 ones
2	13 twos + 0 units
2	6 fours + 1 two
2	3 eights + 0 fours
2	1 sixteen + 1 eight
	0 thirty-twos + 1 sixteen

So $26_{ten} = 11010_{two}$, from the remainders.

Exercise 2A

Convert each of the following from decimal to binary form:

1 3 *2* 5 *3* 6 *4* 9 *5* 11 *6* 14

7 18 *8* 24 *9* 30 *10* 37 *11* 64 *12* 71

Exercise 2B

Convert each of the following from decimal to binary form:

1 19 *2* 16 *3* 20 *4* 29 *5* 32 *6* 50

7 63 *8* 65 *9* 78 *10* 94 *11* 129 *12* 260

2 *Models that show numbers in binary form*

Perhaps you would like to make a model that can show numbers in binary form. Here are three suggestions:

1 Figure 2 shows ends to be fixed to a wooden base. A wire runs from end to end and passes through a number of wooden (or card) rectangles. Each rectangle has 0 on one side of it and 1 on the other side. By turning them round the rectangles can be made to indicate a given binary number.

If your neighbour gives you a decimal number you can convert it into a binary number and show the answer on the model, and vice versa.

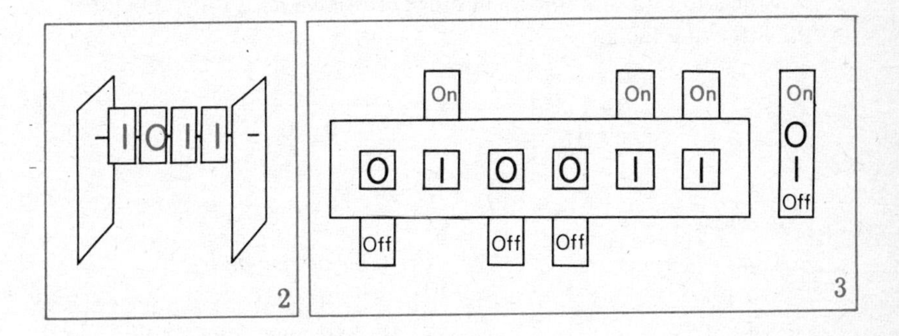

2 Figure 3 shows a cardboard model. Windows are cut in a rectangular piece of cardboard which is glued to a similar back-board, slits being left for the tabs to slide in. Each tab would be as shown, the ON, OFF indicating whether a lamp in an electric circuit would be ON or OFF in this position.

3 Figure 4 shows an electrical model which could be built on a suitable base-board; each light is controlled by a separate switch. In the diagram, flash lights L_2 and L_4 are on, indicating the binary number 101.

Make a list of all the binary numbers that this model could show, and give the corresponding decimal numbers.

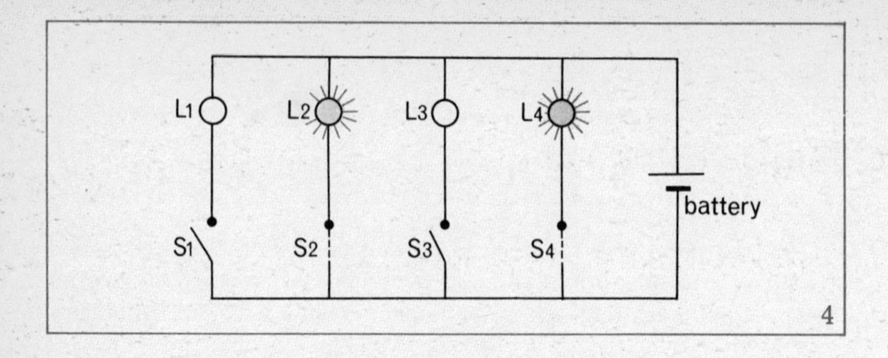

3 Addition

For the addition of numbers in binary form we need only remember the following facts:

$$0+0=0$$
$$0+1=1$$
$$1+0=1$$
$$1+1=10_{two}$$

or

+	0	1
0	0	1
1	1	10

Examples:

(i) (*no carrying*)

$$\begin{array}{r} 101 \\ +\ 10 \\ \hline 111 \\ \hline \end{array}$$

(ii) (*one carried digit*)

$$\begin{array}{r} 101 \\ +1001 \\ \hline 1110 \\ \hline \end{array}$$

(Read as 'one and one, one zero'; put down zero, carry one, etc.)

(iii) (*separate carried digits*)

$$\begin{array}{r} 101 \\ +101 \\ \hline 1010 \\ \hline \end{array}$$

(iv) (*adjacent carried digits*)

$$\begin{array}{r} 101 \\ +11 \\ \hline 1000 \\ \hline \end{array}$$

(v) (*with decimal check*)

$$\begin{array}{r} 10001 \\ +1011 \\ \hline 11100_{two} \\ \hline \end{array} \qquad \textit{Check} \quad \begin{array}{r} 17 \\ +11 \\ \hline 28_{ten} \\ \hline \end{array}$$

(vi) (*addition of three units*)

$$\begin{array}{r} 111101 \\ +\ \ 1111 \\ \hline 1001100_{two} \\ \hline \end{array} \qquad \textit{Check} \quad \begin{array}{r} 61 \\ +15 \\ \hline 76_{ten} \\ \hline \end{array}$$

A simple adding machine

Figure 5 shows a circuit for a very simple adding machine. B is the battery, S a 1-pole 1-way switch, L_1 and L_2 are lamps, S_1 and S_2 are 2-pole 2-way switches. The dotted lines show both switches in the

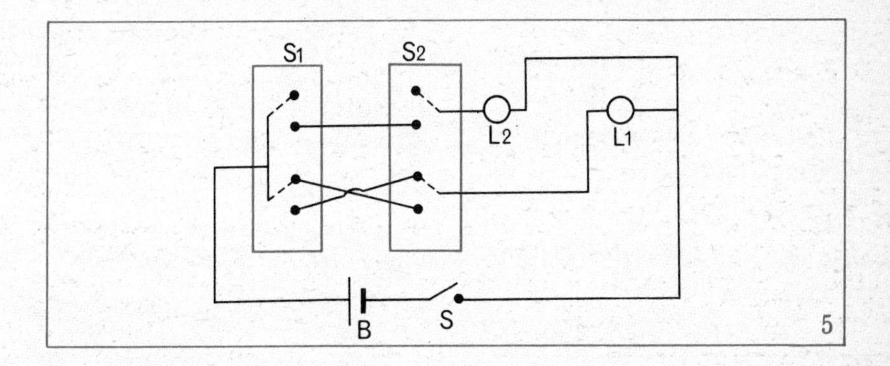

OFF position (i.e. $0+0$). Follow the circuit and see how both lamps are OFF showing the binary number 0. Now when we press switch S_1 ON and leave S_2 OFF (i.e. $1+0$) you can see that L_1 goes ON and L_2 stays OFF, showing the binary number 1. Now notice that S_1 OFF and S_2 ON (i.e. $0+1$) also shows the answer 1. Finally with S_1 and S_2 both ON (i.e. $1+1$), observe that L_2 is ON and L_1 is OFF, showing the binary number 10.

This model illustrates the basic results of the addition of numbers in the binary system.

Exercise 3A

Add the following pairs of numbers which are in binary form.

1 110 + 1 = ____ *2* 1010 + 100 = ____ *3* 1001 + 110 = ____ *4* 10001 + 1010 = ____

5 1010 + 10 = ____ *6* 1011 + 10 = ____ *7* 10101 + 110 = ____ *8* 10101 + 100 = ____

9	1101 1001	*10*	10001 10101	*11*	1001 1001	*12*	10101 10101
13	11 1	*14*	110 10	*15*	1001 11	*16*	10010 111
17	1011 1	*18*	10011 1001	*19*	10111 1010	*20*	1011 101

Add the following numbers in binary form, and then check your work by converting both the given numbers and your answers to decimal form.

21	111 100	*22*	10101 1001	*23*	11011 1001	*24*	11101 101

Exercise 3B

Add the following, all numbers being given in binary form.

1	100 11	*2*	1011 100	*3*	10101 1010	*4*	101 1
5	10001 101	*6*	11010 10	*7*	1010 1010	*8*	101010 1011
9	11010 10011	*10*	101011 11010	*11*	11011 11011	*12*	111111 11111
13	1010 101 10001	*14*	1101 110 1001	*15*	10011 1010 111	*16*	11011 1101 110

Add the following numbers in binary form, and then check your work by converting both the given numbers and your answers to decimal form.

17
```
 10011
  1101
 _____

 _____
```

18
```
 111111
  10011
 ______

 ______
```

19
```
 111011
 101010
 ______

 ______
```

20
```
 101111
 111101
 ______

 ______
```

---

4 Subtraction

For the subtraction of numbers in binary form we need only remember the following facts:

$$0 - 0 = 0$$
$$1 - 0 = 1$$
$$1 - 1 = 0$$
$$10_{two} - 1 = 1$$

Examples

(i)
```
   11
 -  1
 ----
   10
```

(ii)
```
   10
 -  1
 ----
    1
```

(iii)
```
  101
 - 10
 ----
   11
```

(iv)
```
  101
 - 11
 ----
   10
```

(v)
```
  100
 - 11
 ----
    1
```

Exercise 4A

Carry out the following subtractions:

1
```
   110
 -  10
 -----

 -----
```

2
```
   111
 -  10
 -----

 -----
```

3
```
   1011
 -   11
 ------

 ------
```

4
```
   11011
 -  1010
 -------

 -------
```

5
```
   110
 -   1
 -----

 -----
```

6
```
   101
 -  10
 -----

 -----
```

7
```
   101
 -  11
 -----

 -----
```

8
```
   10110
 -  1001
 -------

 -------
```

9
```
   11001
 -    11
 -------

 -------
```

10
```
   10101
 -  1010
 -------

 -------
```

11
```
   10100
 -  1001
 -------

 -------
```

12
```
   11000
 - 10010
 -------

 -------
```

13	100 − 11	*14*	1001 − 110	*15*	1001 − 111	*16*	10110 − 1101

Exercise 4B

Carry out the following subtractions:

1	101 − 100	*2*	11010 − 1010	*3*	10111 − 10001	*4*	110010 − 10000
5	1101 − 11	*6*	10101 − 1011	*7*	11001 − 101	*8*	110010 − 101001
9	11011 − 1010	*10*	11001 − 1110	*11*	101010 − 10101	*12*	100000 − 1111
13	100011 − 101	*14*	1000010 − 11001	*15*	1001000 − 110111	*16*	1010101 − 111111

5 *Multiplication*

For the multiplication of numbers in binary form we need only remember the following facts:

$$0 \times 0 = 0$$
$$0 \times 1 = 0$$
$$1 \times 0 = 0$$
$$1 \times 1 = 1$$

or

×	0	1
0	0	0
1	0	1

Examples (all 'long' multiplication).

(i)
$$\begin{array}{r} 101 \\ 11 \\ \hline 101 \\ 1010 \\ \hline 1111 \\ \hline \end{array} \qquad \text{Check} \qquad \begin{array}{r} 5 \\ \times 3 \\ \hline 15 \\ \hline \end{array}$$

(ii)
$$\begin{array}{r} 101 \\ 101 \\ \hline 101 \\ 10100 \\ \hline 11001 \\ \hline \end{array} \qquad \text{Check} \qquad \begin{array}{r} 5 \\ \times 5 \\ \hline 25 \\ \hline \end{array}$$

Exercise 5A

Carry out the following multiplications, and check your work by converting both of the given numbers and your answer to decimal form.

1 1001×11 *2* 1001×10 *3* 1010×11

4 1101×10 *5* 1001×110 *6* 1011×1001

7 10101×11 *8* 1001×101 *9* 111×11

10 1011×1100 *11* 10101×101 *12* 1111×101

Exercise 5B

Carry out the following multiplications:

1 111×10 *2* 1101×11 *3* 1011×101

4 1101×110 *5* 1101×1001 *6* 10111×101

7 101010×1001 *8* 11111×110 *9* 1010×111

6 Division

Examples (essentially long division).

(i) $1010 \div 10_{two}$

$$\begin{array}{r} 101 \\ 10\overline{)1010} \\ 10 \\ \hline 10 \\ 10 \\ \hline \end{array} \qquad \text{Check} \qquad \begin{array}{r} 5 \\ 2\overline{)10} \\ 10 \\ \hline \end{array}$$

(ii) $1011 \div 11_{two}$

```
      11                  3
  11)1011     Check     3)11
     11                   9
     ---                  -
      101                 2
       11
       --
       10 (remainder)
```

(iii) $11110 \div 101_{two}$

```
       110                6
  101)11110   Check     5)30
      101                 30
      ---                 --
       101
       101
       ---
```

(iv) $100000 \div 101_{two}$

```
        110               6
  101)100000  Check     5)32
      101                 30
      ---                 --
       110                 2
       101
       ---
        10 (remainder)
```

Exercise 6A

Carry out the following divisions, and check your work by converting both of the given numbers, your answer and remainder, if any, to decimal form.

1	$110 \div 10$	*2*	$110 \div 11$	*3*	$1110 \div 10$
4	$1000 \div 10$	*5*	$1100 \div 11$	*6*	$1001 \div 11$
7	$1111 \div 101$	*8*	$1010 \div 101$	*9*	$10010 \div 11$
10	$1010 \div 11$	*11*	$1101 \div 11$	*12*	$11010 \div 101$
13	$111111 \div 111$	*14*	$100011 \div 101$	*15*	$100110 \div 101$

Exercise 6B

Carry out the following divisions:

1	$1111 \div 11$	*2*	$11011 \div 11$	*3*	$11001 \div 101$
4	$100101 \div 110$	*5*	$1101101 \div 1011$	*6*	$1001101 \div 1101$
7	$1010100 \div 1100$	*8*	$1010100 \div 111$	*9*	$10011001 \div 10001$

7 Electronic computers

Modern computers have two main functions. One is to store information, and the other is to carry out calculations very rapidly. Often these two functions are combined, as for example in the preparation of gas, electricity and telephone accounts, and in the production of wageslips and salary cheques.

A computer has to be given precise detailed information in order to carry out a task. The instructions for the task are called the *program*, and these may be fed into the computer in various forms, including *punched cards* and *punched tape*.

The details of a computer are complicated to understand, but the main parts can be shown in the form of a diagram, as in Figure 6.

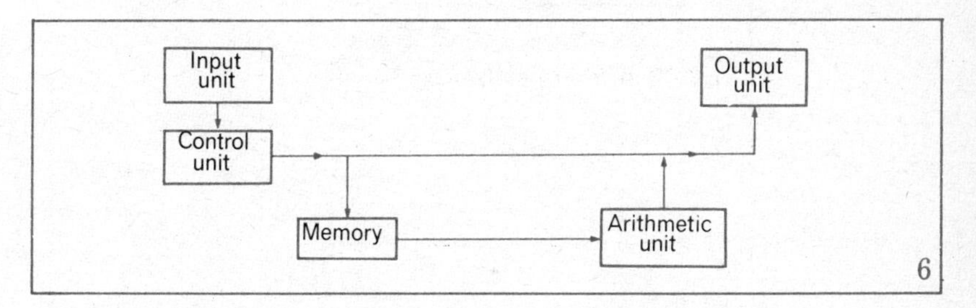

6

The *input unit* of the computer is the part where the pattern of holes on the cards or tape is changed to a series of electrical impulses which flow through the machine.

To carry out a particular instruction the impulses must travel along a specified route. The *control unit* shuts off all parts of the computer except those which are required to carry out the instruction, and it ensures that the instructions are carried out in the correct order.

Computers are constructed so that information and instructions can be stored in the *memory*. Any required number or instruction can be produced on request from the memory.

The *arithmetic unit* of a computer is the part where simple arithmetical operations such as addition, subtraction, multiplication, and division are carried out.

The *output unit* transfers information from the memory to the teleprinter, which prints out the results of the computer's calculations.

8 *Punched cards*

Instructions are entered on the punched cards by punching holes in certain places, and as the card passes through the computer's input unit the pattern of holes is translated into the machine's language.

We will make our own punched-card messages in a slightly different way, using the special card shown in Figure 7.*

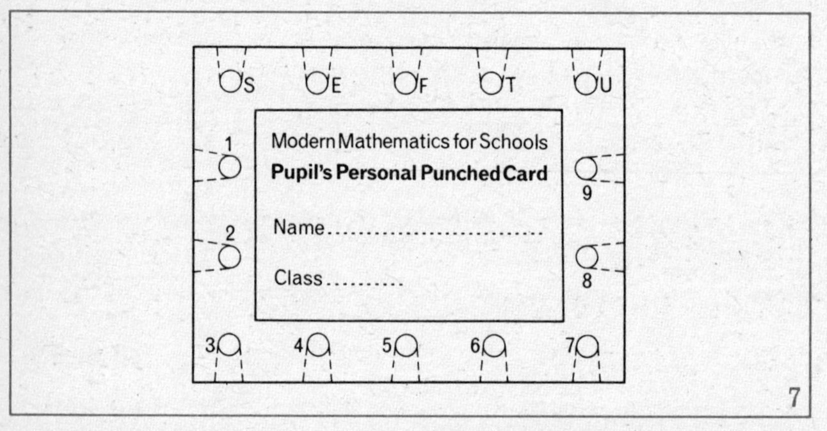

7

First of all we must work out a code for putting your name on the card so that it can be found quickly in a large store of cards.

Exercise 7

1 Copy and complete this binary code for every letter in the alphabet.

Letter	*Code* SEFTU
A	1
B	1 0
C	1 1
D	1 0 0
.	.
.	.
Z	1 1 0 1 0

* These cards are available in packs of 100 (Blackie/Chambers).

We can now show a letter on the card by opening holes corresponding to the code—open for 1, closed for 0.

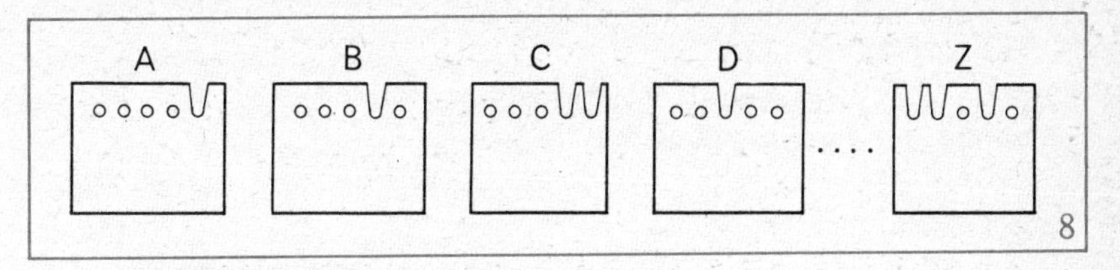

2 On your personal card open the holes that give the pattern of the first letter of your surname.

The cards for all the pupils in the class can now be put together in any order, like a pack of cards.

They can be arranged in alphabetical order quickly as follows.

3 Insert a thin pencil or rod or knitting-needle through all the 'units' holes, and holding the cards loosely above a table, shake them.

All cards 'opened' in the units holes will fall out. Put these at the back of the pack.

Insert the needle in the 'twos' holes, and repeat the process; then in the 'fours', 'eights' and 'sixteens' holes.

The pack of cards will now have your names arranged approximately in alphabetical order. (Why not exactly?)

4 How could you obtain your card very quickly from the pack without looking through all the cards?

5 Take your card again, and open holes 1–9 to provide information about the other matters suggested on the back of the cards, for example:

Hole 1: Boy/girl
Hole 2: Brother or sister in school/No
Hole 3: Take school lunch/No
Hole 4: Use school transport/No, or age 12 and under/over 12
Holes 5, 6, 7: Binary code for favourite subject, or TV programme, or pop star.
Holes 8, 9: Binary code for House in School, etc.

6 When the cards are put together as a pack again how could you find the names of:

a all the boys in the class
b pupils in the class who have a brother or sister in school
c girls who take school lunch and use school transport?

7 Make up some more questions like these, and find the answers.

The above Exercise illustrates how to:

a put information on punched cards
b put the cards in order
c provide a compact store of information
d obtain data from the store very quickly.

9 Punched tape

Computers can also be programmed by means of punched tape. Here numbers in binary form can be represented by patterns of holes punched on a paper tape as shown in Figure 9*a*.

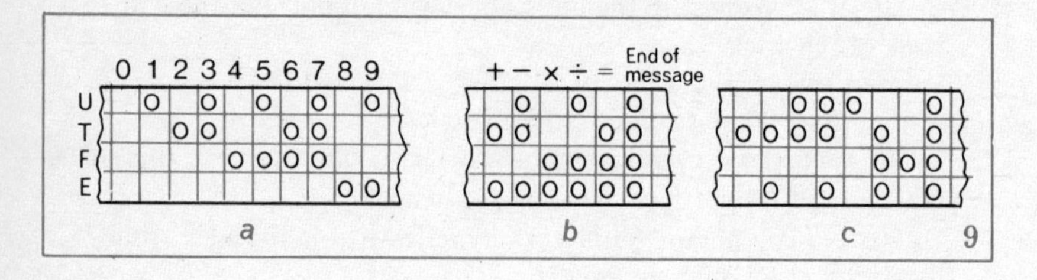

Useful symbols can also be represented by different patterns of holes as shown in Figure 9*b*.

Exercise 8

1 How many different patterns of holes are possible using the punched tape shown in Figure 9?

2 What is the message given on the tape in Figure 9*c*?

3 Draw some paper tape on squared paper, and enter the patterns giving the figures and symbols shown in Figure 9*a* and *b*.

4 Show the following 'messages' on squared paper:

a $5+5 = 10$ *b* $1+2+3+4 = 10$ *c* $8\times9 = 100-28$

5 Show some other calculations of your own.

6 Why will the above punched tape not be able to show all the letters of the alphabet?

7 With paper tape containing five rows for holes we can punch the pattern for every letter as in the table in question *1* of Exercise 7.

To encode or decode a message quickly it might be best to list the letters of the alphabet opposite numbers in decimal form from 1 to 26, and then to change to and from binary form as necessary.

Can you decipher the message in Figure 10?

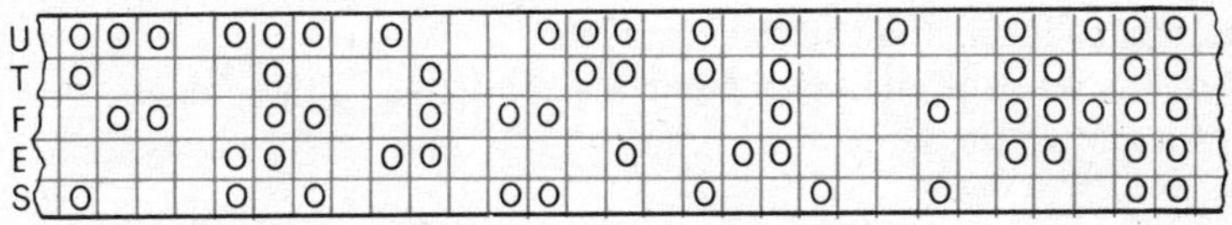

10

8 Make up some messages in punched tape code, and send them to your friends to answer.

10 Miscellaneous questions

Exercise 9A

1 Select the greatest and least binary numeral in each of the following sets:

a {1011, 1101, 1110, 10000}
b {11000, 10100, 10010, 10001}
c {10101, 11010, 11001, 10110}

2 Convert each of the following from binary to decimal form, and say which of them represent prime numbers.

a 1101 *b* 11010 *c* 11111 *d* 10101

3 *Binary Cross-figure.* All numbers are expressed in binary form.

Across

1. $10001-1100$
100. $1+111+1111$
101. A prime number

Down

1. $1000110 \div 101$
10. 101×110
11. The solution of the equation $x+11 = 1010$

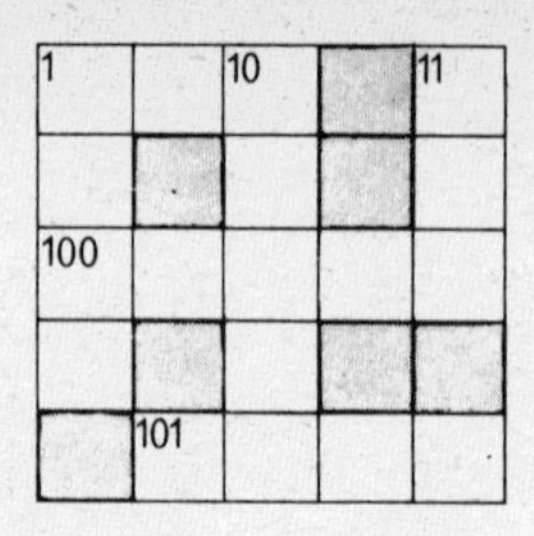

4 Calculate the area and perimeter of a rectangle whose dimensions (in binary form) are 101 cm and 11 cm.

5 a Write down the multiples of 3 which are less than 16. Convert them to binary form, and call this set A.

b List the set B of all binary numerals consisting of three digits.

c Write down the set $A \cap B$.

6 List the first six prime numbers in decimal form, and convert them to binary form.

7 State whether each of the following is true or false:

a $10101 > 11010$ *b* $10^{10} = 100$ *c* $100^{10} = 1000$

d $(110 \times 1010) \div 100 = 1111$ *e* 1010101 represents an odd number.

Exercise 9B

1 Arrange the binary numerals in each of the following in descending order:

a 10101, 10110, 11000, 11001

b 11111, 10111, 11101, 11011

2 State whether each of the following is true or false:

a $11011 < 11101$ *b* $101 \times 100 = 11010$

c $100^{100} = 10000$ *d* $1+11+111+1111 = 11010$

3 *Binary Cross-figure.* All numbers are expressed in binary form.

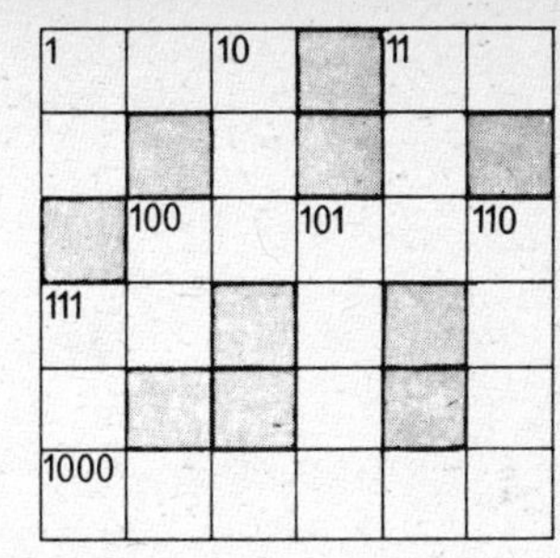

Across

1. A common factor of 1111 and 101101
11. A common factor of 100 and 1010
100. $100000 - 11$
111. The value of x if $x^{10} = 1001$
1000. The number of black squares on a chess-board

Down

1. Half of 11 down
10. $111111 \div 1001$
11. The number of edges on a pyramid with a triangular base
100. The value of y in 111 down
101. 111×10
110. $111 + 11$
111. The value of x if $xy = 1111$

4 How can you tell when a binary numeral represents:

a an even number *b* a number which is divisible by four?

5 Calculate the area and perimeter of a square whose length (in binary form) is 101 cm.

6 When you started to count, did you ever use your fingers to help you? The base of our number system may be ten because long ago people counted on their fingers.

a If an octopus could count, what would the base of its system probably be?

b What would 35 in octopus arithmetic be equal to in our system?

c Can you add 46 and 22 in octopus arithmetic? Then try $56 + 33$.

7 Suppose that the 'inhabitants' of Mars have 3 hands, with 3 fingers on each hand. What is likely to be the base of their system of numbers?

Make up some calculations of the kind that they might have to work out.

Summary

1 a The binary system has only two figures, 0 and 1, and is a *base two* system.

b *Notation*

S E F T U = T U

$1\ 0\ 1\ 0\ 1_{two} = 16+4+1 = 2\ 1_{ten}$

c *Addition*

+	0	1
0	0	1
1	1	10

d *Multiplication*

×	0	1
0	0	0
1	0	1

2 a Computers can:

(i) store information

(ii) carry out calculations very rapidly

b Punched cards and tape can be used to programme a computer by means of binary patterns of holes.

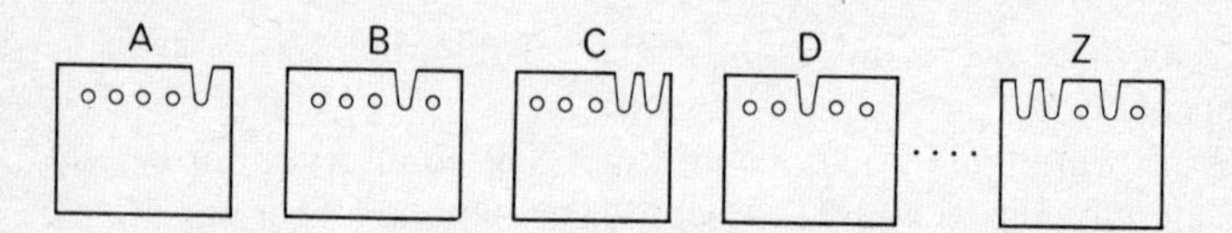

Introduction to Statistics

1 Pictographs

A survey was carried out in a class of 30 sixth-year pupils to find out the number of household pets, and the result was as follows:

Dog	Cat	Budgie	Fish	Pony
6	8	7	3	1

This information can be expressed more effectively pictorially. Figure 1 is a *pictograph*, in which each item is shown by a group of pictorial units.

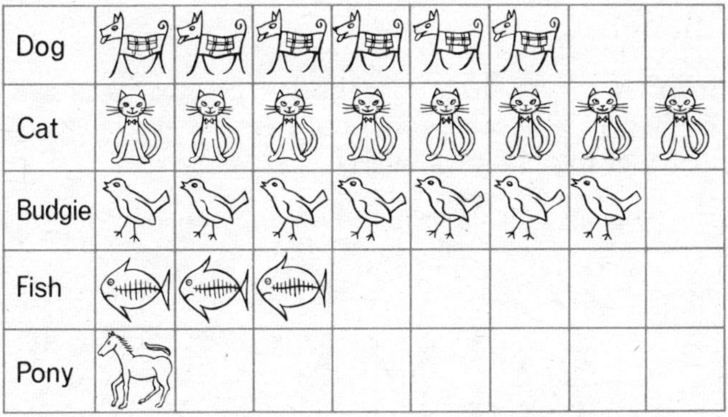

Dog
Cat
Budgie
Fish
Pony

1

a Which is the most popular pet?

b What can you deduce from the fact that the total number of pets is 25?

* * * * *

The same pupils were asked about their methods of travelling to school each day. The results were as follows:

Walk	Cycle	School bus	Service bus	Motor scooter	Train
7	3	9	4	2	5

This information can be shown by means of 'matchstick' figures, as in Figure 2. The *scale* of the diagram is given by the fact that 'one symbol represents 1 pupil'.

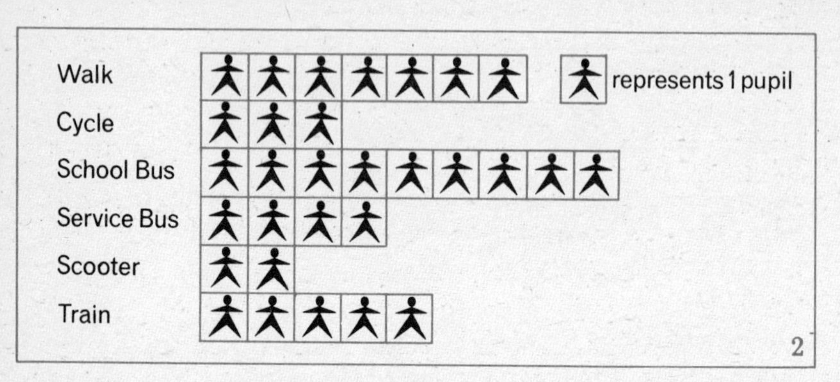

2

a What is the most common method of travelling to school?
b Does this agree with the most common method of travelling to school for your own class?

* * * * *

A survey was carried out among some first-year pupils to find out the most popular school subjects. Here are the results, giving the number of votes for each subject:

English	History	Geography	French	Mathematics	Science
30	20	20	27	45	41

English
History
Geography
French
Mathematics
Science
represents 5 votes

3

In Figure 3 one 'matchstick' figure represents 5 votes; notice that this means that we will have some incomplete symbols to represent part units. For example, the 41 votes for science are represented by 8 complete symbols (8×5) and a head (1).

a Which is the most popular subject?
b Which two subjects got an equal number of votes?

* * * * *

There are two difficulties in drawing pictographs:

(i) It is not easy to draw repeated symbols which are exactly the same. Stencils or tracing paper may be used, or the symbols can be drawn inside equal squares.

(ii) When dealing with large numbers we must let one symbol represent a number of objects or units, and so it may be difficult to interpret parts of symbols.

Exercise 1A

Remember to label your diagrams clearly and to show the scale used.

1 Figure 4 shows, in pictograph form, the cereal production of five European countries in 1967.

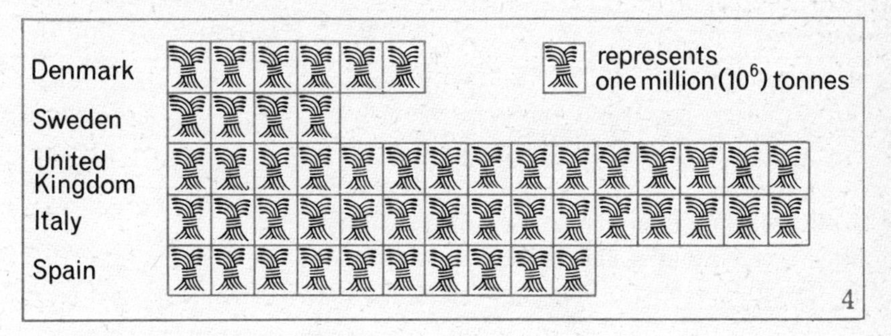

a Write down the number of tonnes produced by each of the five countries, first in millions, and then in standard form ($a \times 10^n$).
b Which country produced one fifth of the total production of the five countries?

2 In the same year (1967) the five countries in question *1* produced 3 million, 2 million, 12 million, 10 million and 4 million head of cattle, respectively.

Draw a pictograph on 5-mm squared paper to illustrate these

figures. Use the symbol  to represent one million cattle.

3 Figure 5 shows the birth-months of the first-year pupils in a school.

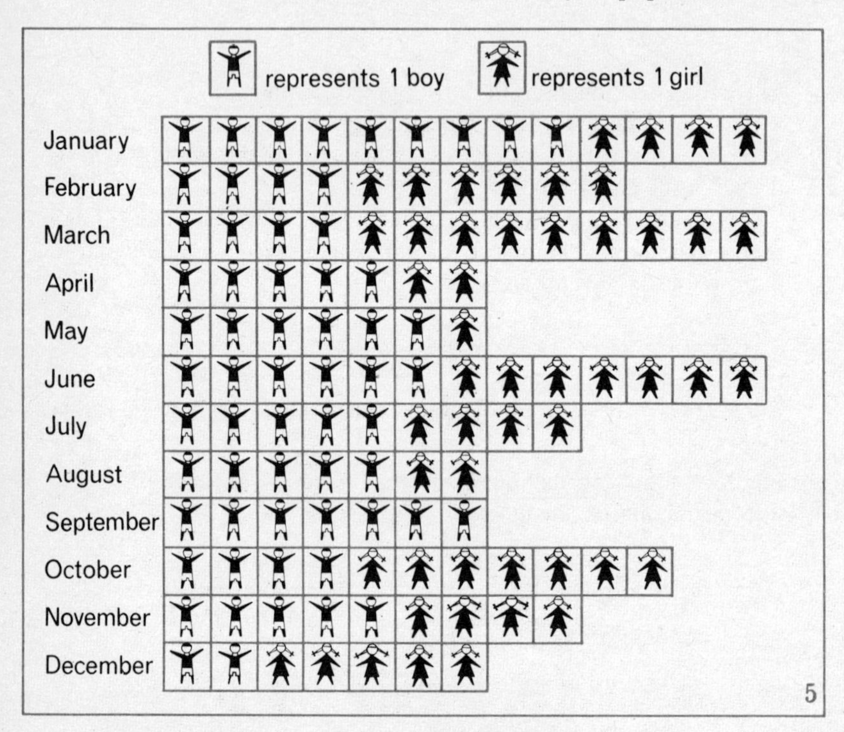

5

a In which months were there most birthdays?
b In which months were there fewest birthdays?
c Give the number of boys born in each month.
d Give the number of girls born in each month.
e What percentage of the total number of boys were born in May, June, July and August taken together?
f Compare your answer to ***e*** with the percentage of girls born in these months.

4 Investigate the popularity of various sports amongst the pupils in your class, e.g. football, rugby, cricket, tennis, hockey, skiing, riding, etc.

Show the results in the form of a pictograph.

Exercise 1B

1 The numbers of life-boat stations in England, Scotland, Ireland, and Wales are 75, 31, 21 and 18 respectively. Draw a pictograph to show this information, using the symbol 

to represent a suitable number of stations.

2 The number of pits in operation in the "traditional" coalfields of Scotland in 1968 can be obtained from the pictograph of Figure 6.

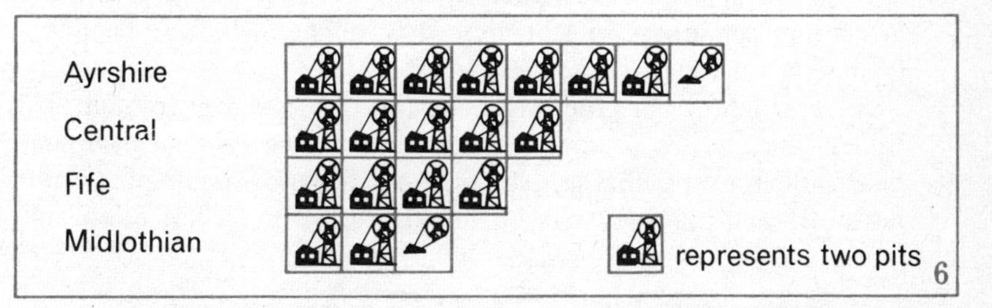

6

The approximate number of men employed was: Ayrshire—6000, Central—8500, Fife—8000, Midlothian—7000. Calculate the average number of men per pit in each coalfield.

3 The membership of a certain club is as follows:

Boys	Girls	Men	Women
250	400	1100	750

Express this information in the form of a pictograph.

4 The first letters of the Christian names of the 32 pupils in a first-year class are as follows:

Letter	A	C	D	J	L	P	R	S	T	V
Boys	4	2	3	2	0	2	2	2	2	0
Girls	1	1	0	3	2	0	0	2	0	4

Draw a pictograph to illustrate this information.

5 Investigate the popularity of certain television programmes: records, or 'groups'; football teams; magazines, or home-readers; foods; etc. Illustrate the results by means of pictographs.

2 Bar charts

The main advantage of a well-drawn diagram is that it can convey, almost at a glance, the general meaning of some statistical information. But it is important to study the scales and titles before drawing any firm conclusions from the diagram.

Effective statistical diagrams can be seen in newspapers, advertisements, and Government pamphlets. Perhaps your class could arrange a display of newspaper cuttings showing the various pictorial representations of statistics; they might include some particularly good and bad examples.

A *bar chart* (or bar graph) is easier to draw than a pictograph; if we can imagine the squares without the symbols we shall have an idea of the form of a bar graph. Note that the bars (or columns) may not touch and that they may be interchanged without loss of effect.

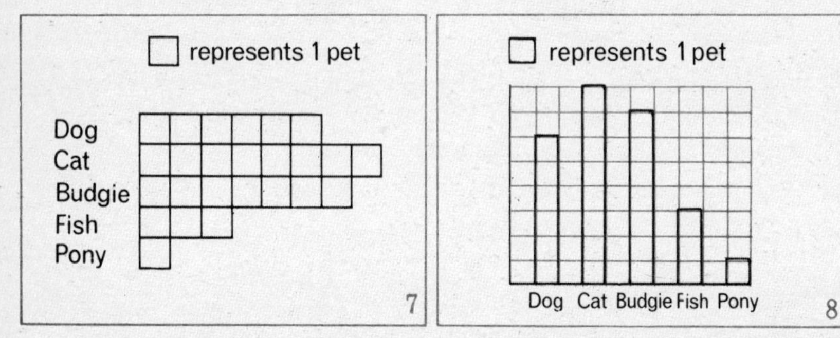

Bar charts may be horizontal as in Figure 7 or (more usually) in vertical columns as in Figure 8. These two figures show bar charts for the table on page 207 giving the number of household pets belonging to the 30 pupils.

We can see at a glance which is the most popular household pet.

Exercise 2A

Illustrate by bar charts (questions *1–4*):

1 & 2 The information supplied in questions *1* and *2* of Exercise 1A.

3 The results of some of the surveys you carried out in question *4* of Exercise 1A.

4 The methods used by pupils in your class to travel to school.

5 The 'composite' bar chart of Figure 9 shows the allocation of periods to subjects for a typical 40-period week in the first year of Scottish and Russian schools.

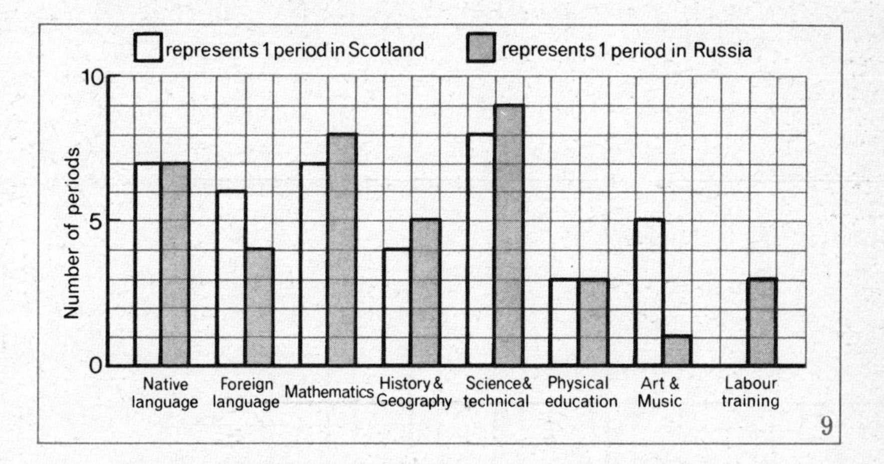

9

a How many periods are given to mathematics in each country?
b Which subjects have most periods both in Scotland and Russia?
c Which subjects have equal times in both countries?
d Make a table showing the number of periods for each subject in each country. Fill in the number of periods you have for each subject.

6 The table shows the points scored for academic and athletic activities by the four 'houses' in the competition for their school championship trophy.

House	EXE	WYE	TEE	VEE
Academic	85	75	100	95
Athletic	80	100	70	90

a Draw separate bar charts showing:
(*1*) academic achievement (*2*) athletic achievement
b Draw a third bar chart showing the 'athletic' column immediately above the 'academic' column. From this chart read off the overall winner of the trophy.

Exercise 2B

Illustrate by bar charts (questions *1–3*):

1 & 2 The information supplied in questions *1* and *2* of Exercise 1B.

3 The results of some of the surveys you carried out in question *5* of Exercise 1B.

4 The bar chart in Figure 10 shows the average number of hours of sunshine per day in the months October to April at Costa del Sol.

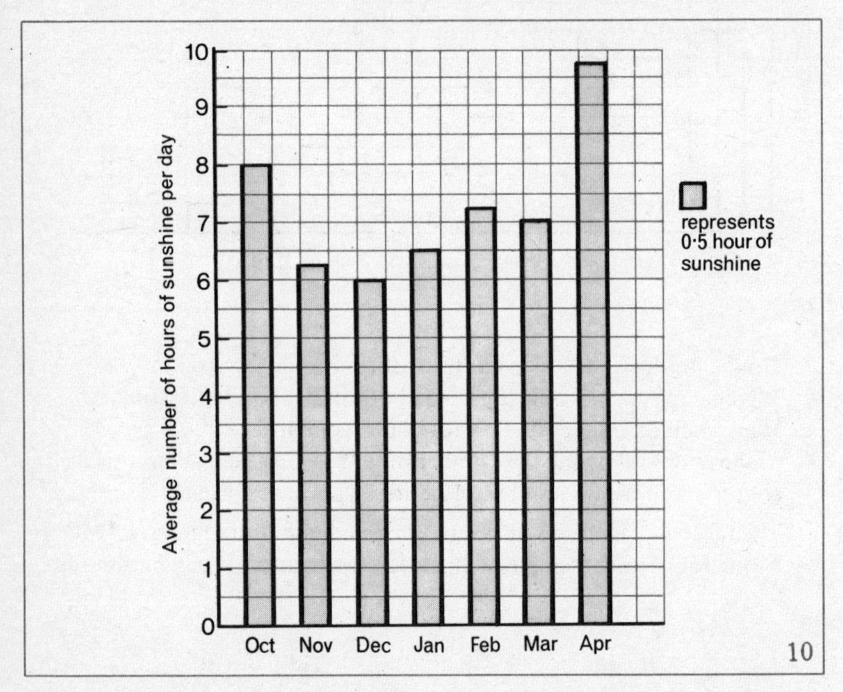

10

a What is the average number of hours of sunshine per day in (*1*) December (*2*) February?

b Which month has the greatest, and which the least, number of hours of sunshine?

c List the average number of hours of sunshine per day in each of the months, and hence calculate the total number for each month.

5 In 1966 the world production of olive oil was 1320 thousand tonnes. The chief producers were Spain (460 thousand), Italy (360), Greece (180), Turkey (150), and Portugal (40).

Choosing a suitable scale illustrate this information by means of a bar chart, remembering to include a column labelled 'others' to denote producers other than the five quoted.

6 Figure 11 shows the approximate number of cars sold by various manufacturers in the first three months of 1970 and 1971.

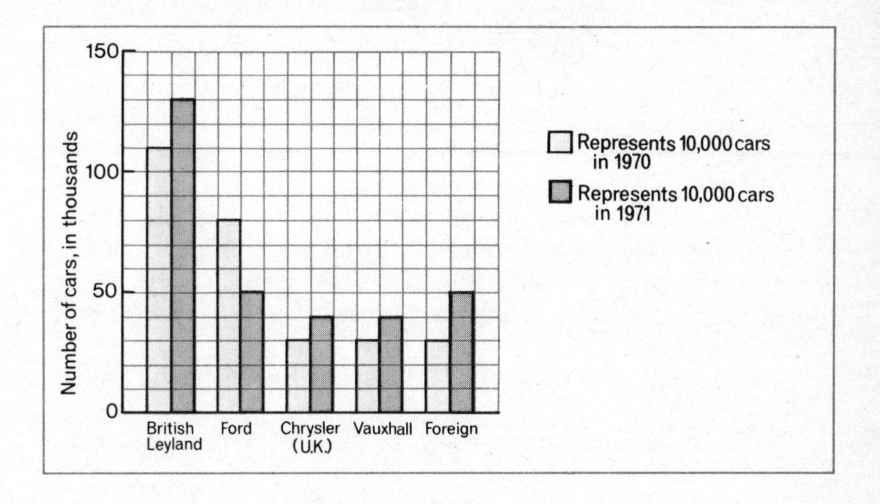

a Find the total number of British cars sold in each period.
b One company lost production due to a strike. Which?
c Find the increase in the number of foreign cars sold.

3 *Pie charts*

Another diagram which shows statistics pictorially is the *pie chart*. This may be thought of as a circular pie divided into slices or 'sectors' which correspond in size to the various items being shown.

It is particularly useful when the actual number of items is not so important as is the relation of the items to each other. Thus in Figure 12 you can see at a glance the relative fractions of sixth-year groups of pupils taking courses based on English, Mathematics and Science, Classics, Modern Languages. Can you arrange these fractions in order? Estimate the actual fractions.

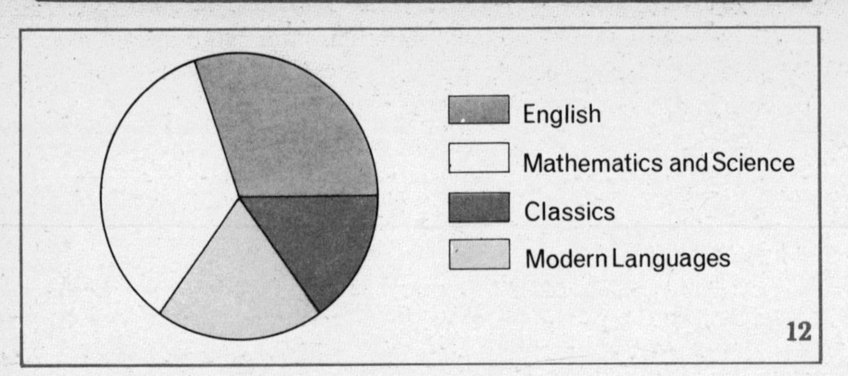

12

To construct the pie chart in Figure 12

We can only construct a pie chart when all the statistical information about the subject being considered is available. It is known that in this sixth-year class of 40 pupils, 12 take the English course, 14 the Mathematics-Science course, 6 the Classics course, and 8 the Modern Languages course.

We can calculate the sizes of the angles at the centre as follows:
For English, the angle is $\frac{12}{40} \times 360° = 108°$.
For Mathematics-Science, the angle is $\frac{14}{40} \times 360° = 126°$.
For Classics, the angle is $\frac{6}{40} \times 360° = 54°$.
For Modern Languages, the angle is $\frac{8}{40} \times 360° = 72°$.

To construct the pie chart we use compasses to draw a circle with radius about 5 cm, then measure the angles at the centre with a protractor and finally draw in the lines to form the 'slices' of the pie chart.

Note. Percentages are sometimes marked in the diagram. Here, for example, the percentage of pupils taking the English course is $\frac{12}{40} \times 100$, i.e. 30%.

Exercise 3A

1 Construct the pie chart shown in Figure 12, and mark in the percentages of pupils who take the various courses.

2 The pie chart in Figure 13 illustrates the balance of time given to the programmes presented by the Independent Television Authority in 1970, and is reproduced with their permission.

a List the programmes in the order of time given to them.
b Do you agree with this order?

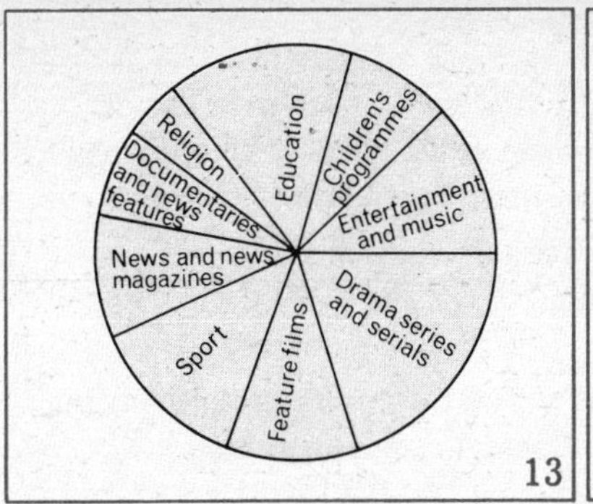

13

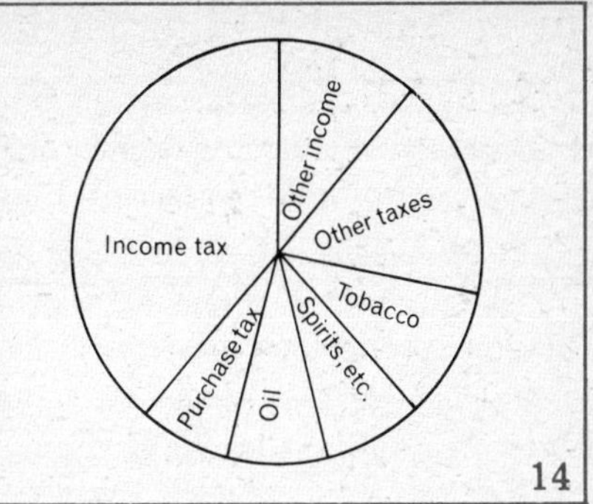

14

3 Figure 14 shows the principal sources of revenue estimated by the Chancellor of the Exchequer in the 1970 Budget.

a What is the largest single source of national revenue?

b What fraction of the total revenue is obtained by 'other taxes', approximately?

c List the sources of revenue in order.

d If the total revenue expected was £12000 million, give an estimate to the nearest £100 million of the revenue from income tax.

4 Make a pie chart showing the various methods of transport to school of pupils in your class.

5 A household spends 35% of the family income on food, 20% on clothes, 10% on rent, 15% on gas and electricity, and puts 5% into the Savings Bank. Show these data on a pie chart.

6 A survey provided the following information about the categories of the first 100 vehicles passing a school: private cars 60, small vans 15, larger vans and lorries 10, motor cycles 10, bicycles 5. Illustrate by a pie chart.

7 In 1967–68 the percentages of electricity produced in Scotland by various means were as follows: steam turbine 56%; hydro 18%; nuclear 19%; others 7%.

Draw a pie chart to illustrate this information. If the total electricity produced was 20000 million units, how much was produced by each means?

Exercise 3B

1 During a short observation period the colours of cars passing a school were noted, with the following results: black 4, blue 10, green 16, red 14, white 9, others 7. Illustrate by a pie chart.

2 Draw two pie charts to illustrate the information given in Exercise 1B, question *2*, about: ***a*** the number of pits ***b*** the personnel employed in the pits.

3 The percentages of the total 1967 crude oil imports refined in the main refining areas of the United Kingdom were as follows:

Thames Estuary 31%; West Coast of England 16%; Scotland 7%; Southampton 19%; Wales 23%; Other areas 4%

a Illustrate these figures by a pie chart.

b If the total crude oil imported was 73 000 000 tonnes, how much was refined in each area?

4 The North of Scotland Hydro-Electric Board issued the following percentages about the number of consumers connected, and the amount of electricity supplied to the consumers, who were of three types: domestic, commercial, industrial.

	Percentage of consumers	Percentage of units supplied
Domestic	82	53
Commercial	16	27
Industrial	2	20

Draw two pie charts to illustrate:

a the percentages of the different types of consumer,

b the percentages of units supplied to these consumers.

5 Collect some examples of pie charts from newspapers, magazines, etc., if you have not already done so, and discuss the information they depict.

6 ***a*** Of 100 students, all the men are more than twenty years old. There are 50 men in the group. If there are 60 persons in the group who are over twenty years old, make pie charts showing 'men–women' and 'over twenty–under twenty'; hence construct a pie chart showing men and women in respective age groups.

b In the same group, there are 25 married women of whom 10 are over twenty years old; 15 married students are over twenty. Using the charts in ***a*** obtain a new pie chart which distinguishes between twenty and under, and over twenty; men and women; and married and unmarried students.

4 Line graphs

Data which refer to information obtained over a period of time may be shown by plotting points, using the horizontal axis as the time scale. Statistics about temperatures, populations, sales, etc., are often shown in this way and, when the various points are joined up, a certain pattern or *trend* may be seen.

It is sometimes possible to estimate intermediate values; this is called *interpolation*. Occasionally it is reasonable to extend the line to the right and estimate future statistics; this is called *extrapolation*. Such estimates must be regarded with caution.

It is important to show the scale clearly in all cases. Why?

Example 1. The temperature in degrees Celsius was measured at a certain place every two hours from midnight to midnight on a day in summer, and the results are shown by the solid line in the graph in Figure 15.

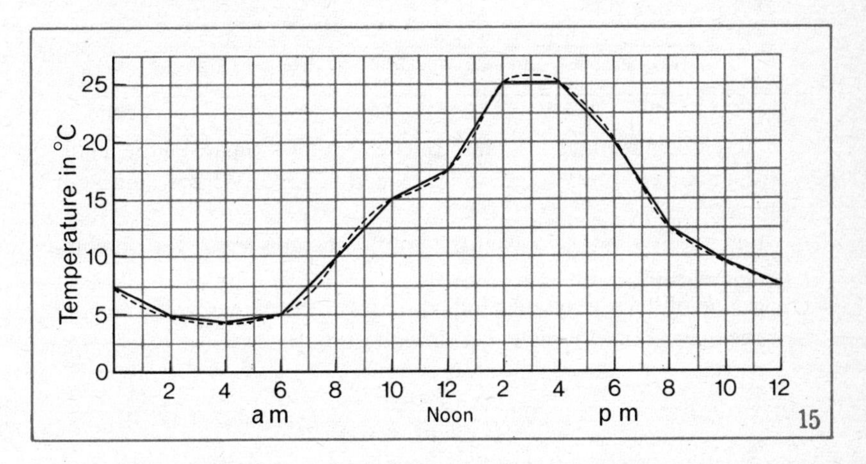

15

Since it is reasonable to assume that the temperature changed gradually, it is safe to interpolate. At 1 pm the temperature would be about 21 or 22°C. We must be careful about an estimate at 3 pm; the trend of the graph suggests that the temperature then was higher than 25°C.

Estimate the temperature at 3 am, 7 am, 11 am, 7 pm, 11 pm.

It is unwise to extrapolate on this graph. Why?

Note.—The dotted line shows a more accurate representation of the temperature throughout the period.

Example 2. Figure 16 shows the quantity of oil refined by an oil company each year from 1954 to 1964.

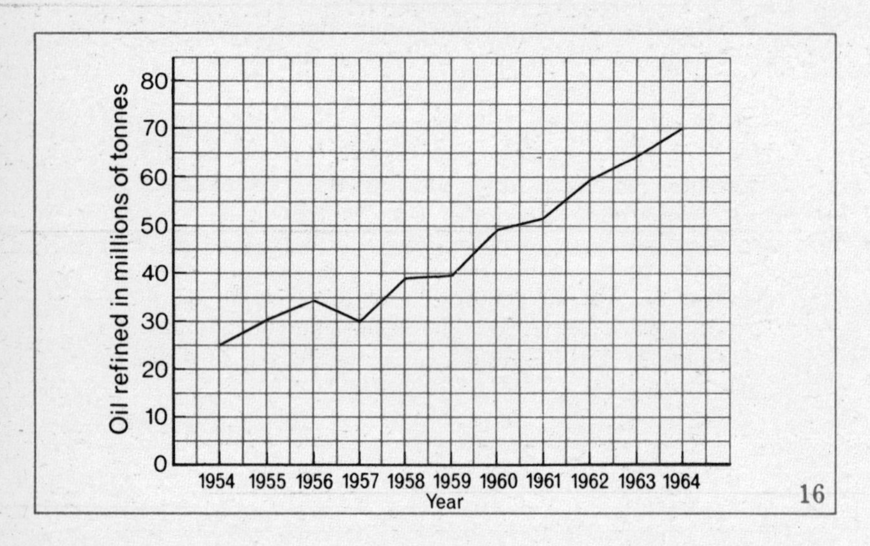

16

Here it is meaningless to interpolate, but an estimate *can* be made of the quantity of oil likely to be refined in 1965. What would your estimate be?

Explain in a sentence what the graph indicates about the oil company's output.

In one of these years oil production was interrupted by the closing of the Suez Canal. Which year do you think this was?

Exercise 4A

1 The quarterly sales of ice-cream (in £) in a school shop over a period of three years were as follows:

	Sep–Nov	Dec–Feb	Mar–May	June–Aug
1968	20	10	30	40
1969	40	10	35	45
1970	40	15	20	35

Show the sales on a line graph drawn on 5-mm squared paper. Take a scale of 1 cm to represent 1 quarter (Sep–Nov, etc.) horizontally, and 1 cm to represent £5 vertically.

a Estimate the probable sales in Sep–Nov 1971.

b Explain the figures for Sep–Nov 1968, and for Jun–Aug 1969.

2 A boy's height was measured each birthday and the following data obtained:

Age in years	9	10	11	12	13	14	15	16	17	18	19
Height in cm	125	130	132	137	144	152	158	164	173	178	178

Draw a line graph, taking a scale of 5 mm to represent 1 year of age horizontally, and 5 mm to represent 10 cm of height vertically. (Do you need to start at zero?) Use your graph to estimate:

a the boy's height at $12\frac{1}{2}$ years of age

b his age when he was 160 cm in height

c the year during which he was growing fastest

d the year when he stopped growing.

3 The population of the county of Fife for the years 1851–1931 was as follows (to the nearest 1000 persons):

Year	1851	1861	1871	1881	1891	1901	1911	1921	1931
Population, in thousands	150	150	160	170	190	220	270	290	270

Draw a line graph to show the population each year.

a Was the growth of population fairly regular?

b Estimate the population in 1906 and in 1926. Would you expect these to be good estimates?

c Suggest an explanation for the 1931 figure.

d Can you estimate the population in 1961?

4 The two line graphs in Figure 17 show the average monthly rainfall in millimetres at Inverness and Eskdalemuir weather stations.

a Which is the driest and which is the wettest month at each station?

b Estimate the rainfall for February and for September at each place.

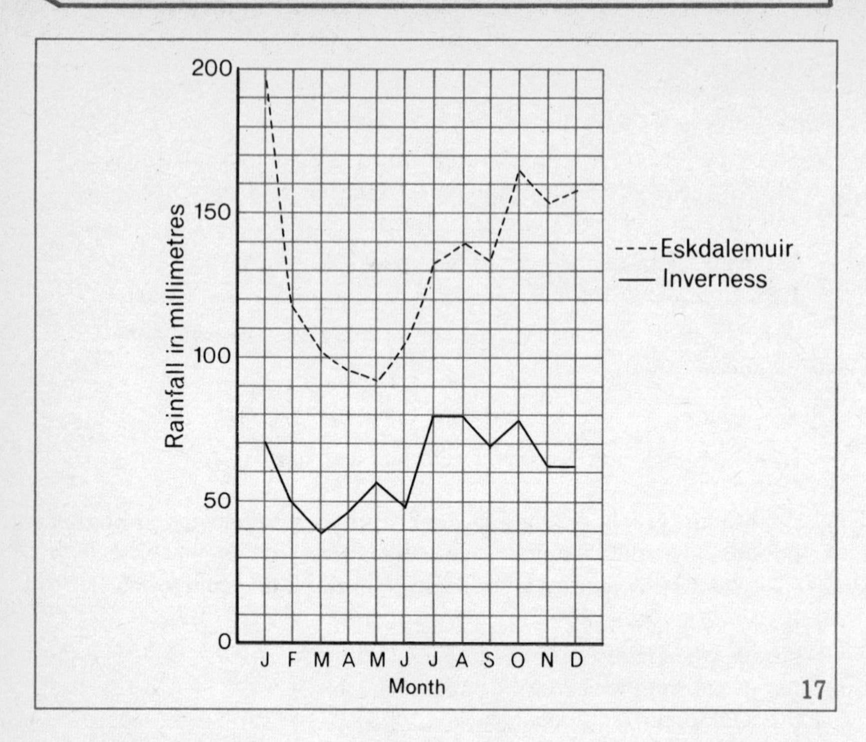

c What is the difference between the August rainfalls in the two places?

d Between which two consecutive months was a decrease in rainfall at Inverness accompanied by an increase at Eskdalemuir?

5 The following table gives the mean monthly temperatures in degrees Celsius at Inverness and Eskdalemuir.

Temperature in °C	Jan	Feb	Mar	Apr	May	Jun	Jul	Aug	Sep	Oct	Nov	Dec
Inverness	3·9	3·9	5	6·7	9·5	12·3	14	14	11·8	9	5·6	4·5
Eskdalemuir	1·7	2·2	3·4	5·5	8·3	11·1	13·3	12·8	10·0	7·8	4·4	2·2

Draw two line graphs on the same diagram to illustrate this information. From your graphs answer the following:

a In what month was there the biggest difference between the mean temperatures at the two places?

b For what period of three consecutive months was there a constant difference between the temperatures at the two stations?

Exercise 4B

1 The table shows the estimated net migration of population from Scotland for the years 1962 to 1970. The figures are in thousands.

Year	1962	1963	1964	1965	1966	1967	1968	1969	1970
Migration, in thousands	30	34	47	43	47	45	33	25	21

Draw a line graph to show the above information.
Write a sentence or two about the trends of the graph during these years.

2 Figure 18 supplies some information about the coal industry in Scotland. The *black* line graph gives the number of pits operating from 1958–68. The *broken* line graph gives the output per man-year in tonnes for the same period.

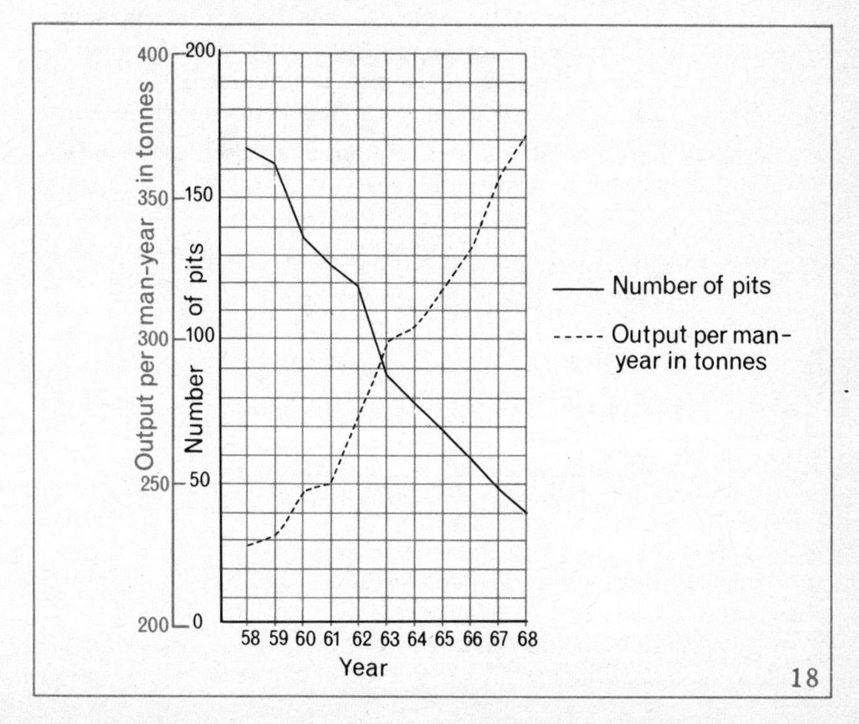

18

From the graph estimate:

a the number of pits which closed during the period.

b the increase in output per man-year over the period. Can you suggest a reason for this increase, and also say why a decrease in the number of pits should be accompanied by increased output?

c the average number of pits closed per year between 1963 and 1968.

d the average increase per year in output per man-year between 1961 and 1963.

3 The maximum and minimum mean monthly temperatures in degrees Celsius at Braemar (in the Scottish Highlands) are given below.

Month	Jan	Feb	Mar	Apr	May	Jun	Jul	Aug	Sep	Oct	Nov	Dec
Max temp in °C	3·9	4·5	7·3	9·0	12·9	16·2	17·9	16·8	14·6	10·6	6·7	5·0
Min temp in °C	−1·6	−1·6	−1·0	1·1	2·8	6·2	8·4	7·8	5·6	3·4	0·0	−1·0

Draw two line graphs on the same diagram showing the maximum and minimum monthly temperatures. Notice that you will require to make provision for the negative temperatures recorded.

Write a sentence or two about interesting features of the graph.

4 Figure 19 shows the profits after taxation of an electrical manufacturing company from 1964 until 1969.

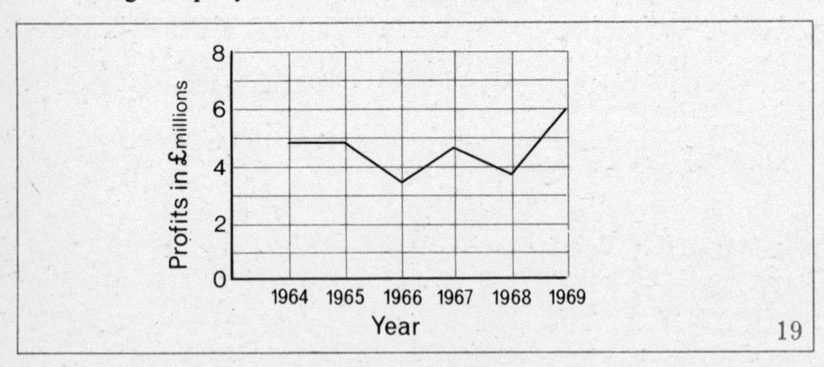

19

a What was the profit in 1967? In 1969?

b In what year was the profit smallest? By how much had the profit fallen that year from the previous year?

c Can you estimate the probable 1970 profit?

5 The following table gives the barometric pressure in metric units as recorded at noon each day for a fortnight at the same place.

Day	Sun	Mon	Tue	Wed	Thu	Fri	Sat
First week	$1{\cdot}02\times10^5$	$1{\cdot}01\times10^5$	$1{\cdot}006\times10^5$	$9{\cdot}94\times10^4$	$9{\cdot}8\times10^4$	$1{\cdot}003\times10^5$	$9{\cdot}98\times10^4$
Second week	$9{\cdot}9\times10^4$	$9{\cdot}96\times10^4$	$9{\cdot}78\times10^4$	$1{\cdot}004\times10^5$	$9{\cdot}98\times10^4$	$1{\cdot}015\times10^5$	$9{\cdot}93\times10^4$

Draw a line graph to illustrate these pressures.

Summary

Statistical data are shown in this chapter by means of:

1 *Pictographs*, e.g.

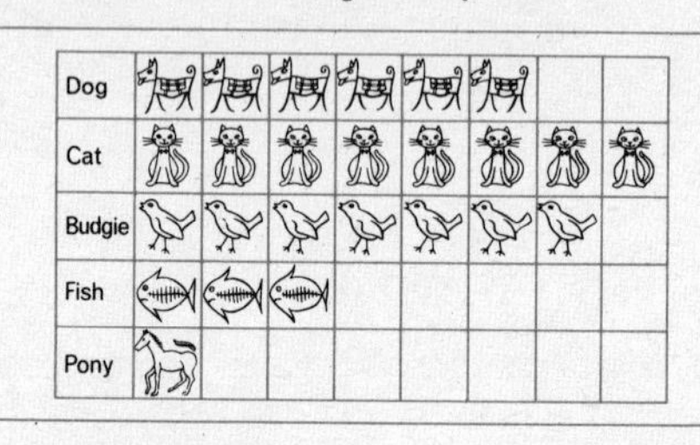

2 *Bar charts*, e.g.

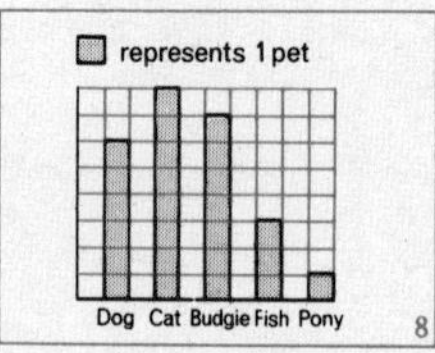

3 *Pie charts*, e.g.

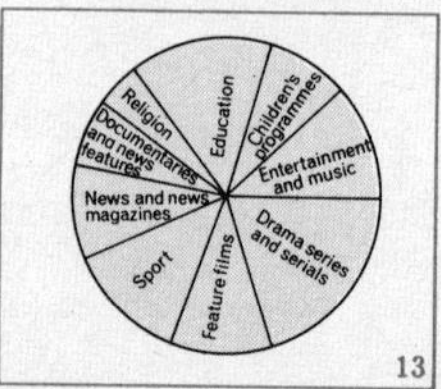

4 *Line graphs*, e.g.

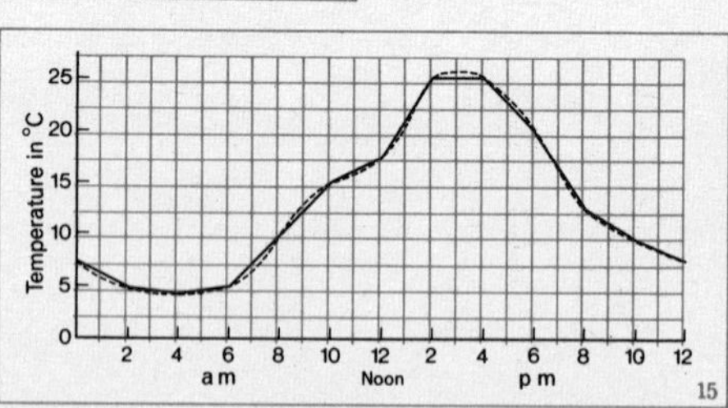

Revision Exercises

Revision Exercises on Chapter 1
Decimals and the Metric System

Revision Exercise 1A

1 Express each of the numbers shown as a decimal, and also as a mixed number:

H	T	U	t	h
		4	3	
		6	7	2
	1	0	4	
2	3	4	5	6

2 Round off each of the following to 2 decimal places:

3·141, 8·247, 12·009, 0·123, 0·345,
15·101, 8·556, 10·107, 0·052, 0·009

3 Round off each of the following to 3 significant figures:

28·37, 4·962, 0·9815, 0·7028, 3698,
0·005692, 0·04097, 69·94, 0·09996

4 Write down an estimate for the answer to each of the following, using 1 significant figure in each number:

$12 \times 8{\cdot}4$, $18{\cdot}9 \times 11{\cdot}1$, $6{\cdot}5 \times 8{\cdot}2$, $2{\cdot}8 \times 0{\cdot}85$,
$15{\cdot}7 \times 0{\cdot}04$, $0{\cdot}8 \times 0{\cdot}07$, $34 \div 2{\cdot}7$, $23{\cdot}8 \div 6{\cdot}8$,
$0{\cdot}76 \div 5{\cdot}1$, $0{\cdot}0092 \div 0{\cdot}23$, $0{\cdot}065 \div 0{\cdot}037$

5 Calculate the perimeters and areas of the shapes in Figure 1 on page 228, rounding off your answers to 2 significant figures. (All angles are right angles.)

6 Find the values of:

a $26{\cdot}9 + 31{\cdot}2 - 19{\cdot}6$ *b* $(5{\cdot}5 \times 2{\cdot}8) \div 0{\cdot}14$

7 Express each of the following as a decimal fraction:

$\frac{9}{10}$, $\frac{1}{2}$, $\frac{3}{4}$, $\frac{4}{5}$, $\frac{1}{8}$, $\frac{7}{8}$

8 Express each of the following as a decimal, rounded off to 2 decimal places: $\frac{2}{3}$, $\frac{5}{7}$, $\frac{4}{9}$, $\frac{2}{13}$

9 Write the following as whole numbers in their usual form:

5×10^2, 15×10^3, $2{\cdot}8 \times 10^2$, $3{\cdot}14 \times 10^3$

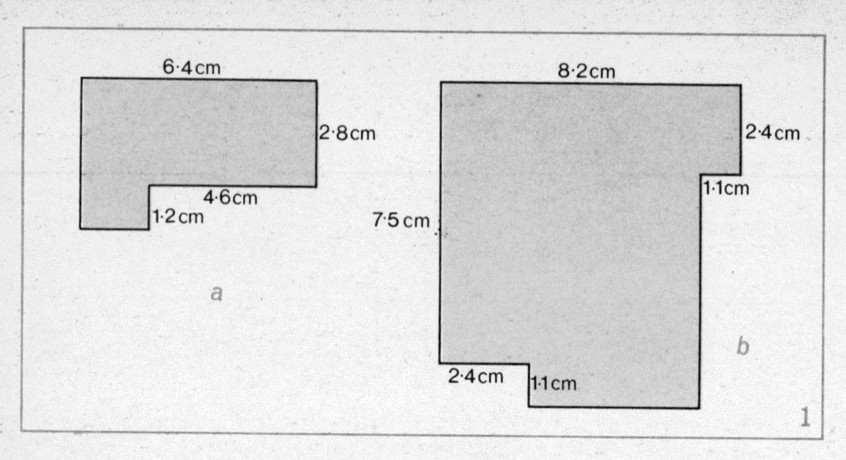

10 Express in standard form (i.e. $a \times 10^n$):

53·7, 376, 5042, 3000000,
562·1, 2760, 246000, 17·65

11 The height of Mont Blanc is $4{\cdot}8 \times 10^3$ metres. How many metres is this?

12 A carton contains 144 tins of soup, each weighing 425 grammes. What is their total mass in kilogrammes?

13 The petrol tank of a car holds 35 litres. If the car travels 500 kilometres on a full tank of petrol, how far will it travel on 1 litre?

14 On a plan a door is marked 3150 mm high and 1250 mm wide.

a What are its dimensions in metres?

b What is the area of the door in square metres (to 3 significant figures)?

15 Calculate the following, to the nearest penny where necessary:

a 25% of £18·60 *b* 20% of £6·35

c 9% of £3·42 *d* $3\frac{1}{2}$% of £75·24

16 A man on holiday in Germany stays 12 days in a hotel which charges 24·50 marks per day, plus $12\frac{1}{2}$% service charge. What is the total charge he has to pay?

Revision Exercise 1B

1 Write down an estimate of the answer to each of the following:

$9{\cdot}37 \times 4{\cdot}83$, $23{\cdot}76 \times 31{\cdot}42$, $5{\cdot}81 \times 0{\cdot}063$, $0{\cdot}0184 \times 0{\cdot}772$,

$158 \div 42{\cdot}9$, $3{\cdot}74 \div 8{\cdot}91$, $0{\cdot}856 \div 22{\cdot}7$, $0{\cdot}236 \div 0{\cdot}619$,
$0{\cdot}0742 \div 5{\cdot}4$, $0{\cdot}0426 \div 0{\cdot}00439$, $218000 \div 38$ million

2 Round off each of the following to 3 significant figures, and also to 1 decimal place:

64·32, 8·729, 373·5, 206·51, 0·06047, 0·08818

3 Calculate the perimeters and the areas of the diagrams shown in Figure 2, giving your answers to 2 significant figures. (All angles are right angles.)

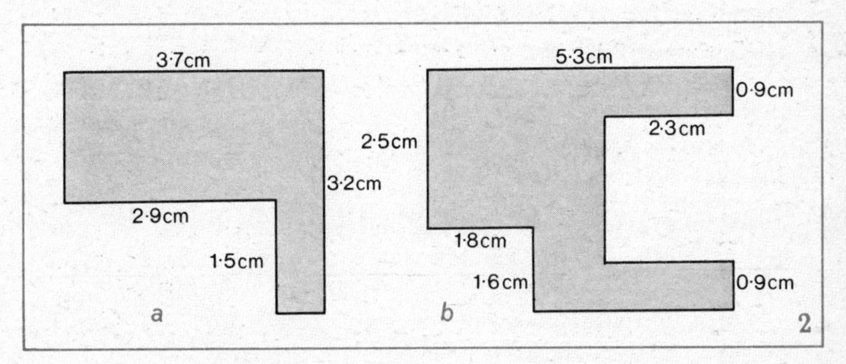

2

4 Evaluate $\dfrac{(0{\cdot}6)^2 \times 0{\cdot}5}{(0{\cdot}2)^2 \times 0{\cdot}45}$

5 Express in standard form (i.e. $a \times 10^n$):

a 285 g *b* 20·8 litres *c* 3400 m
d 280000 km *e* 156700 cm *f* 186 million

6 Express each of the following as a decimal fraction:
$\frac{3}{10}$, $\frac{21}{100}$, $\frac{1}{4}$, $\frac{3}{5}$, $\frac{3}{8}$, $\frac{5}{16}$

7 Express each of the following as a decimal, rounded off to 2 decimal places: $\frac{6}{7}$, $\frac{1}{3}$, $\frac{1}{9}$, $\frac{10}{11}$

8 If a wavelength is 0·000032 cm, how many wavelengths are in 1 mm? Give your answer in scientific notation to 2 significant figures.

9 If $(8{\cdot}2 \times 10^4) \times (2{\cdot}5 \times 10^2) = q \times 10^n$ find q and n.

10 The final positions of the two top clubs in the Scottish Football League in 1964–65 are given in the table:

	Played	Won	Lost	Drawn	Goals for	Goals against
Kilmarnock	34	22	6	6	62	33
Hearts	34	22	6	6	90	49

a The championship was decided on goal average, that is 'goals for' divided by 'goals against'. Work out each goal average to 3 significant figures and find by how much Kilmarnock won.

b Some people think a better way of deciding the championship would be to calculate the average number of goals scored *per match* for and against, and then to subtract these two averages. Find out which team would have won, and by how much, if this method had been used.

c Calculate to 3 significant figures what percentage of their matches these teams won, lost, and drew.

11 In 1968–69 the average sum paid in taxation in Britain was £183 for every man, woman, and child. In 1969–70 it was expected to be £192. If the total population was 54·6 million, how much money was raised in each of these years (to a reasonable degree of approximation)?

12 For a holiday in France, a man has arranged to stay for 13 days in a hotel which charges 36·50 fr per day, to which has to be added 10% for service. On leaving Dover he changes £75 into francs at the rate £1 = 13·16 fr. How many francs will he have available for other expenses over and above his hotel bill?

Revision Exercises on Chapter 2
Computers and Binary Arithmetic

Revision Exercise 2A

1 Convert the following from decimal to binary form:

a 17 *b* 21 *c* 35 *d* 67

2 Convert the following from binary to decimal form:

a 1011 *b* 10111 *c* 101101

3 Calculate:

a 11011 + 1001 *b* 10101 + 11001 *c* 10111 + 1111
d 1011 − 101 *e* 10101 − 1001 *f* 10101 − 1111

4 Calculate:

a 1101 × 11 *b* 1101 × 101 *c* 1111 × 11
d 11101 ÷ 10 *e* 11101 ÷ 11 *f* 111100 ÷ 110

5 Convert the following from base two to base ten, and find out which one is not prime:
1101, 10001, 11001, 11101

6 Calculate the area and perimeter of a rectangle whose dimensions (in binary form) are 110 cm and 11 cm.

7 Calculate the breadth of a rectangle with area 11110 cm^2 and length 1010 cm.

8 A square has perimeter 1100 m. What number: *a* in decimal form *b* in binary form, must you divide by to get the length of the side of the square? Find this length in binary form.

Revision Exercise 2B

1 Convert the following from decimal to binary form:

a 39 *b* 59 *c* 95 *d* 255 *e* 260

2 Convert the following from binary to decimal form:

a 11011 *b* 11110 *c* 111011 *d* 1111101

3 Calculate:

a 101111 + 1010 *b* 1011110 + 11101 *c* 1000001 − 10110
d 1000100 − 1011 *e* 11011 × 1001 *f* 101010 × 111
g 1100010 ÷ 1110 *h* 10000101 ÷ 11010

4 What multiplication in the binary system is equivalent to multiplication by 2 in the decimal system? By 4 in the decimal system?
The following are all in binary form.

a Write down the answers to 101 × 10, 110 × 10, 11011 × 10, 11 × 100, 1011 × 100.

b Find the perimeters of squares with sides of length 11, 101, 100, and 1101 cm.

c Calculate the perimeters of rectangles with lengths and breadths 11 cm and 10 cm; 101 cm and 11 cm; 1111 cm and 111 cm.

5 Calculate the areas of the squares in question ***4b***.

6 Calculate the areas and perimeters of rectangles with lengths and breadths 110 m and 11 m; 1011 m and 1010 m.

7 Find the lengths of the sides of squares with perimeters:

a 1100 cm *b* 1010100 cm *c* 11000 m

8 The inhabitants of the planet Diabolicus have only one hand with three fingers. So when we write 5 they write 12. How would the Diabolicals write the numbers which we write 8, 19, 27?

Write out the addition and multiplication tables which the lucky Diabolical infants have to learn at school.

A Diabolical child receives pocket money of 1120 Diabolos. How much is that in our money, given that one Diabolo is worth a penny?

Revision Exercises on Chapter 3
Statistics

Revision Exercise 3A

1 100 boys and 100 girls were questioned about the beverage they had for breakfast; the results of the investigation are shown in Figure 3.

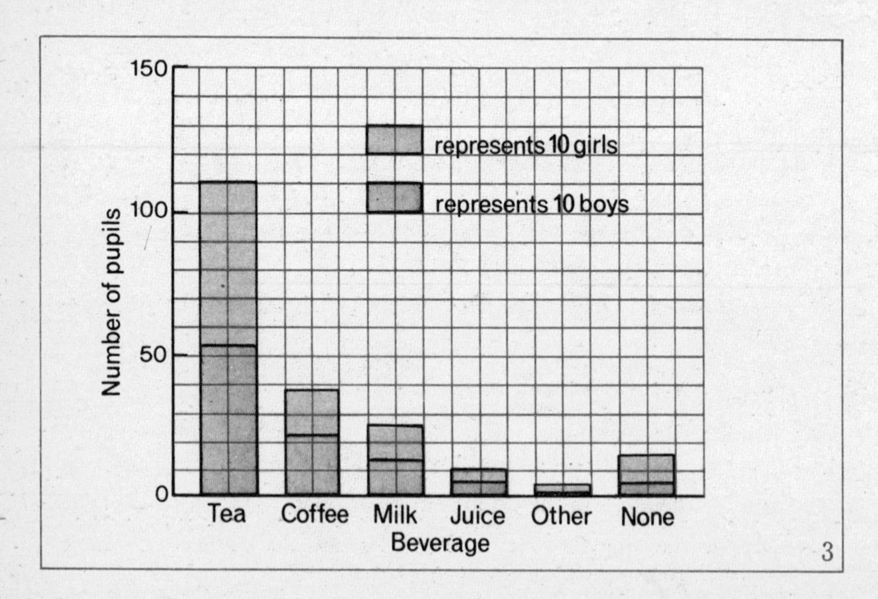

3

From the bar chart estimate the number of boys and the number of girls who had tea, coffee, etc., and the numbers who had none.

2 The following figures (supplied by the Meteorological Office) show the rainfall and the number of rainy days recorded at certain places for the month of March 1968. Draw a bar chart to illustrate each set of figures.

Place	Aberdeen	Braemar	Fort William	Inverness	Wick
Rainfall in mm	51	53	127	38	46
Number of rainy days	16	16	17	14	20

3 The total cost of building a modern school is made up as follows: professional fees (architects, surveyors, etc.) 9%; cost of plant 4%; wages 40%; building materials 47%. Draw a pie chart to illustrate these figures.

4 The number of pupils in a school who passed Ordinary Grade Mathematics in Certificate examinations from 1962 to 1970 was as follows:

Year	1962	1963	1964	1965	1966	1967	1968	1969	1970
Number of pupils	80	107	91	74	59	83	92	86	88

Illustrate these data by a line graph.

5 The graph in Figure 4 shows the level of water in a harbour from noon until midnight on a certain day.

From the graph estimate:

a the depth of the water at 2 pm and at 9 pm
b the time of high water and of low water, and the depths at those times
c the times between which a vessel which requires a depth of 4 metres of water could *not* enter the harbour

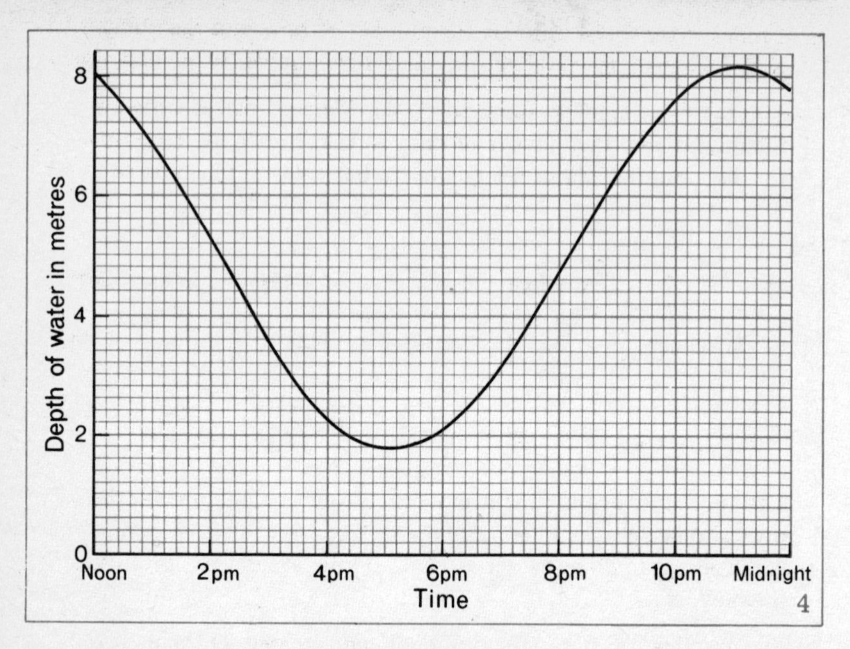

4

6 The relative sizes of the main tribal groups in Nigeria (based on a fairly recent census) are shown in Figure 5.

a List the tribes in their order of size.

b What fraction of the total do you estimate 'others' to be?

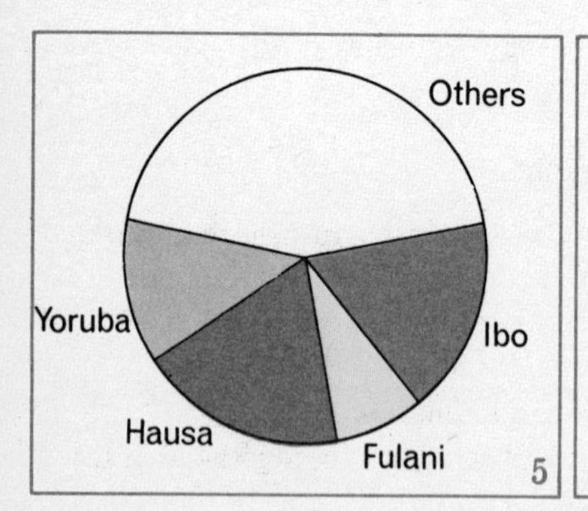

5

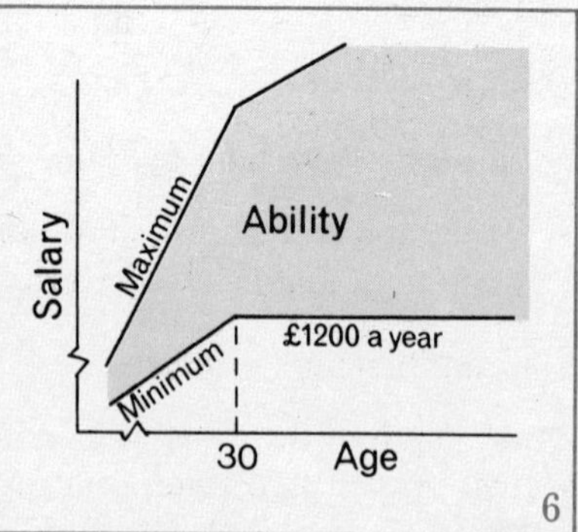

6

7 Figure 6, taken from an advertisement, had the caption, 'Have you plotted your future? If not, you will be interested in the graphic presentation of our salary structure, based on reward for enthusiasm and hard work.'

What information can you get from this graph? Point out any deficiencies in the graph.

Revision Exercise 3B

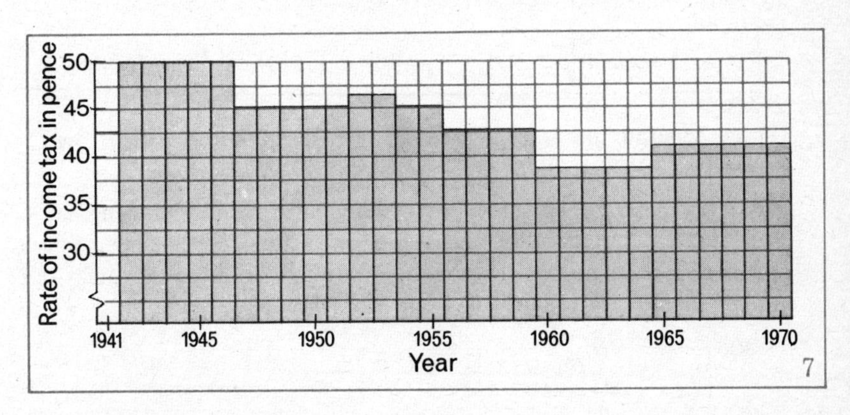

7

1 Figure 7 shows how the standard rate of Income Tax has changed over a certain period.

a Why is there a break shown on the vertical axis?

b What were the highest and lowest rates charged during the period? How long did the tax remain at the highest rate? Give an explanation of why the rate was so high during these years.

c What was the greatest reduction that took place in any year, and what year was that?

2 The table below gives the approximate population and areas of some Scottish counties.

County	Orkney	Aberdeen	Angus	Midlothian	Wigtown
Population (thousands)	18	317	278	592	28
Area (km²)	973	5104	2263	947	1261

Draw a bar chart to illustrate each set of figures.

3 The percentages of the total land areas of Norway and the United Kingdom which fall into various categories are given below.

	Percentage of total land area			
	arable	pasture	forest	wasteland, etc
Norway	3	1	22	74
United Kingdom	31	50	7	12

Draw separate pie charts to show the distributions in the two countries.

4 The graph in Figure 8 shows the cooling curve for naphthalene, i.e. the graph of temperature against time as the naphthalene cools from its boiling-point, when it is liquid, to its normal solid state.

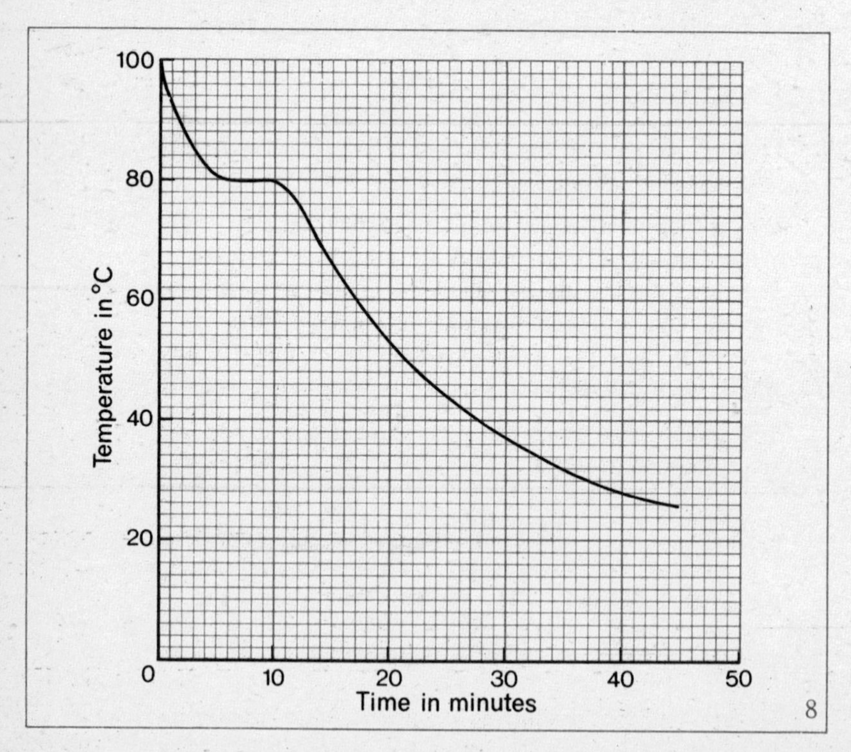

8

From the graph answer the following:

a Read off the temperatures after 3, 6, 10, 25, and 37 minutes.

b Read off the times after which the temperatures were 70°C, 66°C, 40°C, 35°C.

c Make a true statement about the temperature corresponding to the 'flat' portion of the curve. Can you guess (or find out) what is happening during this period?

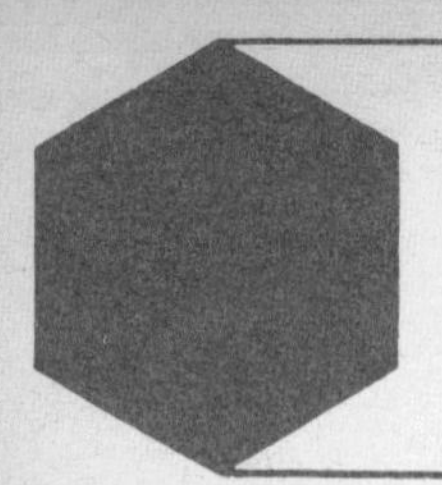

Cumulative Revision Section (Books 1 and 2)

Book 1 Chapter Summaries

Chapter 1 The system of whole numbers

The set of *whole numbers* is {0, 1, 2, 3, 4, ...}
The set of *natural numbers* is {1, 2, 3, 4, ...}
The set of *even numbers* is {0, 2, 4, 6, ...}
The set of *odd numbers* is {1, 3, 5, 7, ...}
The set of *prime numbers* is {2, 3, 5, 7, 11, 13, 17, 19, ...}
Prime numbers are divisible only by themselves and by 1.

1 *The Commutative Laws of addition and multiplication*

$a+b = b+a$ $\qquad$ $a\times b = b\times a$

e.g. $8+24 = 24+8$ $\qquad$ e.g. $8\times 24 = 24\times 8$

2 *The Associative Laws of addition and multiplication*

$(a+b)+c = a+(b+c)$ $\qquad$ $(a\times b)\times c = a\times(b\times c)$

e.g. $(6+4)+3 = 6+(4+3)$ $\qquad$ e.g. $(6\times 4)\times 3 = 6\times(4\times 3)$

3 *The Distributive Law*

$(a\times b)+(a\times c) = a\times(b+c)$ or $a(b+c) = ab+ac$

e.g. $(7\times 3)+(7\times 5) = 7\times(3+5)$

4 *The identity element for addition* (0)

$a+0 = 0+a = a$

e.g. $5+0 = 0+5 = 5$

5 *The identity element for multiplication* (1)

$a\times 1 = 1\times a = a$

e.g. $8\times 1 = 1\times 8 = 8$

6 *Squares and cubes*

a^2 means $a\times a$ $\qquad$ a^3 means $a\times a\times a$

e.g. $9^2 = 9\times 9 = 81$ $\qquad$ e.g. $5^3 = 5\times 5\times 5 = 125$

7 *The principle for multiplying by* 0

$a\times 0 = 0 = 0\times a$

e.g. $7\times 0 = 0 = 0\times 7$

If a, b are whole numbers for which $a\times b = 0$, then at least one of a, b is 0.

Chapter 2 Decimal systems of money, length, area, volume and mass

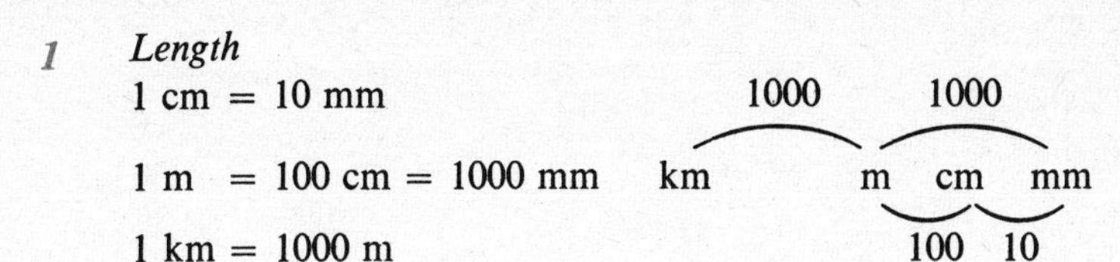

1 *Length*

1 cm = 10 mm

1 m = 100 cm = 1000 mm

1 km = 1000 m

2 *Area*

1 cm^2 = 100 mm^2

1 m^2 = 10000 cm^2 — *For a rectangle*, $A = lb$

1 km^2 = 1000000 m^2 — *For a square*, $A = l^2$

3 *Volume*

1 litre = 1000 ml = 1000 cm^3

1 m^3 = 1000000 cm^3 = 1000 litres

1000
litre ⌒ ml

For a cuboid, $V = lbh$, or $V = Ah$

4 *Mass*

1 g = 1000 mg

1 kg = 1000 g

1000 1000
kg ⌒ g ⌒ mg

Chapter 3 Fractions, ratios and percentages

The meaning of a fraction

1 $\frac{a}{b} = a \div b$ — a is the *numerator* of the fraction; b is the *denominator* of the fraction

2 *Equal fractions*

$$\frac{1}{2} = \frac{1\times 3}{2\times 3} = \frac{3}{6} = \frac{3\times 4}{6\times 4} = \frac{12}{24} = \frac{12\div 2}{24\div 2} = \frac{6}{12} = \ldots$$

3 *Least common multiples*

The LCM of *a* 2, 3, is 6 *b* 2, 3, 4 is 12 *c* 5, 10 is 10

4 *Addition and subtraction of fractions*

$\frac{3}{4}+\frac{5}{6} = \frac{9}{12}+\frac{10}{12} = \frac{19}{12} = 1\frac{7}{12}$

5 *Multiplication of fractions*

$$\frac{3}{4} \text{ of } \frac{5}{6} = \frac{\overset{1}{\cancel{3}}}{4} \times \frac{5}{\underset{2}{\cancel{6}}} = \frac{5}{8}$$

6 *Division of fractions*

$$3\tfrac{1}{3} \div 2\tfrac{1}{2} = \frac{10}{3} \div \frac{5}{2} = \frac{\overset{2}{\cancel{10}}}{3} \times \frac{2}{\underset{1}{\cancel{5}}} = \tfrac{4}{3} = 1\tfrac{1}{3}$$

$$\frac{p}{q} \div \frac{a}{b} = \frac{p}{q} \times \frac{b}{a}$$

7 *Ratios*

$$a : b = \frac{a}{b}$$

8 *Percentages*

$$1 \text{ per cent} = 1\% = \frac{1}{100}$$

$$a\% = \frac{a}{100} \qquad \frac{x}{y} = \frac{x}{y} \times 100\%$$

Cumulative Revision Exercises

Exercise A

In questions *1–8*, say whether each statement is true or false.

1 A football field is about 100 km long.

2 £3·50 ÷ 100 = $3\frac{1}{2}$p.

3 18 square tiles each of side 12 cm will exactly cover a hearth 72 cm long and 36 cm wide.

4 The area of a rectangle 1·25 cm long and 0·48 cm broad is 0·06 cm².

5 3% of £18 = 54p.

6 91 is a prime number.

7 $3\frac{1}{4} + 2\frac{1}{3} = 5\frac{2}{7}$.

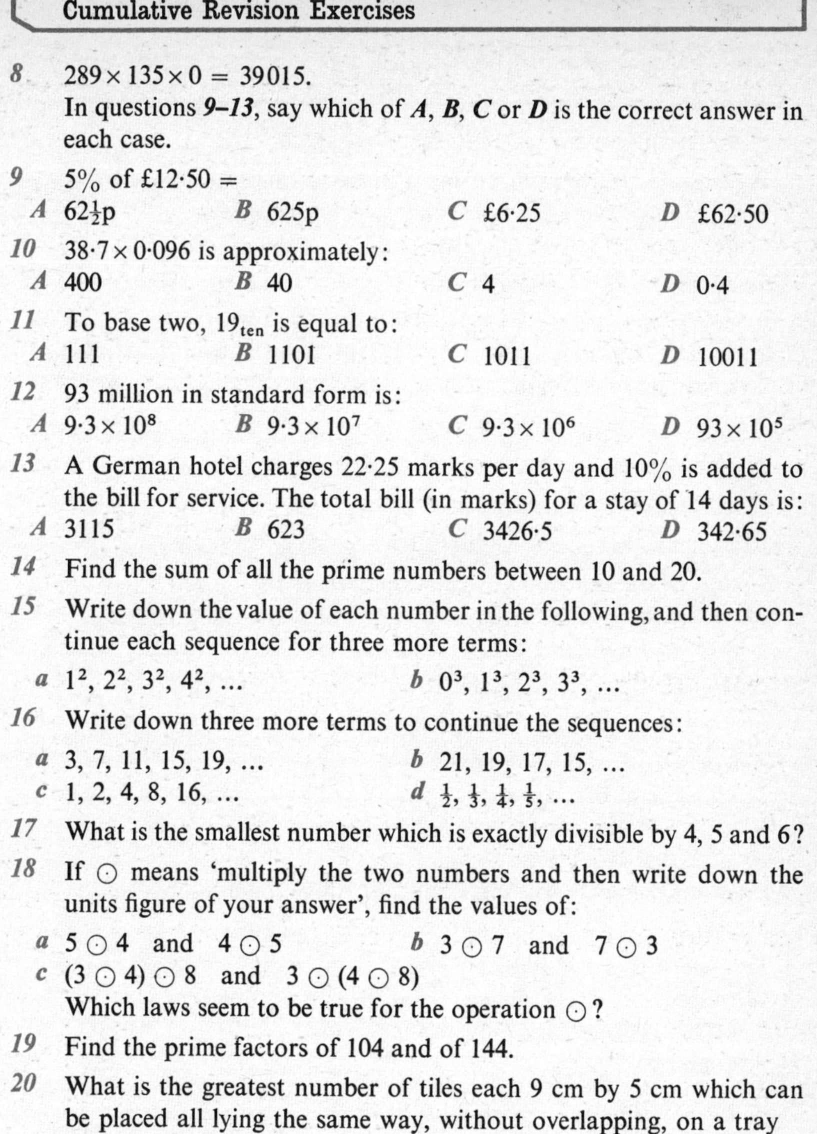

8 $289 \times 135 \times 0 = 39015$.

In questions ***9–13***, say which of ***A***, ***B***, ***C*** or ***D*** is the correct answer in each case.

9 5% of £12·50 =

A $62\frac{1}{2}$p *B* 625p *C* £6·25 *D* £62·50

10 $38{\cdot}7 \times 0{\cdot}096$ is approximately:

A 400 *B* 40 *C* 4 *D* 0·4

11 To base two, 19_{ten} is equal to:

A 111 *B* 1101 *C* 1011 *D* 10011

12 93 million in standard form is:

A $9{\cdot}3 \times 10^8$ *B* $9{\cdot}3 \times 10^7$ *C* $9{\cdot}3 \times 10^6$ *D* 93×10^5

13 A German hotel charges 22·25 marks per day and 10% is added to the bill for service. The total bill (in marks) for a stay of 14 days is:

A 3115 *B* 623 *C* 3426·5 *D* 342·65

14 Find the sum of all the prime numbers between 10 and 20.

15 Write down the value of each number in the following, and then continue each sequence for three more terms:

a $1^2, 2^2, 3^2, 4^2, \ldots$ *b* $0^3, 1^3, 2^3, 3^3, \ldots$

16 Write down three more terms to continue the sequences:

a 3, 7, 11, 15, 19, ... *b* 21, 19, 17, 15, ...
c 1, 2, 4, 8, 16, ... *d* $\frac{1}{2}, \frac{1}{3}, \frac{1}{4}, \frac{1}{5}, \ldots$

17 What is the smallest number which is exactly divisible by 4, 5 and 6?

18 If $\odot$ means 'multiply the two numbers and then write down the units figure of your answer', find the values of:

a $5 \odot 4$ and $4 \odot 5$ *b* $3 \odot 7$ and $7 \odot 3$
c $(3 \odot 4) \odot 8$ and $3 \odot (4 \odot 8)$

Which laws seem to be true for the operation $\odot$?

19 Find the prime factors of 104 and of 144.

20 What is the greatest number of tiles each 9 cm by 5 cm which can be placed all lying the same way, without overlapping, on a tray

a 27 cm by 100 cm *b* 45 cm by 64 cm?

21 Which hockey pitch is bigger, one measuring 75 m by 45 m, the other 80 m by 42 m?

22 Find the areas of rectangles with lengths and breadths as follows:

a 13 m, $4\frac{1}{2}$ m *b* 3·5 cm, 2·8 cm *c* 2 km, 12 m

23 Find the volume of a box 6 cm long, 3 cm broad and 1·5 cm deep. How many such boxes could be filled from a tank full of liquid which is 1 m long, 72 cm broad and 15 cm deep?

24 Write down rough estimates for each of the following; then calculate each, rounding off the answer to 2 significant figures:

a $30{\cdot}9 \times 1{\cdot}45$ *b* $0{\cdot}82 \times 0{\cdot}64$ *c* $57 \times 0{\cdot}063$
d $98{\cdot}4 \div 21{\cdot}5$ *e* $1{\cdot}76 \div 8{\cdot}1$ *f* $0{\cdot}269 \div 0{\cdot}017$

25 A rectangular sheep enclosure in Australia is 7·82 km long and 3·94 km broad. By how much is the length greater than the breadth? What is the total length of a wire fence round the perimeter?

26 A carton weighing 1·25 kg contains 72 tins of meat, each weighing 225 g. What is the total mass of the full carton?

27 Calculate to the nearest penny:

a 25% of £1·25 *b* 5% of £36·40 *c* $2\frac{1}{2}$% of £7·50

28 Carry out the following calculations, where the numbers have base two. Check your answers by expressing the numbers and the answers to base ten.

a $1101 + 1011$ *b* $1011 - 101$ *c* 1110×101
d $1111 \div 101$ *e* $(111 \times 11) - (111 + 11)$

29 Calculate the total cost of:
$2\frac{1}{2}$ kg steak at 65p per kg
$\frac{3}{4}$ kg ham at 46p per kg
15 eggs at 18p per dozen
How much change will there be from a five-pound note?

30 Find the total cost of:
200 daffodils at £3·20 for 50
48 tulips at £1·80 per dozen
6 autumn crocuses at $17\frac{1}{2}$p each
Postage 15p.

31 Express in the usual whole-number form:

a $1{\cdot}76 \times 10^3$ *b* $9{\cdot}3 \times 10^6$ *c* $4{\cdot}8 \times 10^4$ *d* 8×10^7

32 Express in standard form ($a \times 10^n$):

a 3000 *b* 286000 *c* 75000000 *d* 9 million million

33 The table gives the number of units manufactured in a factory each year from 1965 to 1970.

Year	1965	1966	1967	1968	1969	1970
Number of units	120	150	180	180	240	210

Illustrate the data by means of:

a a bar chart *b* a pie chart *c* a line graph

Exercise B

In questions *1–8*, say whether each statement is true or false.

1 A medium-sized motor-car weighs about 1000 kg.

2 The operation of division is commutative.

3 The capacity of a rectangular tank measuring 75 cm by 45 cm by 12 cm is 40·5 litres.

4 £512 increased by $2\frac{3}{4}\%$ is £526·08.

5 $0{\cdot}773 \div 0{\cdot}00429$ is approximately 20.

6 $3\frac{1}{4} \times 2\frac{1}{3} = 6\frac{1}{2}$.

7 3^6 is greater than 6^3.

8 $10101_{two} \times 10101_{two} = 100010001_{two}$

In questions *9–13*, say which of ***A***, ***B***, ***C*** or ***D*** is the correct answer in each case.

9 $1{\cdot}5 \times 0{\cdot}7 \times 0{\cdot}04 =$

A 42 *B* 0·42 *C* 0·042 *D* 0·0042

10 How many tins, each a cube of edge 5 cm, can be filled from a bottle containing 1 litre of liquid?

A 8 *B* 40 *C* 50 *D* 200

11 $(1{\cdot}3 \times 10^6) + (3{\cdot}1 \times 10^5) =$

A $4{\cdot}4 \times 10^{11}$ *B* $4{\cdot}4 \times 10^6$ *C* $1{\cdot}61 \times 10^5$ *D* $1{\cdot}61 \times 10^6$

12 A rectangular lawn is 20 m long and 12 m broad. If the length is increased by 50% and the breadth is increased by 25%, by what percentage is the area of the lawn increased?

A $87\frac{1}{2}\%$ *B* 75% *C* 50% *D* $37\frac{1}{2}\%$

13 The perimeter of a rectangle is 101000_{two} units and one side is 1101_{two} units. The length of the other side is:

A 11011 *B* 11010 *C* 1110 *D* 111

14 Find the sum of all the prime numbers between 40 and 50.

15 Write down three numbers to continue each of the following sequences:

a 99, 92, 85, 78, ...

b 0, 1, 8, 27, 64, ...

c 2, 5, 10, 17, 26, ...

d 2, 3, 5, 7; 11, 13, 17, ...

16a Find the LCM of 6, 8, and 9.

b I have a bag of caramels. When I try to share them equally among 2, 3, 4, 5 or 6 children, I always have one left over for myself. What is the smallest number of caramels which can be in the bag?

17 In the operation of 'modular multiplication' on the set {0, 2, 4, 6, 8} we write down the units figure in the product of two elements. For example, 8 multiplied by 8 gives 64, so in modular multiplication $8 \times 8 = 4$. Complete this table for modular multiplication:

		Second Number				
	×	0	2	4	6	8
	0	0				
First Number	2					
	4					
	6					
	8					4

a Is the table symmetrical about the main diagonal? What does this tell you?

b By trying several examples such as $(2 \times 4) \times 6$ and $2 \times (4 \times 6)$ decide whether the associative law appears to be true or not for this multiplication.

c There is an identity element. What is it?

18 Find the prime factors of 196, and of 1152.

19 A rectangle is 10 cm long and 8 cm broad. What is its area? If the length is increased by 15% and the breadth by 25%, what is the new area? What is the ratio of the new area to the original area? What percentage is this equivalent to?

20 What is the greatest number of boxes each 4 cm by 6 cm by 10 cm which can be packed, all lying the same way, into a carton 1 m long, $\frac{1}{2}$ m broad, and 15 cm deep? What will be the volume of the remaining space in this case?

21 Write down rough estimates for each of the following calculations; then work each one, rounding off the answer to 3 significant figures:

a $219 \times 0{\cdot}135$ *b* $0{\cdot}076 \times 0{\cdot}238$ *c* $47{\cdot}8 \div 0{\cdot}56$

d $0{\cdot}056 \div 0{\cdot}43$ *e* $\dfrac{50{\cdot}2 \times 0{\cdot}69}{3{\cdot}14 \times 10{\cdot}7}$

22 How many pieces of wire each 38·5 m long can be cut from a coil of length 1 km? What length will be left over?

23 A patchwork quilt has a plain border and a centre composed of squares of material each of side 7·5 cm sewn together. The border is 17·5 cm wide all round, and the finished quilt measures 2 m by 1·4 m. How many square patches are required?

24 A man whose pay is £28 per week gets a rise of $11\frac{1}{2}\%$. What is his new pay?

25 After one year a certain car is worth only 65% of what it cost when new. What is the value of a one-year-old car whose new price was £860?

26 Find to base two the perimeter and area of a rectangle whose length and breadth are 10111 cm and 101 cm. Check by converting all numbers to base ten.

27 Work out the total cost of:
$10\frac{1}{2}$ m brocade at £1·12 per m
$3\frac{3}{4}$ m nylon net at 38p per m
$2\frac{1}{2}$ m curtain tape at 9p per m

28 Work out my hotel bill at Linz on the Rhine:

9 days room	at	11·50 marks per day
18 breakfasts	at	2·25 marks each
6 lunches	at	6·25 marks each
18 dinners	at	8·80 marks each

10% is added to the bill for service.

If £1 is nearly equal to 9 marks find the cost of the hotel bill to the nearest £1.

29 Express in the usual whole-number form:

a $1{\cdot}3 \times 10^5$ *b* $8{\cdot}4 \times 10^8$ *c* 7×10^{10}

30 Express in standard form ($a \times 10^n$):

a 2240   *b* 87000000000   *c* 2 thousand million
d the number of mg in 1000 kg

Answers

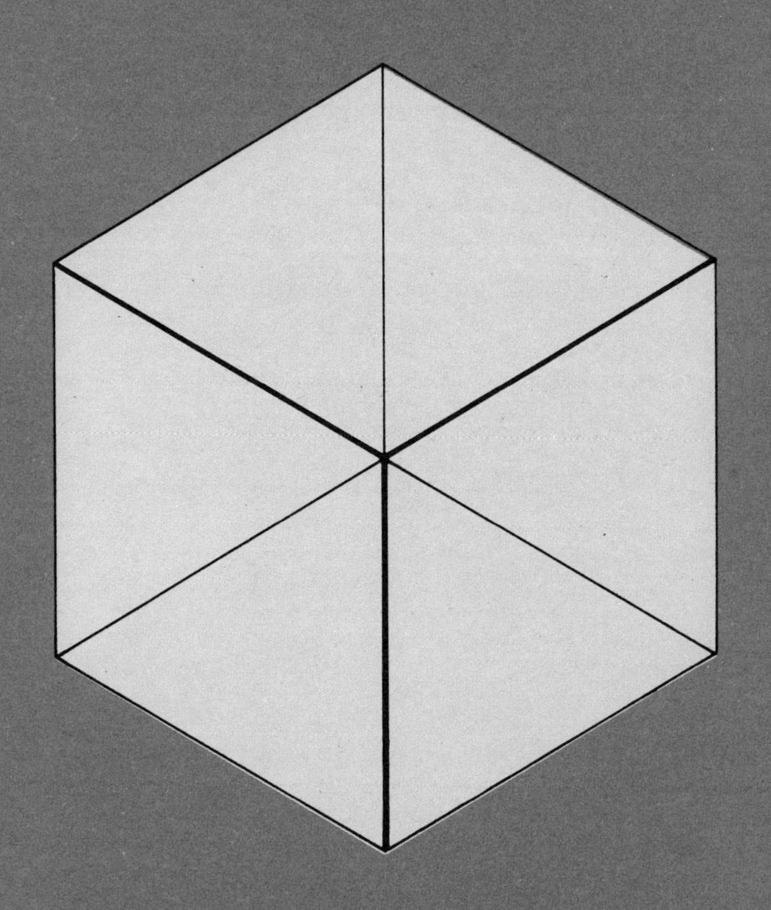

Answers

Answers

Algebra—Answers to Chapter 1

Page 3 Exercise 1A

1 a 5 *b* 1 *c* 15 *d* 5 *e* 11 *f* 4
g 9 *h* 27

2 a 10 *b* 20 *c* 25 *d* 250 *e* 125 *f* 250

3 a 17 *b* 11 *c* 8 *d* 132 *e* 6 *f* 0
g 31 *h* 18

4 a 9 *b* 31 *c* 0 *d* 15 *e* 90 *f* 23

5 a 9 *b* 13 *c* 1 *d* 5 *e* 25 *f* 45
g 16 *h* 30

6 a T *b* F *c* T *d* T *e* T *f* F

7 {3, 6, 9, 12, 15}. Set of multiples of 3 from 3 to 15.

8 {5, 11, 17, 23, 29}; prime numbers. No. $x = 6$ gives $n = 35$ which is not a prime number.

9 {7, 13, 19, 25, 31}; prime numbers except 25. No. $x = 4, 8, 9$ do not give prime numbers.

10 {10, 12, 18, 28, 42, 60}

Page 4 Exercise 1B

1 T *2* T *3* F *4* F *5* T

6 T *7* T *8* F *9* T *10* F

11a {2, 4, 6, 8, 10, 12} *b* {5, 7, 9, 11, 13, 15}
c {1, 4, 9, 16, 25, 36} *d* {2, 5, 10, 17, 26, 37}

12a {4, 7, 10, 13} *b* {0, 1, 4, 9} *c* {0, 5, 20, 45} *d* {0, 1, 8, 27}
e {0, 1, 2, 3, 4, 5, 6} (*1*) 6 (*2*) 0

13 4, 6, 8, 10 *14a* $v = 1$ *b* $v = 35{\cdot}3$ *15* 6400 and 2400

16a 400 *b* 48 km/h *17a* 3; 11 *b* 12 *18* 31 metres

Page 6 Exercise 2

1 a 48 cm *b* 60 cm *2a* 100 cm^2 *b* 484 cm^2

3 a 120 cm^2 *b* 3·75 cm^2 *4a* 240 m^3 *b* 1000 m^3

5 80°, 35°, 11° *6* 80°, 125°, 57°

7 a 260 *b* 50 *8a* 14 *b* 30 *c* 7·1

9 a £8·30 *b* £66·40 *10a* 8 pm *b* 8 am

Page 7 Para. 3

a (*1*) 10 cm (*2*) 6 cm (*3*) 16 cm
b (*1*) $2x$ cm (*2*) $2y$ cm (*3*) $2x+2y$ cm
c $2(x+y)$ cm

Page 7 Exercise 3

1 10, 26, 40, 18, $8+2b$, $2l+2b$ or $2(l+b)$

2 a $P = 4x$ *b* $P = x+2y$ *c* $P = 4p+2q$ *d* $P = 2a+2b$
e $P = 2a+2b+2c$ *f* $P = 8x+4t$

3 a $P = 3m$ *b* (*1*) 24 (*2*) 16·5 (*3*) 31·2

4 a $d = x+y$ *b* $d = 1300$ *5a* $d = 3x+3y$ or $3(x+y)$ *b* 1515

6 $W = x+y$ *a* $W = 5{\cdot}25$ *b* $y = 2\frac{1}{2}$

7 $L = p-q$; $L = 3{\cdot}5$

8 $P = 2x+2y+z$ *a* $P = 112$ *b* $z = 20$ *c* $x = 7$

9 a $S = 4a+40$ *b* 100 *10a* $P = 10k$ *b* $A = 4k^2$ *c* 75 cm; 225 cm²

Page 9 Exercise 3B

1 a (*2*) x^2, y^2, x^2-y^2, $A = x^2-y^2$
(*3*) x^2, yz, x^2-yz, $A = x^2-yz$
b 60 cm², 84 cm², 72 cm²

2 a (*1*) ab, cd, $ab-cd$, $A = ab-cd$
(*2*) ab, $2ef$, $ab-2ef$, $A = ab-2ef$
(*3*) ab, $4g^2$, $ab-4g^2$, $A = ab-4g^2$
b 120 mm², 72 mm², 116 mm²

3 a (*1*) $2f$ tonnes (*2*) $3f$ tonnes (*3*) hf tonnes
b $W = t-hf$ *c* $W = 10\frac{3}{4}$

4 a (*1*) 30°C (*2*) 10°C (*3*) 0°C *b* $C = \frac{5}{9}(F-32)$; $C = 100$; $C = 540$

5 a 19, 23, 27 *b* 399 *c* 399, 3999

6 a 12 *b* 20 *c* (*1*) $S = xy$ (*2*) $P = (x+1)(y+1)$ (*3*) 870; 930

7 5 *a* 14 *c* 338 350

8 a 2, 5 *b* 9 *d* 170 *e* 12

Algebra—Answers to Chapter 2

Page 14 Exercise 1A

1 a $7 > 4$ *b* $2 < 3$ *c* $5 > 1$ *d* $10 < 100$

2 a 2 is less than 5 *b* 10 is greater than 6
c 0 is less than 4 *d* 8 is not equal to 3

3 ***a, c, d***

4 a $3 < 4$ *b* $10 = 10$ *c* $8 > 7$ *d* $1 > 0$
e $0 < 10$ *f* $4+1 > 5-3$ *g* $99 = 99$ *h* $99 < 101$

5 ***c*** *6a* 3, 4 *b* 1, 2, 3, 4, 5, 6, 7

7 a $5 < 6 < 7$ *b* $2 < 4 < 6$
c $6 < 7 < 8$ *d* $1 < 5 < 9$

Page 15 Exercise 1B

1 a $199 < 200$ *b* $2\times7 = 13+1$ *c* $2^3 < 3^2$
d $0 < 1$ *e* $91 > 19$ *f* $35 > 34$
g $0{\cdot}1 > (0{\cdot}1)^2$ *h* $\frac{1}{4} < \frac{1}{2}$ *i* $3+0 > 3\times0$

2 *a* F *b* T *c* T *d* T *e* T *f* F

3 *a* $1 < 2 < 5$ *b* $0 < 5 < 10$ *c* $2 < 3 < 5$ *d* $5 < 10 < 20$

4 *a* $3 < 6 < 8$ *b* $1 < 10 < 12$ *c* $1 < x < 5$ *d* $a < x < b$

5 *a* $5 > 2 > 1$ *b* $10 > 5 > 0$ *c* $5 > 3 > 2$ *d* $20 > 10 > 5$

6 *a* $1 < 3 < 5$, or $5 > 3 > 1$ *b* $0 < 1 < 2$, or $2 > 1 > 0$
c 1 mm < 1 cm < 1 m, or 1 m > 1 cm > 1 mm
d 2 mg < 2 g < 2 kg, or 2 kg > 2 g > 2 mg
e $0{\cdot}02 < 0{\cdot}2 < 2{\cdot}0$, or $2{\cdot}0 > 0{\cdot}2 > 0{\cdot}02$
f 5p < 10p < 50p, or 50p > 10p > 5p

Page 17 Exercise 2

1

2

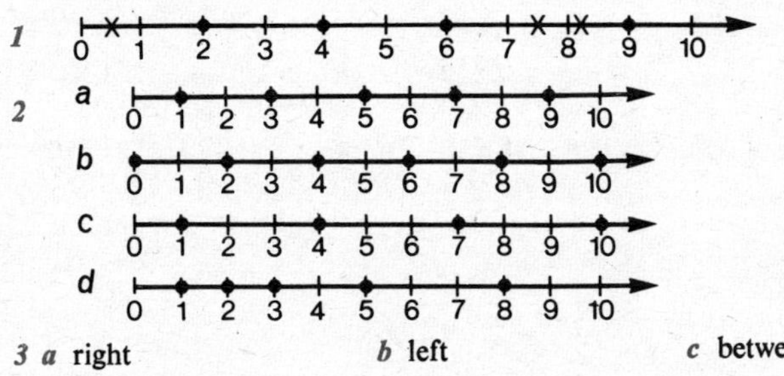

3 *a* right *b* left *c* between

4

5

6 *a* $\{p \in E: 0 \leqslant p \leqslant 3\}$ *b* $\{p \in E: 3 \leqslant p \leqslant 6\}$

Page 19 Exercise 3A

1 {3, 4, 5, 6} *2* {0, 1, 2} *3* {0, 1, 2, 3} *4* {0}

5 ø *6* {0, 1, 2, ..., 6} *7* {2, 3, 4} *8* {3, 4, 5}

9 {2, 3, 4, 5, 6} *10* {3} *11* {2, 3, 4} *12* {1, 2, 3, 4, 5, 6}

13 {0, 1, 2, 3}

14 {0, 1}

15 {0, 1, 2, 3, 4}

16 {4, 5, 6, ..., 10}

17 {1, 2, 3, ..., 10}

18 {9, 10}

19 {0, 1, 2, 3, 4, 5, 6, 7, 8}

20 {0, 1, 2, 3}

21 {8, 9, 10}

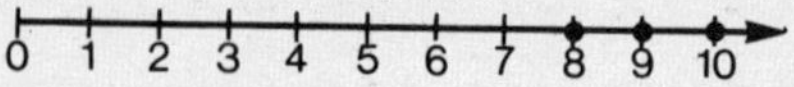

22 {5, 6, 7, 8, 9, 10} *23* {1, 2, 3, 4} *24* {1, 2, 3}

25 {5, 6, 7, 8, 9, 10} *26* {1, 2, 3} *27* {1, 2, 3, 4}

28a $p > 5$ *b* $n < 10$ *c* $x+10 > 20$
d $c-1 < 13$ *e* $2y < 30$ *f* $p+q \geqslant r$

Page 20 Exercise 3B

1 a {3, 4, 5} *b* {0, 1, 2} *c* {0, 1, 2, 3, 4}
d {0} *e* {0, 1, 2, 3} *f* {2, 3, 4, 5}

2 a {1, 2, 3, 4, 5} *b* {1, 2} *c* {1, 2, 3, ..., 10}
d ø *e* {3, 4, 5, ..., 10} *f* {1, 2, 3}

3 a {0, 1} *b* {0} *c* {0, 1, 2} *d* {0, 1, 2}
e {0, 1, 2} *f* {3}

4 a {4, 5, 6, ..., 10}, {0, 1, 2, ..., 7}; {4, 5, 6, 7}
b {0, 1, 2, 3, 4, 5, 6}, {5, 6, 7, 8, 9, 10}; {5, 6}
c {6, 7, 8, 9, 10}, {0, 1, 2, 3}; ø
d {0, 1, 2, 3}, {3, 4, 5, 6, 7, 8, 9, 10}; {3}

5 a $p > 50$ *b* $g \leqslant 10$ *c* $3t < 25$
d $x+2 < y$ *e* $5 < y < 15$ or $15 > y > 5$ *f* $0 < h < 10$ or $10 > h > 0$

Page 21 Exercise 4

1 {5, 6, 7, ...} *2* W *3* {19, 20, 21, ...}

4 {0, 1, 2, ..., 17} *5* {5, 6, 7, ...} *6* {0, 1, 2, ..., 7}

7 ø *8* {3, 4, 5, ...} *9* {12, 13, 14, ...}

10 {0, 1, 2, 3} *11* {0, 1, 2, ..., 7} *12* {12, 13, 14, ...}

13 {0} *14* ø *15* W

16 ø *17* {0, 1, 2} *18* {2, 3, 4, ...}

19 {0, 1} *20* {8, 9, 10, ...} *21* {0, 1, 2, ..., 6}

22 {0, 1, 2, ..., 15} *23* {0, 1, 2, ..., 5} *24* {4, 5, 6, ...}

Page 22 Exercise 4B

1 a 5 *b* 19+15 *c* 7+4 *d* $9 = 5+4$ *e* $1 = 0+1$

2 a $>$ *b* $=$ *c* $<$ *d* $<$ *e* $<$ *f* $<$

3 a $<$ *b* $>$ *c* $<$ *d* $>$ *e* $>$ *f* $>$

4 $c = 5$ *5 a* {0, 1, 2, 3} *b* {0, 3, 4, 5} *c* {0, 1, 3, 4, 5}

6 a 6, 3 *b* 6, 3 *c* 4, 1 *d* 4, 1
e 18, 6 *f* 18, 6 *g* 8, 4 *h* 8, 4

7 8, 6, 4, 3, 2, 1; 10, 25

8

p	1	2	3	4	6	9	12	18	36
q	36	18	12	9	6	4	3	2	1

; 12, 37

9 *a* 16 *b* 8 *c* 8 *d* 5, 4, 3, 2, 1 *e* 7 *f* 16

10*a* 12 *b* 6 *c* 6 *d*

p	1	2	3	4	5
q	5	4	3	2	1

e 5 *f* 9

11*a* $t > 4$ *b* $t < 6$ 12*a* 32, 12, no box *b* $12 \leqslant n \leqslant 36$

Page 24 Exercise 5

1 $S \geqslant 5$ 2 $H \geqslant 1{\cdot}72$ 3 $W \geqslant 326$ 4 $L \leqslant 130\,000$

5 $S \leqslant 275$ 6 $N \leqslant 74$ 7 $180 \leqslant m \leqslant 1250$ 8 $2{\cdot}2 < d < 2{\cdot}4$

9 *a* $88 \leqslant v \leqslant 101$, or $87 < v < 102$ *b* $v > 121$

10 $7{\cdot}998 < d < 8{\cdot}002$

11*a* {2, 3, 4, ..., 12} *b* $x \geqslant 7$; {7, 8, 9, 10, 11, 12}

12 $2x > 10$, so $x > 5$

Algebra—Answers to Chapter 3

Page 28 Exercise 1A

1 *a* 0, −3, −6 *b* −1, −3, −5 *c* −6, −4, −2 *d* −1, 3, 7

2 −1°C *a* (*1*) −5°C (*2*) +5°C, or 5°C
b (*1*) 10° below zero (*2*) freezing-point
(*3*) 20° above zero (*4*) 8° below zero
c (*1*) 3° (*2*) −10°

3 7, 6, 5, 4, 3, 2, 1, 0, −1, −2, −3, −4

4 *b* (*1*) 10 to 15 m above high water (*2*) 5 to 10 m below high water

5 *a* 10 km south *b* 15 newtons to the left *c* 25 m/s downwards
d (*1*) +1 minute (*2*) −2 minutes

6 −2°C; 0°C 7*c* (−4, −4), (−5, −5), (−6, −6)

8 A′(−3, 3), B′(−2, 2), C′(−1, 1) 9 P′(−1, 1), Q′(−2, 2), R′(−3, 3)

10*a* (−4, 2), (1, 1), (3, −5), (−6, 0) *b* (4, −2), (−1, −1), (−3, 5), (6, 0)

Page 30 Exercise 1B

1 *a* 0, −4, −8 *b* −5, −8, −11 *c* −3, −1, 1

2 *a* 4°C *b* 0°C *c* −7°C

3 *c* (−1, 1), (−3, 1), (−2, 4) 4 (1, −1), (3, −1), (2, −4)

5 *b* a rectangle *c* (−3, −1), (1, −1), (1, −2), (−3, −2)
d (3, 1), (−1, 1), (−1, 2), (3, 2)

6 *a* £2 *b* −£2 7 lose 1 minute

Page 32 Exercise 2A

1 a 4 *b* 1 *c* 0 *d* -4

2 a 15°C *b* 3°C *c* -2°C *d* 1°C

3 a 200 m *b* 50 m *c* -30 m *d* 0 m

4 a $5 > 3$ *b* $0 > -1$ *c* $3 > 1$ *d* $1 > 0$ *e* $\frac{1}{2} > -2$
f $2 > -2$ *g* $-3 > -4$ *h* $-20 > -30$ *i* $0 > -2$ *j* $-1 > -5$
k $\frac{3}{4} > \frac{1}{4}$ *l* $10 > -1$

5 a $5 < 12$ *b* $-9 < 9$ *c* $3 < 6$ *d* $-\frac{1}{2} < \frac{1}{4}$
e $\frac{1}{3} < \frac{1}{2}$ *f* $-5 < -3$ *g* $-6 < 2$ *h* $-1 < 0$
i $-0{\cdot}01 < 0$ *j* $-100 < -99$ *k* $9{\cdot}01 < 9{\cdot}11$ *l* $0 < 1$

6 ***a, c, d, g, h***

7 a $5 > -7$ *b* $10 > -10$ *c* $-2 < -1$ *d* $-7 > -100$

8 a $\{-4, -3, -2, -1\}$ *b* $\{-4, -3\}$ *c* $\{3, 4\}$
d $\{0, 1, 2, 3, 4\}$ *e* $\{4\}$ *f* $\{-3, -4\}$ *g* $\{0, 1, 2, 3, 4\}$ *h* ø

Page 33 Exercise 2B

1 a $-3 < 3$ *b* $0 > -2$ *c* $7 < 17$ *d* $-1 > -2$
e $1 > -10$ *f* $-\frac{1}{2} < -\frac{1}{4}$ *g* $-1{\cdot}0 < 0{\cdot}1$ *h* $0{\cdot}07 > -100$

2 a $3 < 5 < 7; 7 > 5 > 3$ *b* $-6 < 2 < 4; 4 > 2 > -6$
c $-2 < -1 < 2; 2 > -1 > -2$ *d* $-100 < 0 < 10; 10 > 0 > -100$

3 a $\{2, 3, 4, 5\}$ *b* $\{-5, -4, -3\}$ *c* $\{0, 1, 2, ..., 5\}$
d $\{-5, -4, -3, -2, -1\}$ *e* $\{-5, -4\}$ *f* $\{-3, -2, -1, 0, 1, 2\}$

4 a £80 *b* £2 *c* £(-5)

5 b 'Every replacement which makes the open sentence $x > 0$ a true sentence is a positive number.'

6 a $-2, -1$ *b* -3 *c* $-2, -1, 0, 1, 2$
d 1, 2, 3 *e* $-5, -4$ *f* 0

Page 34 Exercise 3

1 5 *2* 6 *3* 9 *4* 9 *5* 9 *6* 12
7 5 *8* 6 *9* 13 *10* 11 *11* 13 *12* 13

Page 35 Exercise 4

1 8 *2* -4 *3* 1 *4* 0 *5* -5 *6* -7
7 -9 *8* 0 *9* 5 *10* 0 *11* 0 *12* -9
13 2 *14* -4 *15* 4 *16* 0 *17* -6 *18* 3
19 2 *20* -2 *21* -7 *22* 0 *23* -9 *24* -8
25 -11 *26* -10 *27* -15 *28* -15 *29* 8 *30* 22
31 -7 *32* 0 *33* -2 *34* 6 *35* -20 *36* 0
37 3 *38* -15 *39* -2 *40* -1 *41* 3 *42* -100
43 5 *44* 10

45a -2 *b* 10 *c* 14 *d* 8 *e* -2
f between 9 am and 12 noon, and between 6 pm and 9 pm

Page 36 Exercise 4B

1 3 *2* −3 *3* −9 *4* −2 *5* 1 *6* −4
7 $-\frac{1}{3}$ *8* −1 *9* −4·7 *10* −0·4 *11* −2·5 *12* −1

13 *Rows:* −6, −5, −4, −3, −2, −1, 0; −5, −4, −3, −2, −1, 0, 1;
−4, −3, −2, −1, 0, 1, 2; −3, −2, −1, 0, 1, 2, 3;
−2, −1, 0, 1, 2, 3, 4; −1, 0, 1, 2, 3, 4, 5;
0, 1, 2, 3, 4, 5, 6

a main diagonal *c* numbers increase *d* numbers increase
e −3, 3; −2, 2; −1, 1; 0, 0

14 $a+b = b+a$. Yes.

15 −21, −14, −7; 0, 7, 14, 21; same

16 −21, −14, −7; 0, 7, 14, 21; same *17* $(a+b)+c = a+(b+c)$

19a 19 *b* −1 *c* 3 *d* 0 *e* −16 *f* 4

20a 3 *b* −2 *c* −4 *d* 1 *e* −2 *f* 3

21a *Rows:* −1, 10, 6; 12, 5, −2; 4, 0, 11
b *Rows:* −3, 1, 2; 5, 0, −5; −2, −1, 3

22a 133 *b* 50 *23* −30 *24* $y = x + 2$. −3, −2, −1, 0, 1, 2, 3, 4

Page 39 Exercise 5

1 a −3 *b* −5 *c* 6 *d* 10 *e* −10 *f* −12
g −100 *h* 100 *i* −1 *j* 9 *k* −4 *l* 4

2
-4 -3 -2 -1 0 1 2 3 4

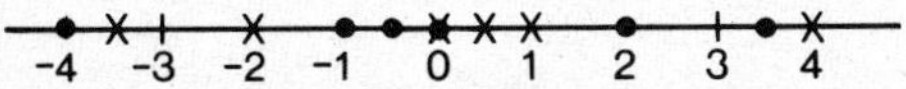

3 a $7+(-7) = 0$ *b* $2+(-2) = 0$ *c* $-9+9 = 0$
d $-5+5 = 0$ *e* $-1{\cdot}5+1{\cdot}5 = 0$ *f* $8+(-8) = 0$

4 a 6 *b* −8 *c* 5 *d* 0 *e* −4 *f* 3

Page 40 Exercise 6

1 a 3 *b* 3 *2a* −2 *b* −2 *3a* −5 *b* −5
4 a −3 *b* −3 *5a* 2 *b* 2 *6a* 5 *b* 5
7 a 4 *b* 14 *c* −14 *d* −4

8 a $14+(-8) = 6$ *b* $5+(+3) = 8$ *c* $2+(-6) = -4$
d $-3+(-3) = -6$ *e* $6+(+4) = 10$ *f* $-3+(+3) = 0$
g $0+(-4) = -4$ *h* $8+(-24) = -16$

9 a 2 *b* −5 *c* −16 *d* 5 *e* 30 *f* −24
g −10 *h* 4 *i* 4 *j* −4 *k* 2 *l* 8

10a 9 *b* −8 *c* 3 *d* 11 *e* −13 *f* 5

11a $4c$ *b* $3x$ *c* $5y$ *d* $-10h$ *e* $-2n$ *f* $-6p$
g $2n$ *h* 0 *i* $-2x$ *j* $2m^2$ *k* $-8y^2$ *l* $3t^2$

12a $7p$ *b* $6a$ *c* $-9a$ *d* 0 *e* $-3t$ *f* $-9g$
g $-5m$ *h* $-6x$ *i* $-y$ *j* a^2 *k* $-9w$ *l* $6v$

13a 3 *b* −1 *c* 5 *d* 4 *e* 9 *f* −5 *g* −4 *h* 7

Page 42 Exercise 7

1

	6	−8	0·1	−5	$\frac{3}{4}$	100	0	−3·14
N	✓	.	.	.	.	✓	.	.
W	✓	.	.	.	.	✓	✓	.
Z	✓	✓	.	✓	.	✓	✓	.
Q	✓	✓	✓	✓	✓	✓	✓	✓

2 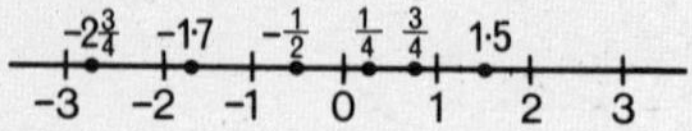

3 a $\frac{3}{4} < 1\cdot5$ *b* $-\frac{1}{2} < \frac{1}{4}$ *c* $-\frac{1}{2} > -2\frac{3}{4}$
d $-1\cdot7 < 2$ *e* $0 > -\frac{1}{2}$ *f* $-2\frac{3}{4} < -1$

4 a $-\frac{1}{2}$ *b* $\frac{3}{4}$ *c* 1·5 *d* $\frac{11}{4}$ *e* −1·7 *f* 9·81

5 a {2} *b* ø *c* {10} *d* ø *e* $\{2\frac{1}{2}\}$
f {9} *g* {−4} *h* ø *i* {0}

6 a {0, 1, 2, 3, 4} *b* {8, 9, 10, ...} *c* {−2, −3, −4, ...}
d $\{x \in Q : x > 1\frac{1}{2}\}$ *e* ø *f* $\{x \in Q : x > 2\frac{1}{2}\}$

7 a 1 *b* $\frac{2}{5}$ *c* $-\frac{1}{2}$ *d* $-\frac{1}{2}$ *e* $-\frac{1}{4}$ *f* $\frac{3}{4}$
g $-\frac{1}{4}$ *h* $-1\frac{5}{8}$ *i* $\frac{1}{6}$ *j* $1\frac{5}{12}$ *k* 1 *l* $\frac{2}{5}$

Page 45 Exercise 8

1 a $10^0, 10^{-1}, 10^{-2}, 10^{-3}, 10^{-4}, 10^{-5}, 10^{-6}$
b $1, \frac{1}{10^1}, \frac{1}{10^2}, \frac{1}{10^3}, \frac{1}{10^4}, \frac{1}{10^5}, \frac{1}{10^6}$

2 a 10^{-2} *b* 10^{-4} *c* 10^{-5} *d* 10^{-1} *e* 10^{-3}

3 3×10^{-1} *4* $4\cdot7\times10^{-1}$ *5* 5×10^{-2} *6* $2\cdot78\times10^{-2}$

7 6×10^{-4} *8* $2\cdot73\times10^{-3}$ *9* $6\cdot49\times10^{-4}$ *10* $4\cdot56\times10^{-2}$

11 $1\cdot73\times10^{-7}$ *12* $4\cdot5\times10^{-3}$ *13* $7\cdot004\times10^{-1}$ *14* $1\cdot05\times10^{-2}$

15 $7\cdot5\times10^{-2}$ *16* $4\cdot6\times10^{-5}$ *17* $1\cdot2\times10^{-1}$ *18* $9\cdot9\times10^{-3}$

19 $1\cdot1\times10^{-1}$ *20* $3\cdot2\times10^{-6}$ *21* $4\cdot0\times10^{-1}$ *22* $7\cdot8\times10^{-4}$

23 $8\cdot8\times10^{-7}$ *24* 5×10^{-8} mm *25* 3×10^{-23} g

Page 46 Exercise 9A

1 a A square *b* A(1, 3), B(−2, 3), C(−2, 0), D(1, 0)

2 A(3, 1), B(0, 1), C(0, −2), D(3, −2)

3 $k = 5$. A(5, 0), B(3, −4), C(7, −4) *4* A(0, 5), B(−2, 1), C(2, 1)

5 b $p = -2$ *6a* (6, −1) *b* (6, −5) *c* (0, −2) *d* (7, −9)

7 $\begin{pmatrix}0\\7\end{pmatrix}$, or 'sail 7 km north'. *8* $\begin{pmatrix}-4\\2\end{pmatrix}$ *9* $\begin{pmatrix}-4\\2\end{pmatrix}$

Page 47 Exercise 9B

1 a (−3, −1), (−6, −2), (−9, −3) *b* (15, 5), (−15, −5), (−30, −10)
c $p = 3q, q = \frac{1}{3}p$

2 *a* (1, 13) *b* S(1, −15), T(14, −15), U(14, 0), V(1, 0)

3 *a* T *b* F *c* T *d* T *e* F *f* T

4 $\begin{pmatrix} 8 \\ -1 \end{pmatrix}$ 5*a* $\begin{pmatrix} 1 \\ -1 \end{pmatrix}$ *b* $\begin{pmatrix} -3 \\ 1 \end{pmatrix}$ 7 $\begin{pmatrix} -a \\ -b \end{pmatrix}$ 8 *d*; *b*

Algebra—Answers to Chapter 4

Page 50 Exercise 1

1 *Rows:* 2, 5, 3, 8, 16, 10, 6, 16; 4, 3, 6, 9, 36, 12, 24, 36;
0, 3, 8, 11, 0, 0, 0, 0; 8, 1, 5, 6, 48, 8, 40, 48;
10, 5, 7, 12, 120, 50, 70, 120; 4, $\frac{1}{2}$, $\frac{1}{4}$, $\frac{3}{4}$, 3, 2, 1, 3

2 *Rows:* 3, 5, 2, 3, 9, 15, 6, 9; 1, 8, 3, 5, 5, 8, 3, 5;
0, 7, 5, 2, 0, 0, 0, 0; 4, 9, 2, 7, 28, 36, 8, 28;
5, 10, 6, 4, 20, 50, 30, 20; 3, $\frac{1}{2}$, $\frac{1}{4}$, $\frac{1}{4}$, $\frac{3}{4}$, $\frac{3}{2}$, $\frac{3}{4}$, $\frac{3}{4}$

Page 51 Exercise 2

1	$2a+2$	2	$7c+14$	3	$5+5p$	4	$8m+8n$
5	$a+b$	6	$2n+16$	7	$3x^2+3$	8	$4p-8q$
9	$10-10x$	10	$21u+28v$	11	$12x-6y$	12	$9c+6d$
13	$3y^3-3$	14	$4p+pq$	15	$5ah+2ak$	16	$10m-50n$
17	a^2+ab	18	$2hl+2hb$	19	$4a^2-4ab$	20	$2x-3x^2$
21	$2x^2+2xy$	22	$2p^2-3pq$	23	$abc+2ab$	24	$rs-r^2s$
25	$3a+3b+3c$	26	$2a-2b+2c$	27	$a^2+ab+ac$	28	$x^2+2xy-xz$
29	$6p-9q+12r$	30	$xy-x^2y+xy^2$				

Page 51 Exercise 2B

1	$15p+10q$	2	$4px-2py$	3	$9a^2+15a$	4	r^3-r^2s
5	a^2b+ab^2	6	x^2-7x^3	7	$2xy+2yz+2zx$		
8	$2a^2+3ab+a$	9	$2p^2+2pq-4pr$			10	$\frac{1}{2}m+\frac{1}{2}n$
11	$2x+4y$	12	$a-3b$	13	$9x+3y$	14	$2h-4$
15	$6u+8{\cdot}4v$						

16*a* (*1*) *Rows:* 0, 1, 2, 3, 4; 1, 2, 3, 4, 0; 2, 3, 4, 0, 1; 3, 4, 0, 1, 2; 4, 0, 1, 2 3
(*2*) *Rows:* 0, 0, 0, 0, 0; 0, 1, 2, 3, 4; 0, 2, 4, 1, 3; 0, 3, 1, 4, 2; 0, 4, 3, 2, 1

b *Rows:* 4, 2, 1, 3, 2, 3, 4, 2; 1, 4, 3, 2, 2, 4, 3, 2;
3, 3, 3, 1, 3, 4, 4, 3; 4, 2, 3, 0, 0, 3, 2, 0

c ‘Multiplication’ is distributive over ‘addition’.

Page 53 Exercise 3A

1 *a* $p+q$ *b* $p+5$ *c* $c+3$ *d* $x-y$ *e* $2m-3$
f $2a+3$ *g* $n-2m$ *h* $2+b$ *i* $x-1$

2 *a* $2(x+2)$ *b* $3(a+b)$ *c* $4(c+3d)$ *d* $3(y+2z)$
e $4(p+3)$ *f* $a(x+y)$ *g* $4(x-8)$ *h* $7(m-7)$
i $8(2t-3)$ *j* $p(q-s)$ *k* $q(p-r)$ *l* $5x(3+4y)$
m $6(a+2b)$ *n* $x(x+1)$ *o* $y(y-1)$ *p* $y(x+z)$

3 *a* $7(11+9) = 7\times 20 = 140$ *b* $13(14+16) = 13\times 30 = 390$
c $14(38+12) = 14\times 50 = 700$ *d* $15(23-19) = 15\times 4 = 60$
e $16(16-11) = 16\times 5 = 80$ *f* $\frac{3}{4}(27-23) = \frac{3}{4}\times 4 = 3$

4 *a* $s = 3(u+v)$; $s = 3(10+160) = 3\times 170 = 510$
b $Q = 15(x+y)$; $Q = 15(17{\cdot}5+2{\cdot}5) = 15\times 20 = 300$
c $A = 2h(a+b)$; $A = 2\times 4\times (15+10) = 8\times 25 = 200$

5 $A = ab+ac = a(b+c)$; 26 m^2

6 $A = xy-yz = y(x-z)$; 9600 mm^2

Page 54 Exercise 3B

1 *a* $5(x+3y)$ *b* $2(4y+9)$ *c* $3(x^2+2)$ *d* $p(p+r)$
e $a(a-2b)$ *f* $p(p-q)$ *g* $x(xy+1)$ *h* $5x(3x+1)$
i $3p(3p-5q)$ *j* $a(x+y+z)$ *k* $2(2p+3q-4r)$
l $a(a^2+a+1)$ *m* $x(1-x^2)$ *n* $x^2(y+z)$ *o* $2ab(1+2ab)$

2 *a* 3; 300 *b* 11; 33 *c* 29; 290 *d* 13; 390 *e* 3·4; 17 *f* 360; 90

3 *a* $s = 5t(4-t)$; $s = 19{\cdot}2$ *b* $I = m(v-u)$; $I = 160$
c $A = \frac{1}{2}h(a+b)$; $A = 75$

4 *a* $(\frac{1}{2}\times 8a)+(\frac{1}{2}\times 4b) = 4a+2b$
b $\frac{4(2a+b)}{2} = 2(2a+b) = 4a+2b$. Results are the same.

5 *a* $\frac{2x+2y}{2} = \frac{2(x+y)}{2} = x+y$; $\frac{1}{2}(2x+2y) = (\frac{1}{2}\times 2x)+(\frac{1}{2}\times 2y) = x+y$
b $2p+2q$ *c* $2a+3b$ *d* $m-2n$ *e* $3x-5$ *f* $2c-3d$ *g* $a+2b$
h $2x-2y$ *i* $c+d$ *j* $x+y$ *k* $2t+3$ *l* $p-4$

6 $P = 2a+2b+2c = 2(a+b+c)$ 7*a* 16 *b* 13 *c* 10 *d* 9

8 *a* $a+2b$ *b* $3a-6b$ *c* $a-4$ *d* x^2+x+1 *e* y^2-y *f* $2z^2+3z+4$

Page 56 Exercise 4

1 *a* $(2+3)m = 5m$ *b* $9a$ *c* $3y$ *d* $5p$ *e* $7x$ *f* $8n$

2	$7a$	3	$4c$	4	$17m$	5	$6y$	6	$10x$
7	$13h$	8	$23p$	9	$2w$	10	$6n$	11	$4a$
12	$7c$	13	0	14	$7d$	15	b	16	$\frac{1}{2}p$
17	$6x$	18	$11q$	19	m	20	$6ab$	21	$10x^2$
22	xy	23	$14pq$	24	$6cd$	25	$6a$	26	$13a$
27	$6y$	28	none	29	$9x$	30	none	31	none
32	$7pq$	33	$4xy$	34	$9a$	35	$9c$	36	$5m$
37	$15r+s$	38	$9n-3$	39	$11y$	40	none	41	$10x+2y$

Answers

Page 57 Exercise 4B

1 $5x+3$ *2* $6x+6$ *3* $8x-3$ *4* $15m+10$ *5* $24m+4n$
6 $2a+b$ *7* $6x$ *8* $3y+6$ *9* $5a+k$ *10* $3x$
11 $2x^2+x$ *12* $6q$ *13* $x+2y$ *14* a^2 *15* $5x^2+8x$
16 $5a+7$ *17* $17a$ *18* $5a+1$ *19* $15x$ *20* $24c+9d$
21 $17p$ *22* $4a^3+2a-a^2$ *23* $5a+8b+11c$
24 $7x+7y+14z$ *25* $5a$ *26* $12a+7ab+5bc+9b+5c$
27 $5a^2+6b$ *28* a^3+b^3

Page 58 Exercise 5

1 $x = 3$ *2* $y = 7$ *3* $z = 6$ *4* $p = 1$ *5* $t = 9$
6 $x = 3$ *7* $x = 15$ *8* $y = 12$ *9* $m = 2$ *10* $p = 2$
11 $q = 3$ *12* $r = 6$ *13* $x = 1$ *14* $x = 5$ *15* $x = 3$
16 $y = 1$ *17* $y = 5$ *18* $y = 1$ *19* $x = 1$ *20* $x = 2$
21 $y = 2$ *22* $y = 9$ *23* $z = 2$ *24* $z = 1$ *25* {1, 2}
26 {1, 2, 3, 4} *27* {4, 5, 6, ...} *28* {1, 2, 3} *29* {2, 3, 4, ...}
30 {1, 2, 3, 4, 5} *31* {1, 2, 3, 4} *32* {1, 2} *33* {4, 5, 6, ...}
34 {2, 3, 4, ...} *35* { } or ø *36* {1, 2, 3, 4} *37* {1, 2, 3, ...}

Page 58 Exercise 5B

1 a $x = 2$ *b* $x = 1, 2, 3, 4, 5$ *c* $x = 0, 1, 2$
d $x = 3, 4, 5$ *e* $x = 0$ *f* $x = 4$
2 {5} *3* {3} *4* {4, 5, 6, ...}
5 {3, 4, 5, ...} *6* {0, 1, 2, 3, 4} *7* {0, 1, 2, 3}
8 {7} *9* {1, 2, 3, ...} *10* {0, 1, 2, 3, 4}
11 {0, 1, 2} *12* {2, 3, 4, ...} *13* {2}
14 {0, 1, 2} *15* {2, 3, 4, ...} *16* {1}
17 {0, 1} *18* {0, 1, 3} *19* {0, 1, 3}
20 ø *21* {3, 5}

Page 59 Exercise 6A

1 $p = 2, q = 6, r = 5, s = 16$ *2* $2(x+2)-4 = 2x$. Even. *3* 25
4 $3(n+5) > 33$. {7, 8, 9, ...}
5 a • + •• = ••• b ••• + •••• = •••••••
      •   ••   •••   •••   ••••   •••••••
      •   ••   •••   •••   ••••   •••••••
6 $(3\times 10)+(3\times 5)$; $(10\times 8)-(7\times 5)$. 45 cm²
7 a $2x+10$, or $2(x+5)$ cm *b* $4(x+1)$, or $4x+4$ cm². $x > 1$. Area >8 cm²
8 a 77 *b* 385 *c* 946 *d* 10857
9 'Multiply by 10, then subtract the number'. *a* 324 *b* 666 *c* 4140
d 4842. $99\times 57 = (100-1)57 = 5700-57 = 5643$

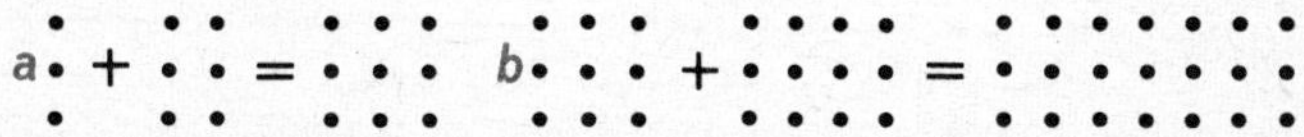

10a $9+x$ *b* $9+y$ *c* x^2+3 *d* y^2+3
e $9+x+y$ *(1)* No *(2)* * is not distributive over +

Page 61 Exercise 6B

1 a $6(x+5)$ cm² *b* $x = 10$. 15 cm
2 a $4x+4y+4z$, or $4(x+y+z)$ *b* $2xy+2yz+2zx$, or $2(xy+yz+zx)$
3 $3n+5 = 200$. $n = 65$. £1·30 and £0·70
4 $4(x-2) < 40$. {2, 4, 6, 8, 10}
6 a $n+1$ *b* $n-1$ *c* $n = 48$. 47, 48, 49
7 a $6x+30$ *b* $x = 25$ *c* 25°, 100°, 55°
8 a $36x+36$ pence *b* $x = 5$ *c* 12
9 a $75x-280$ pence *b* $x = 8$ *c* 36
10a $4x+18$, or $2(2x+9)$ cm
b $x(x+9)$, or x^2+9x cm². $4x+18 > 30$, i.e. $x > 3$, and $4x+18 < 42$, i.e. $x < 6$. $3 < x < 6$

Algebra—Answers to Revision Exercises

Page 64 Exercise 1A

1 a 14 *b* 48 *c* 35 *d* 4 *e* 32 *f* 8
2 *Rows:* 10, 0, 12, 5, 9; 12, 36, 7, 29, 25; 5, $\frac{1}{2}$, 1, $5\frac{1}{3}$, $\frac{13}{36}$; 10, 27, 3, 23, 18
3 {1, 4, 7, 10, 13, 16} *4a* 81 *b* 8 *5a* $3a$ cm *b* $9a^2$ cm²
6 a F *b* T *c* T *d* F *7* $A = 7x^2$; 63 cm²
8 a $6x$ cm², $6y$ cm² *b* $A = 6x+6y$, or $A = 6(x+y)$ *c* $A = 84$
9 a 2100 km, 3150 km, 5250 km, $1050x$ km *b* $D = 1050x$
c $D = 3675$ *10* y: 5, 7, 9, 11, 13, 15

Page 65 Exercise 1B

1 a 32 *b* 31 *c* 38 *d* 30 *e* 11 *f* 13
2 a 25 *b* 41 *c* 81 *d* 9 *e* 1
3 a 2 *b* $\frac{1}{2}$ *c* 5 *d* 2 *4a* 60 *b* 8
5 a No *b* 28; yes
6 a $S = 4x+8y$ *b* $A = 4xy+2y^2$ *c* $V = xy^2$
d $S = 88$; $A = 290$; $V = 300$
7 a *Rows:* 30, £15, £9, £6; 35, £17·50, £9, £8·50; 40, £20, £9, £11; n, £$0{\cdot}50n$, £9, £$(0{\cdot}50n-9)$
b $P = 0{\cdot}5n-9$, or $P = \frac{1}{2}n-9$ *c* $\frac{1}{2}n-9 = 0$ and so $n = 18$
8 a 1 *b* 0 *c* 2
9 a $P = 4x+2y$, or $2(2x+y)$ *b (1)* $P = 6x+2y$, or $2(3x+y)$
(2) $P = 8x+2y$, or $2(4x+y)$ *(3)* $P = 10x+2y$, or $2(5x+y)$
(4) $P = 12x+2y$, or $2(6x+y)$ *(5)* $P = 2nx+2y$, or $2(nx+y)$

Page 67 Revision Exercise 2A

1 a $>$ *b* $=$ *c* $<$ *d* $\leqslant$

2 a $3 < 9$ *b* $3^2 = 9$ *c* $\frac{1}{2} < 1$ *d* $1 > 0$ *e* $5{\cdot}80 < 5{\cdot}81$ *f* $\frac{1}{5} > \frac{1}{10}$

3 ***a, b, c, d*** and ***e*** are true

4 a $p < 5$ *b* $q > 12$ *c* $4 < r < 10$ *d* $15 < n < 17$

5 a $x+8 < 25$ *b* $5y > 30$ *c* $n \neq 0$ *d* $p-9 \leqslant 15$

6 a 0 1 2 3 4 5 b 0 1 2 3 4 5 6 7

c 0 1 2 3 4 5 6 7 8 9 d 0 1 2 3 4 5

7 a $8 > 5 > 3; 3 < 5 < 8$ *b* $10 > 6 > 1; 1 < 6 < 10$
c $\frac{1}{2} > \frac{1}{3} > \frac{1}{4}; \frac{1}{4} < \frac{1}{3} < \frac{1}{2}$

8 a {8, 9, 10} *b* {0} *c* {0, 1, 2, 3, 4, 5, 6}
d {3, 4, 5, ..., 10} *e* ø

9 a {7,9} 0 1 2 3 4 5 6 7 8 9 b {7} 0 1 2 3 4 5 6 7

c {1,3,5,7} 0 1 2 3 4 5 6 7 8 d {7,9} 0 1 2 3 4 5 6 7 8 9

e {1} 0 1 2 f {5,7,9} 0 1 2 3 4 5 6 7 8 9

10 a $n \leqslant 50$ *b* $t \geqslant 5$ *c* $v \leqslant 110$ *d* $6300 < R < 6400$

11 a {0,1,2,3} 0 1 2 3 4 b ϕ

c {1,2,...,6} 0 1 2 3 4 5 6 d {0} 0 1 2

e {5,6} 0 1 2 3 4 5 6 f {4,5,6} 0 1 2 3 4 5 6

12 $4x < 24$, i.e. $x < 6$

Page 69 Revision Exercise 2B

1 a $5 < x < 10$, or $10 > x > 5$ *b* $0 < x < 3$, or $3 > x > 0$
c $y \geqslant 4$, or $4 \leqslant y$ *d* $8 \leqslant x < 10$, or $10 > x \geqslant 8$

2 a $0 \leqslant n < 4$, or $0 \leqslant n \leqslant 3$ *b* $0 \leqslant n \leqslant 7$, or $0 \leqslant n < 8$
c $1 < n < 5$, or $2 \leqslant n \leqslant 4$ *d* $2 < n < 4$

3 a {4, 6} *b* {2, 4} *c* {4, 6}

4 a {4,5} 0 1 2 3 4 5 b {3,4,5} 0 1 2 3 4 5

c {1,4,5} 0 1 2 3 4 5 d {1,2,3} 0 1 2 3 4

e {1,2,3} 0 1 2 3 4 f {5} 0 1 2 3 4 5

5 a $\{0, 1, 2, 3, 4, 5, 6\}$ *b* $\{0, 1\}$ *c* $\{4, 5, 6, ...\}$
d $\{4, 5, 6, 7\}$ *e* $\{1, 2, 3, 4, 5\}$ *f* $\{5, 6, 7, ...\}$

6 a $\{1, 2, 3\}$ *b* $\{6, 7, 8, ...\}$ *c* $\{2, 3, 4, ...\}$
d ø *e* $\{6, 7, 8, ...\}$ *f* $\{1, 2, 3, 4\}$

7 a $4x < 16$, or $x < 4$ *b* $\{1, 2, 3, 4\}$; $\{4\text{ cm}^2, 6\text{ cm}^2\}$

8 a $x \leqslant 50$ *b* $W \geqslant 1{\cdot}6$ *c* $190 \leqslant S \leqslant 1200$ *d* $C \geqslant 50$

9 $250+w < 410$; $w < 160$

10a $180-2x$ *b* $180-2x > 30$; $x < 75$
c $180-2x < 40$; $x > 70$ *d* $70 < x < 75$

11a $x+14 > 20-x$ *b* $\{4, 5, 6, ..., 20\}$

12a $S_5 = 5\times 6$, $S_6 = 6\times 7$ *b* 9 *c* 20 *d* $r(r+1)$

Page 71 Revision Exercise 3A

1 a (*1*) +£25 (*2*) −£30 (*3*) −£*p*
b (*1*) +80 km (*2*) −125 km (*3*) x km$-y$ km

2 a $6 > 4$ *b* $0 > -1$ *c* $-2 = (-2)$ *d* $-3 < 3$

3 a 5 *b* −7 *c* $n-3$ *d* $-x+1$ *e* 0
f 0 *g* $-2m$ *h* −1

4 ***a, c, d, e, f, g***

6 a −13 *b* 7 *c* −6 *d* 5 *e* −32 *f* −28
g −4 *h* 4 *i* −7 *j* $7a$ *k* $-7a$ *l* $36m$

7 a 3 *b* 13 *c* −13 *d* −8 *e* 11 *f* 0 *g* 6 *h* 10

8 a $\{-1, 0, 1, 2\}$ *b* $\{-4, -3, -2, -1, 0, 1\}$ *c* $\{-4, -3, -2, -1\}$
d no solution *e* 2 *f* −2

9 5

10a 2 *b* $\frac{1}{4}$ *c* $\frac{1}{8}$ *d* $1\frac{1}{2}$ *e* 0·75 *f* −0·75

11b 1 *12a* $\{6\}$ *b* $\{0, 1, 2\}$ *c* ø *d* $\{0, -1, -2, ...\}$ *e* $\{-\frac{1}{2}\}$ *f* ø

13 $(1, -2), (2, -4), (3, -6)$

14a $4{\cdot}7\times 10^{-1}$ *b* $6{\cdot}4\times 10^{-2}$ *c* $1{\cdot}3\times 10^{-6}$ *d* $2{\cdot}3\times 10^{3}$

Page 72 Revision Exercise 3B

1 a < *b* > *c* > *d* = *e* = *f* > *g* < *h* <

2 *Rows:* 7, −1, 0, −6; 5, −3, 2, −4; 3, −5, 4, −2;
2, −6, 5, −1; 1, −7, 6, 0; −6, −14, 13, 7

3 a $\{-5, -4, -3\}$ *b* $\{1, 2, 3, 4, 5\}$ *c* $\{5\}$ *d* ø *e* $\{-1, 0, 1\}$

4 a 2 *b* −4 *c* 5 *d* −6 *e* 4 *f* 9 *g* 8 *h* −2

5 a −3 *b* −5 *c* −1·4 *d* $-\frac{3}{4}$ *6* 65

7 a $6p$ *b* $2m$ *c* $-3y$ *d* $5x$ *e* $-6t$
f n *g* $7p$ *h* $-4a^2$ *i* 0

8 a 11° *b* −19° *9* −10, −9, −8

10 *Rows:* 0, 1, 2, 3, 4; 1, 2, 3, 4, 0; 2, 3, 4, 0, 1; 3, 4, 0, 1, 2; 4, 0, 1, 2, 3
a 3, 1, 4 *b* −2, −4, −1. 3 = −2, 1 = −4, 4 = −1

11b (*1*)T (*2*)T (*3*)T (*4*)F (*5*)F (*6*)F

12a $9{\cdot}6\times10^{-1}$ *b* $1{\cdot}1\times10^{-3}$ *c* $3{\cdot}3\times10^{3}$ *d* $1{\cdot}8\times10^{-6}$

13a A square *c* A(2, −1), B(5, −5), C(1, −8), D(−2, −4). A square

14a 9 *b* at any number

Page 75 Revision Exercise 4A

1 $3a+3b$ and $3(a+b)$; $3ab$ and $a\times3b$

2 a $13n$ *b* $8x^2$ *c* $6p$ *d* ab *e* z *f* $5m+n$

3 Fig. 6: $A = 3x+12 = 3(x+4)$; Fig. 7: $A = 5x+5y = 5(x+y)$

4 a $4c+36$ *b* $3c-6$ *c* c^2+9c *d* c^2-2c
e $ac+3c$ *f* $ac-3c$ *g* $6a+21$ *h* $12x-3y$

5 a $p+2r$ pence; $5(p+2r)$ pence *b* $5p+10r$ pence
c $P = x(p+2r)$ or $P = px+2rx$

6 a $5(x+6)$ *b* $3(2a+5)$ *c* $5(2m-5n)$ *d* $x(3+y)$
e $5y(y+6)$ *f* $c(c-1)$ *g* $4q(3p+5q)$
h $2(2x^2+2x-1)$

7 a {3} *b* {3} *c* {1, 2, 3, 4} *d* {1, 2, 3} *e* {8}
f {4} *g* {7, 8, 9, ...} *h* ø *i* {1, 2, 3, 4}

8 a $7a+3b$ *b* $14x-3y$ *c* $2p+9q$ *d* $7y$
e $7m-3n$ *f* $21u+2v$

9 a $101P$ pence *b* $19m$ *c* $4x$ *d* $15k$ cm *e* $7x$ cm

10a $x+x+80 = 180$; $x = 50$ *b* $x+x+20+x+40 = 180$; $x = 40$
c $x+2x+3x+3x = 360$; $x = 40$

Page 76 Revision Exercise 4B

1 a $6x+6y+15z$ *b* $5p$ *c* $xy+x^2$ *d* $2x^3+x$

2 a $5a(a+2)$ *b* $3n(2m+5n)$ *c* $xy(x+y)$
d $2x(2x^2+2x+3)$ *e* $a(a+2b-c)$ *f* $2x(2x-3y+4z)$

3 a $n+2$, $n+4$ *b* $n+n+2+n+4 = 180$; $n = 58$

4 $\frac{1}{2}(a+b) = \frac{a}{2}+\frac{b}{2}$

5 a $x+3y$ pence *b* $c = n(x+3y)$, or $c = nx+3ny$
c $c = 6(95+15) = 660$, or $c = (6\times95)+(3\times6\times5) = 660$. Cost is £6·60. Formula $c = n(x+3y)$ is easier to use.

6 a Value is $(5\times19^2)+(5\times19) = (5\times19\times19)+95 = 1805+95 = 1900$
b Value is $5\times19\times(19+1) = 5\times19\times20 = 19\times100 = 1900$

7 a $a+b$ *b* $x-2y$ *c* $2x+3y$ *d* $3a-4b$

8 a $9x+8y$ cm
b $(9\times2y)+8y = 52$ i.e. $26y = 52$, so $y = 2$ cm and $x = 4$ cm

9 $A = ac+(a+b)c+(a+2b)c = ac+ac+bc+ac+2bc = 3ac+3bc$
$= 3c(a+b)$

10a {4} *b* {0, 1, 2, 3} *c* {6} *d* {8}
e {4} *f* {0, 1, 2, 3, 4, 5} *g* {8, 9, 10, ...}

11 $42x+6x = 24$; $x = \frac{1}{2}$. 21 km

12a $ac+ad = a(c+d)$ and $bc+bd = b(c+d)$. Thus, $ac+ad+bc+bd = a(c+d)+b(c+d) = (c+d)(a+b) = (a+b)(c+d)$

b $xp+xq+2p+2q = x(p+q)+2(p+q) = (p+q)(x+2)$

Page 80 Cumulative Revision Exercise A

1 a The first six natural numbers, or whole numbers from 1 to 6; 6
b $A = \{1, 3, 5\}$ *c* $B = \{2, 3, 5\}$ *d* $\{3, 5\}$

2 $V = \{P, Q, R, S\}$, $S = \{PQ, QR, RS, SP\}$, $T = \{POQ, QOR, ROS, SOP\}$

3 b *4* $A = G, K = B$

5 a (*1*) (*2*) (*3*)

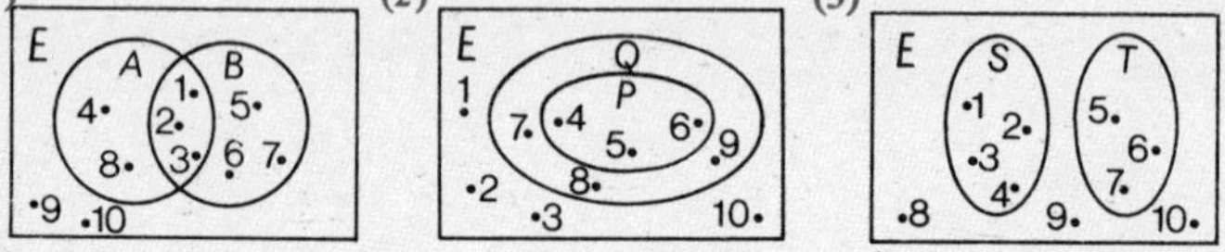

b $A \cap B = \{1, 2, 3\}$, $P \cap Q = \{4, 5, 6\} = P$, $S \cap T = \varnothing$ *c* $B \subset Q$

6 a 8; 5 etc. *b* 10; 9 etc. *c* 9 etc., 10 etc.
d 2 etc., 5 etc. *e* -4, -6 etc.

7 a $\{4\}$ *b* $\{0, 1, 2, 3\}$ *c* $\{3\}$ *d* $\{0, 1, 2\}$
e $\{0, 1, 2, 4, 5\}$ *f* $\varnothing$ *g* $\{1, 2, 3, 4\}$ *h* $\{0, 2, 4\}$

8 a 11 *b* 30 *c* 14 *d* 11 *e* 8 *f* 17
g 6 *h* 12 *i* 3 *j* 6 *k* 9 *l* 60

9 a 0 *b* 6 *c* 2 *d* 5 *e* 20 *f* 9

10 ab, $3c$, $8p$, m^2, $2pq$, $12pq$, abc

11a $p+q$ or $q+p$ *b* $y-x$ *c* mn *d* $4ab$ *e* $10(a+b)$

12a 18 *b* 21 *c* 18 *d* 1250 *e* 20 *f* 6

13a $x+200 = 450$; $x = 250$ *b* $4p+3p+5p = 84$; $p = 7$

14a 150p *b* $C = 2nx$

15 *Rows:* 3, 6, 3, 6; 7, 10, 5, 10; 11, 14, 7, 14; 15, 18, 9, 18; $S+3=2J$; $J=13$

16a $3 \times 4w$ *b* $4c \times 6$ *c* $5p \times 3q$ *d* $35a$
e $36m$ *f* $72mn$ *g* t^2 *h* $3p^3$

17 n, $n+1$, $n+2$; 74

18a 9, 13, 17, 21, 25 *b* 12, 8, 4, 0, -4
c 0, 2, 8, 18, 32 *d* 15, 18, 21, 24, 27

19a $\{0, 1, 2, 3, 4\}$ *b* $\{8, 9, 10\}$ *c* $\{8, 9, 10\}$ *d* $\{0, 1, 2\}$
e $\{8, 9, 10\}$ *f* $\{0, 1, 2, 3, 4\}$ *g* $\{5, 6, \ldots, 10\}$ *h* $\{2, 3, \ldots, 7\}$

20a 37 *b* 20 *c* -19 *d* 0 *e* $-38y$ *f* $-5z$

21 c

22a 1 *b* 5 *c* -8 *d* $-7a$
e $-23c$ *f* 0 *g* $-4p$ *h* $-12h$

23a $a = 6$ *b* (6, 0), (3, -5), (9, -5), *c* (0, 7), (-3, 2), (3, 2)

24a 10 *b* 11 *c* -7 *d* -10 *e* -7 *f* -3

25a $3x+6$ *b* $10x+20$ *c* $20a+15b$ *d* $2pq+pr$
e $5a+5b+10c$ *f* $9x-24$ *g* $2ac-7c$ *h* $29a+10b$

26a $5(p+q)$ *b* $4(n+2)$ *c* $a(b+1)$ *d* $a(a-4)$
e $3(2p-7q)$ *f* $x(x-8)$

27a $2(2p+q)$ *b* $5(5m+3)$ *c* $5(x+8)$ *d* $2x(x+5)$

28a $3(a+b)$ *b* $5a+4b$ *c* $4a+5b$ *d* $10a+11b$

29 $s = t(u+5t)$; $s = 150$

30 ax | bx | cx | x
a b c

Page 84 Cumulative Revision Exercise B

1 a T *b* F *c* T *d* F *e* T *f* F

2 $A = \{2, 3, 5, 7\}$, $B = \{1, 3, 5, 7, 9\}$, $C = \{3, 5\}$
a $\{3, 5, 7\}$ *b* $\{3, 5\}$ *c* $\{3, 5\}$
d $\{3, 5\}$. $C \subset B$, $C \subset A$, $B \cap C = A \cap C$, etc.

3 Integers. $\varnothing$, $\{2\}$

4 $S = \{1, 2, 3, \ldots, 12\}$ *a* $\{1, 3, 5, 7, 9, 11\}$ *b* $\{3, 6, 9, 12\}$
c $\{1, 2, 3, 4, 5, 6\}$ *d* $\{1, 2, 3\}$. $\{3, 9\}$, $\{1, 2, 3\}$

5 a $\{1, 2, 3, 4\}$ *b* $\{5, 6\}$ *c* $\{1, 2, 3, 4, 5\}$
d $\{1, 2, 3\}$ *e* $\{2, 3, 4, 5, 6\}$ *f* $\{1, 2, 3, 4\}$

6 a e.g. $x = 4, y = 3$; $x = 5, y = 2$ *b* none
c e.g. $x = 2, y = 1$; $x = 4, y = 2$ *d* e.g. $x = 1, y = 2$; $x = 1, y = 3$
e e.g. $x = 4, y = 1$; $x = 1, y = 5$ *f* $x \in \{1, 2, 3, 4, 5\}$

7 a 373, 173 *b* 227, −273

8 a 5250 *b* 2500

9 a $2h+w = 75$ *b* 55, 15

10a +28 *b* −2, −5, −8 *c* 13°C *d* $-6a$ *e* 7

11a 5 *b* −9 *c* −7 *d* 12 *e* 2

12a −3 *b* 6 *c* 7 *d* −15 *e* −14 *f* 8
g 0 *h* 5

13 *Rows:* −3, 1, 2; 5, 0, −5; −2, −1, 3 *14* $3x$

15 64; 25

16a $\{0, 1, 2, \ldots, 8\}$ *b* $\{0, 1, 2, \ldots, 6\}$ *c* $\{0, 1, 2, 3, 4, 5\}$

17a $>$ *b* $<$ *c* $>$ *d* $<$ *e* $=$ *f* $>$

18a $A = 100-4x$ *b* $100-4x < 80$, so $4x > 20$ and $x > 5$
c $x \leqslant 10$. Hence $5 < x \leqslant 10$.

19a $10x^2$ *b* $7m+9$ *c* $6a+4b+2c$
d $4a-5b$ *e* $4a^2+15a$ *f* $8ab+8c$

20 *Rows:* $2a^2$, $6ab$, $4ac$; ab, $3b^2$, $2bc$; $4ac$, $12bc$, $8c^2$

21a $2(2n+3)$ *b* $2h(a+b+c)$ *c* $2p(3p+2)$

22a (*1*) $29-x$ (*2*) $15-x$
b $29-x+x+15-x = 35$, so $x = 9$

23a $a \triangle b * a \triangle c$ *b* $a * b \triangle a * c$

24a $\{-5, -4, -3, -2, -1\}$ *b* $\{-3, -2, -1, 0, 1, 2, 3\}$, $\{-3, -2, -1\}$

25a A′(−5, 0), B′(−3, −4), C′(3, −4) *b* A hexagon. 64 square units.

26a 0 *b* −3 *c* $-3a+2$. E.g. 2, −4; −5, 3; 0, −2

Geometry—Answers to Chapter 1

Page 91 Exercise 2

1 ***a*** and ***d***

Page 93 Exercise 3

2 The shapes fit *3a* south *b* west, south west

4 P goes to Q, Q goes to P, and the lines change places.

5 a X → X *b* R → S *c* XR → XS
d the parallel lines change places

Page 95 Exercise 4

1 2, 1, 2, 4, 2, 8, 4, 2, 8

2 B *3* A, C, E, H *4* D, G *5* F, I

Page 97 Exercise 5

1 a PQ, SR; PS, QR *b* PS = 3 cm, SR = 5 cm

2 a (*1*) HG (*2*) EH *b* HG = 2·8 cm, FG = 4·5 cm

3 a VW = YX, WX = VY *b* VW = YX = 17 mm, WX = VY = 8 mm

4 b Doors, windows, pages of books, etc.

5 K ⟶ L, N ⟶ M, KN ⟶ LM, KN = LM;
K ⟶ N, L ⟶ M, KL ⟶ NM, KL = NM

6 You would require two pairs of equal strips.

Page 98 Exercise 6

1 c PR, FH, WY *2b* SU and TV *c* SU = TV, ST = VU, SV = TU

Page 99 Exercise 7

1 O ⟶ O, B ⟶ D, OB ⟶ OD

2 Turn over about PQ. O ⟶ O, A ⟶ B, OA ⟶ OB, OA = OB; etc.

3 OA = OB = OC = OD

4 b OP = OQ = OR = OS *c* PQ = SR, PS = QR

5 b OT = OU = OV = OW *c* TU = WV, TW = UV

6 b EG and FH. Yes *c* FZ = GZ = HZ = 2 cm *d* HG, FG, FH

7 c a rectangle *8* rectangles *9* an infinite number

10a no *b* yes, yes, no *c* LJFD, LHFB, LCFI, LAFG

Page 102 Exercise 8

1 a angles P, Q, R, S; angles E, F, G, H; angles V, W, X, Y
b PQ and SR, PS and QR; EF and HG, EH and FG;
VW and YX, WX and VY

3 360° *4a* AB = DC, AD = BC, AC = BD, OA = OB = OC = OD; ∠A = ∠B = ∠C = ∠D = 90° *b* See pages 95 and 96.

5

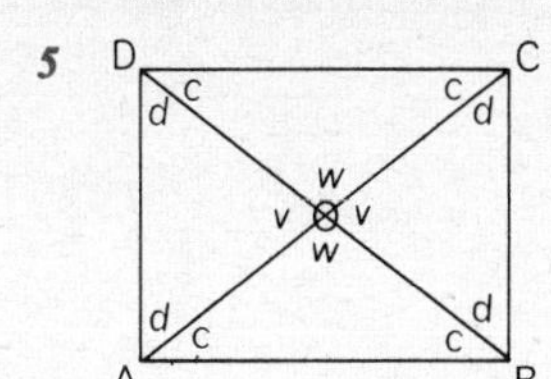

6

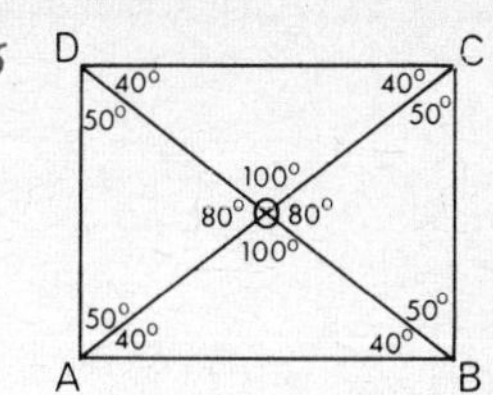

7

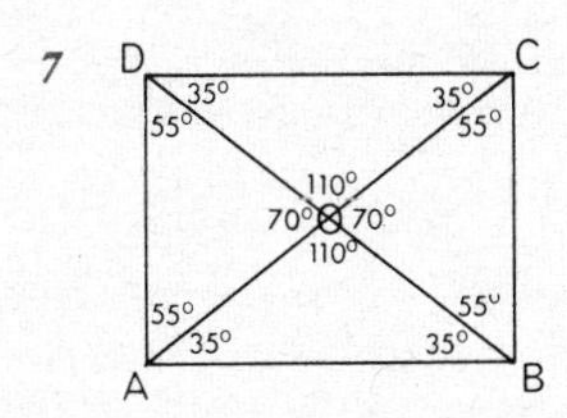

8

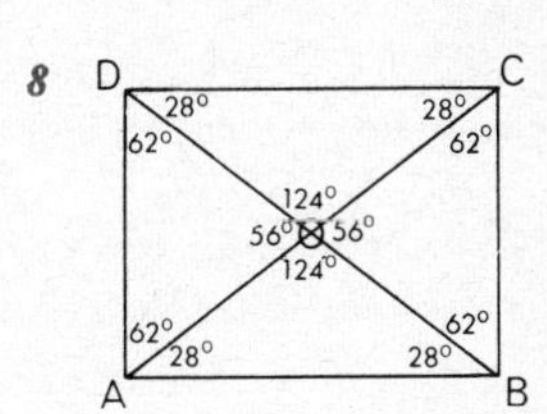

Page 103 Exercise 9A

1 b AB = DC = 4 units; AD = BC = 6 units *c* (3, 5)

2 b EF = HG = 8 units; EH = FG = 5 units *c* $(11, 2\frac{1}{2})$

3 b a rectangle *c* UT, ST, US *d* 60 square units *e* (7, 5)

4 b B(12, 2), D(5, 11) *c* 63 square units

5 b B(25, 5), D(15, 10) *c* 50 square units

6 a N(3, 3) *b* P(8, 5), Q(3, 5)

7 a D(2, 5) *b* F(10, 10) *c* E(10, 0), G(2, 10) *d* 1 : 4

Page 104 Exercise 9B

1 b 32 square units; (4, 5) *2* R(7, 9), S(3, 9)

3 b $b = 4$, $c = 12$, $p = 3$, $q = 10$ *c* (3, 0), (12, 0), (0, 4), (0, 10)

4 a R(12, 8), S(8, 12) *b* (6, 6) *c* U(12, 0), V(0, 8)

5 b F(15, 6) *d* 96 square units; (9, 6) *e* (21, 14), (27, 18), (33, 22)

Page 106 Exercise 10

1 a From Position 5, B ⟶ B, A ⟶ C, BA ⟶ BC, BA = BC; also DA = DC. From Position 6, AB = AD and CB = CD. Hence AB = BC = CD = DA.

b From Position 5, BA ⟶ BC, so ∠DBA ⟶ ∠DBC, and ∠DBA = ∠DBC, etc.

2 From Position 7, ∠AOB ⟶ ∠BOC, so ∠AOB = ∠BOC; each must be a right angle; etc.

3 From Position 2, A ⟶ B, C ⟶ D, AC ⟶ BD, so AC = BD.

4 From Position 4, A ⟶ C, OA ⟶ OC, so OA = OC, etc.

Page 106 Exercise 11

2

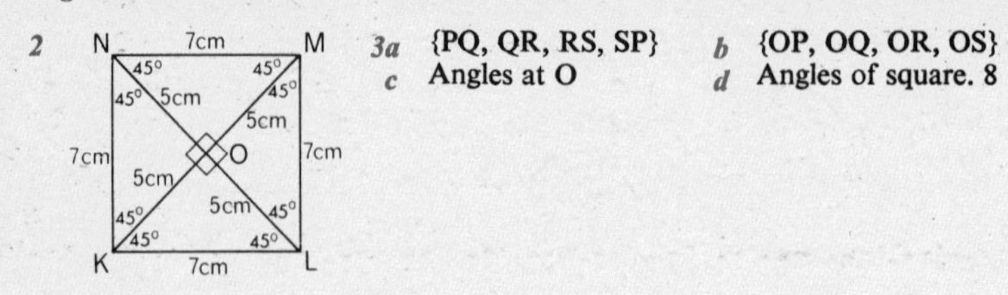

3a {PQ, QR, RS, SP} *b* {OP, OQ, OR, OS}
c Angles at O *d* Angles of square. 8

4 a yes, yes *b* (5, 5) *6* C(10, 7), D(4, 7), (7, 4)
7 G(16, 13). (22, 19), (28, 25), (34, 31).
8 b E(9, 2), F(15, 8), G(9, 14), H(3, 8)
c K(12, 5), L(12, 11), M(6, 11), N(6, 5)

Page 107 Exercise 12

1 T *2* T *3* F *4* T *5* F
6 T *7* T *8* T *9* T *10* T

Page 108 Exercise 13

7 ***a*** and ***d*** *8* ***b***, ***c*** and ***e***

Page 109 Exercise 14

1 12, 25, 7, 15, 20·25 cm^2 *2* The third and fifth are squares.
4 b any number (an infinite number)

Page 110 Exercise 15

2 a AB, BC, CD, DA, EF, FG, GH, HE *b* AE, BF, CG, DH
c ABCD, EFGH *d* ABFE, BCGF, CDHG, ADHE
e {AB, DC, EF, HG}, {BC, AD, FG, EH}, {AE, BF, CG, DH}
3 a PQ, QR, RS, SP, TU, UV, VW, WT *b* PT, QU, RV, SW
c PQRS, TUVW *d* PQUT, QRVU, RSWV, PSWT
e {PQ, SR, TU, WV}, {QR, PS, UV, TW}, {PT, QU, RV, SW}
4 a horizontally *b* 4 cm, 2 cm *5a* vertically *b* 3 cm, 2 cm
6 a vertically *b* 4 cm, 3 cm
7 ADGF *8* EFCD *9* AEGC
10a 6 *b* 6. They are congruent squares.
14a AG, CE, DF *b* They are equal. *16* at the centre

Geometry—Answers to Chapter 2

Page 114 Introduction

a *a, b, g* *b* *c, j* *c* *d, e, f, h, i* *d* none

Page 115 Exercise 1A

2 a ABC, ADC *b* ABD, BCD
3 a ½ *b* 8; 4 *4c* (i) 4, 2 (ii) 4, 2 (iii) 3, 1·5 (iv) 8, 4
5 c 24 square units *6b* (*1*) 10 (*2*) 12·5 (*3*) 42

Page 116 Exercise 1B

1 a ABD, BCD *b* ABC, ADC *2* PQS, PQR, RSP, RSQ
3 a Each is 90° *b* OVW, OWX, OXY, OYV
4 a ½ *b* 24 cm², 12 cm² *5* 64 m², 32 m²
6 100 cm², 50 cm², 25 cm²
7 a 16 cm² *b* 64 cm² *c* 30 cm² *d* 60 cm² *e* 72 cm² *f* 120 cm²
8 a ½ *b* 1 *c* 1 *d* 2 *e* 1½ *f* 3
9 a 2 *b* 5 *c* 6 *d* 5
10a 12 *b* 12·5 *c* 22 *d* 13 square units

Page 118 Exercise 2

3 Because two right angles were placed together to make a straight angle.
4 Turn △AEC over, and place it alongside △ABC as shown.
5 a DB *b* DC *c* ∠DBC
d ∠BDC *e* ∠DCB *f* DBC
6 a (*1*) PMR, PMS (*2*) 15 cm, 14 cm, 28 cm
b (*1*) TZX, TZY (*2*) 70°, 20°, 40°
7 AC, ∠ACB, △ACB

8 a

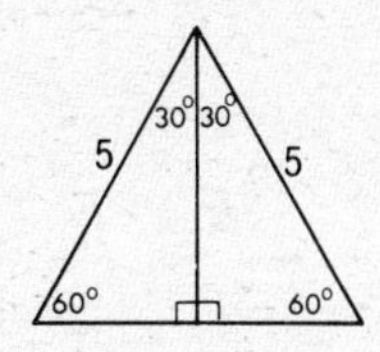

b

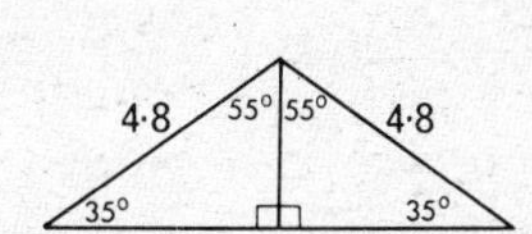

c

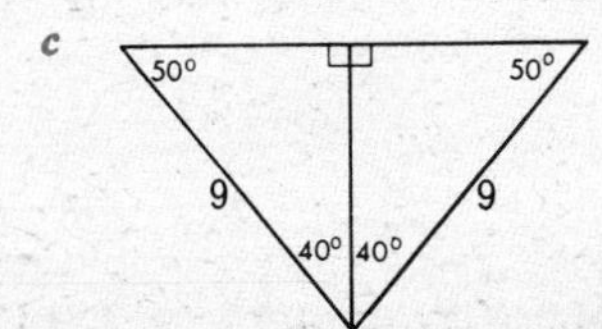

10a ∠D and ∠F *b* ST and SV

11a

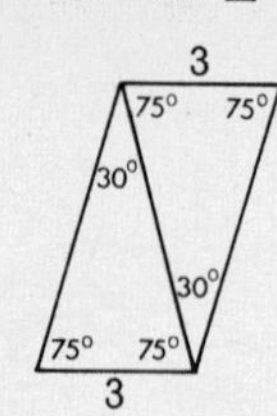

b

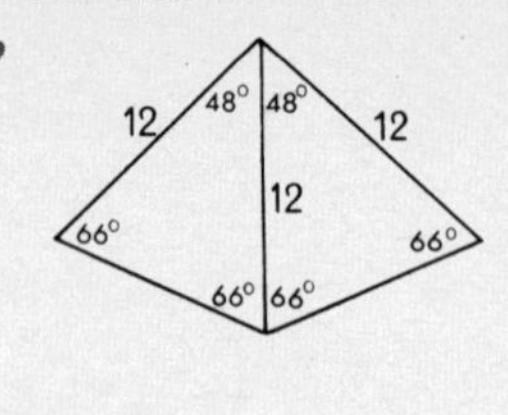

c

13 AOB and DOC, AOD and BOC

14a PMQ, QMR, RMS, SMP *b* PQR, QRS, RSP, SPQ

15 EOF, HOI; FOG, IOJ; GOH, JOE; EOG, HOJ; FOH, IOE; GOI, JOF

Page 123 Exercise 4A

1 a CA, CB *b* ∠BAC, ∠ABC *c* (7, 4)

2 a AC, BC *b* ∠CAB, ∠CBA *c* (2, 16)

3 a $r = 9$ *b* (9, 2) *4b* $p = 6$ *5b* $p = q$

6 b (12, 0). ∠AOB, ∠ABO *7b* (0, 10). ∠AOB, ∠ABO

Page 123 Exercise 4B

1 a (8, 1) *b* (2, 9) *2a* (8, 10) *b* (2, 2)

3 a (3, 1), (5, 1) *b* (23, 10); 4, 16 units *c* (7, 3), (11, 3); 12 square units

4 b (8, 8), (8, 0), (8, 199) *c* $p = 8$ *d* (8, 7)
e (8, 4), (8, 10) *f* $x = 8, 4 < y < 10$

5 b (8, 8), (7, 7) *c* $p = q$ *d* (5, 5)
e (0, 0), (10, 10) *f* $0 < x < 10, 0 < y < 10, x = y$

Page 124 Exercise 5

2 180° *4* 90°; 90°; 180° *5* 90°; 90°; 180°; 180°

Page 125 Exercise 6

1 (i) 80° (ii) 55° (iii) 63° (iv) 108°

2 a (ii) *b* (i) and (iii) *c* (iv)

3 a 90°; right-angled *b* 10°; acute-angled *c* 92°; obtuse-angled
d 60°; acute-angled *e* 36°; obtuse-angled *f* 71°; right-angled
g 59°; acute-angled *h* 48°; acute-angled *4* ***b, d, h***

5 a 60 *b* 70 *c* 60 *d* 30

6

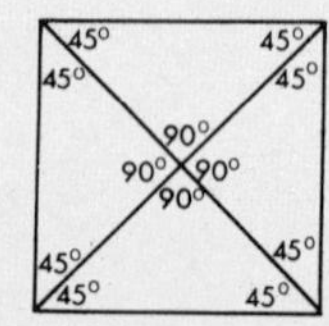

7

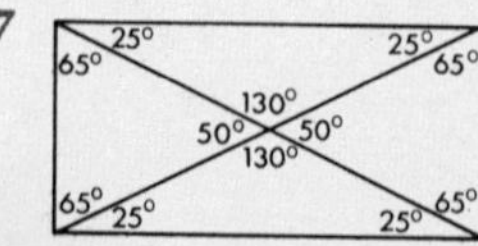

8

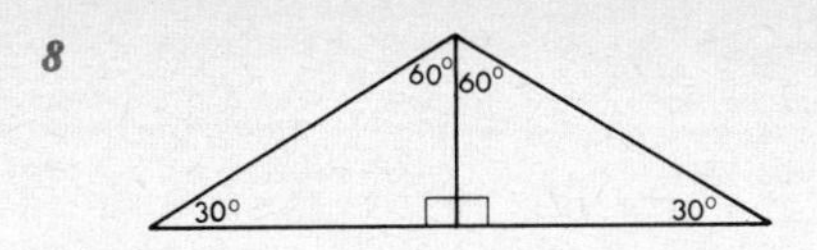

9 *a* T *b* F *c* F *d* T *e* T *f* F

Page 127 Exercise 7

1 *a* AC; ∠ACB *b* BA; ∠CAB

2 Sides equal; angles equal; each angle is 60°.

3 120°; 120°; 120°; 120°; 360°

4 *a* Because six angles of 60° make up a complete turn of 360°
b 180°; BE, CF *c* 12 *d* 120°

5 *a* Because six angles of 60° make up 360°
b Because the three angles of 60° make up 180°, i.e. a straight angle
c 9; 3; 1 6 ***a, c, d*** 7 3 equal sides; ∠ARB = 60°+60°+60°

8 4 faces, 4 corners, 6 edges. 72 cm

Page 130 Exercise 8A

1 *a* 4 *b* 6 *c* 6 *d* 3 square units

2 *a* 24 cm² *b* 48 cm² *c* 56 cm² *d* 75 cm²
e 132 cm² *f* 64 cm² *g* 120 cm² *h* $\frac{1}{2}bh$ cm²

3 *a* 30 cm² *b* 30 cm²

4 72 cm², 55 m², 54 mm², 1275 cm², 2·76 m²

5 AD – BC; BE – AC; CF – AB

6 *a* 38·5 m² *b* 150 m² *c* 78 m²

Page 131 Exercise 8B

1 192 cm², 49·5 m², 216 cm², 14·3 mm²

2 *a* 160 m² *b* 187·5 m² *c* 192 m² *d* 5000 m²

3 *a* 144 cm² *b* 288 cm² 4 Each area is 36 cm².

5 *a* AD, BC *b* 4·5 square units
c PS, QR, 4 square units; EH, GF, 6 square units

Page 133 Exercise 9

3 The sum of two lengths must be greater than the third.

7 The sum of the two angles must be less than 180°.

Page 134 Exercise 10

1 37°, 90°, 53° 2 60°, 60°, 60° 3 7 cm, 80°, 50°

4 9·1 cm, 60°, 40° 5 8·5 cm, 8·5 cm, 90° 6 7·4 cm, 6·5 cm, 60°

8 23°, 67°, 90° 9 500 km, 227° 10 15 km or 93 km

Geometry—Answers to Revision Exercises

Page 137 Revision Exercise 1A

4 (3, 2). CD ⟶ BA, AC ⟶ DB, DB ⟶ AC

5 (3, 2). They bisect each other.

6

7 *b*

8 (4, 1)

9 (7, 5) and (7, 3), or (3, 5) and (3, 3)

10 Fits outline in four ways; opposite sides equal and parallel; all angles right angles; diagonals equal; diagonals bisect each other; two axes of symmetry.

11 Fits outline in eight ways; all sides equal; diagonals bisect angles of square; four axes of symmetry; diagonals at right angles.

12

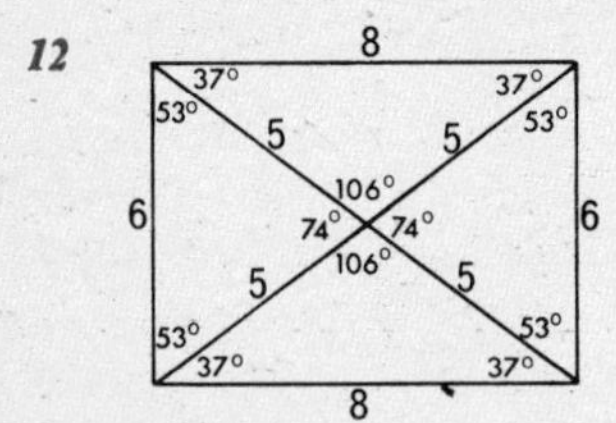

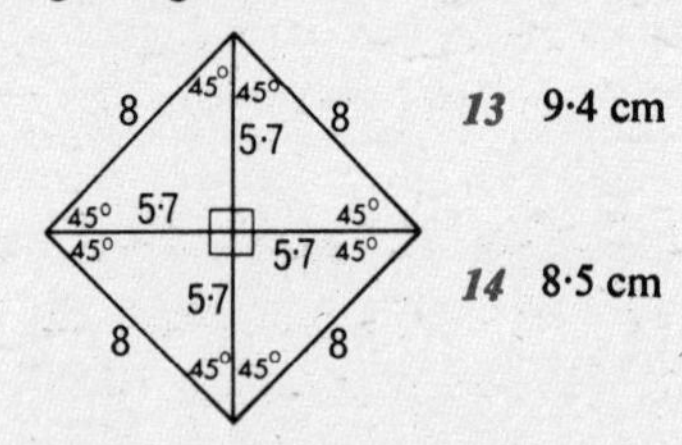

13 9·4 cm

14 8·5 cm

15 3·5 cm

16a T *b* T *c* T *d* F *e* T

17 6 faces 8 corners, 12 edges, 4 space diagonals

18a 512 cm^3 *b* 384 cm^2 *c* 96 cm

19a 3 *b* 3 *c* all the same

Page 139 Revision Exercise 1B

2 b a cube

3 (2, 4), (3, 6), (2, 8), (1, 6); (6, 0), (7, 2), (6, 4), (5, 2)

4 (11, 4)

5 4; 3

6 a rectangle.

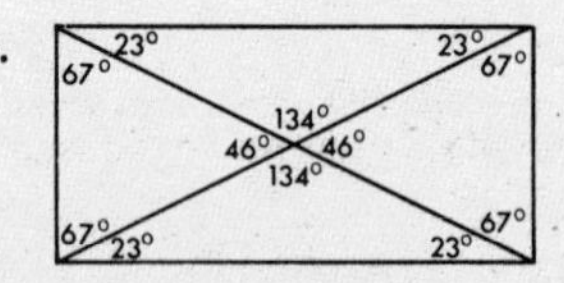

7 Parallel and equal; parallel and equal; bisect each other. ($2\frac{1}{2}$, 3)

8 (2, 3) and (6, 3)

9 Fits outline in eight ways; all sides equal; diagonals bisect angles of square; four axes of symmetry; etc.

11a T *b* F *c* T *d* F

12 144 cm^2, 72 cm^2

13 11·7 cm, 14·1 cm

14 (9, 2), (21, 4), (45, 8), (93, 16)

Page 141 ***Revision Exercise 2A***

1 a OD *b* OA *c* CD *d* B *e* △ODA *f* △CDA

2 a 96 cm^2 *b* 48 cm^2

3 a T *b* F *c* F *d* T

4 a 30 square units *b* 42 square units

6 △AOB, OA = OB, ∠OAB = ∠OBA;
△OBC, OB = OC, ∠OBC = ∠OCB;
△OCD, OC = OD, ∠OCD = ∠ODC;
△ODA, OA = OD, ∠OAD = ∠ODA

7

8 24 cm^2

9 The triangle can be made from two congruent right-angled triangles, etc. AB = AC, ∠ABC = ∠ACB. 10 square units.

10a 12 *b* It is made up of three angles of 60°, i.e. 180° *c* △ABC

11 96 cm; 444 cm^2

12 ***a*** and ***c***

13a ∠C = 70°, ∠A = 40° *b* ∠B = ∠C = 55°

14

15a 80° *b* 116° *c* $180° - p° - q°$

16

17 12 cm^2, 44 m^2, $4x$ mm^2, $6y$ cm^2, $\frac{1}{2}pq$ cm^2

18 28 cm

19 54 cm^2 each

20a ∠A = 57°, ∠B = 79°, ∠C = 44°
b AC = 11·5 cm, ∠A = 41°, ∠C = 29°
c ∠BAC = 65°, AC = 10·4 cm, AB = 7·8 cm

Page 143 ***Revision Exercise 2B***

1 a

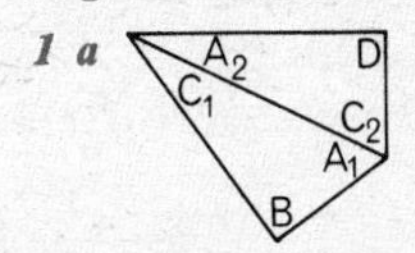

b

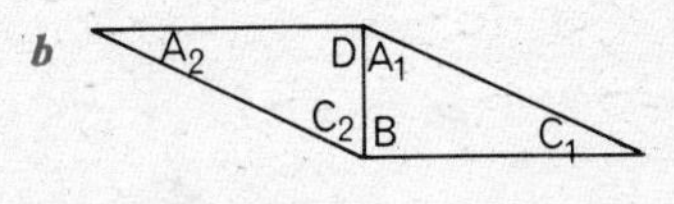

c

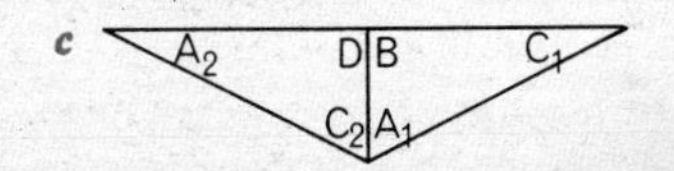

2 8 cm

3 25 square units; M($3\frac{1}{2}$, 5); image of B is (6, 0)
4 6 cm and 8 cm. 16 cm and 12 cm *5* Two
7 C(5, 5·2). 3 cm by 5·2 cm
8 DE, EF and FD. AF and BF, BD and CD, AE and CE.
9 Because all the edges can be seen (e.g. an overhead view)
10a 110° *b* 120° *c* $x° + y°$. The exterior angle is equal to the sum of the two interior opposite angles. *11* 360°
12 6 cm², 2·4 cm *13* 45 cm² *14* 25 m² *15c* yes *16* 8·3 km, 090°
17 51·5 m *18a* yes *b* yes. 10 square units; nearly 4·5 units.

Page 148 Cumulative Revision Exercise A

2 60 cm *3* 3 cm *5a* 90° *b* 36° *c* 24° *d* 15°
6 a 90° *b* 150° *c* 75°
7 a between 0° and 90° *b* 90° *c* between 90° and 180°
d 180° *e* 360°
8 a east *b* south-west *c* west *d* north
9 a supp. *b* neither *c* comp. *d* supp. *e* comp.
10a 32° *b* 58° *c* 148° *d* 328° *11* 62 m
12b (2, 2), (4, 4), (8, 8) *c* $p = q$ *13* (2, 5), (6, 3)
14b Yes. No. *c* (4, 6), (5, 8) *15a* 4 *b* 10 *c* 2
16a angles AOD and BOC, or AOB and DOC
b pairs of angles at A, or B, or C, or D
c pairs of angles at O making up a straight angle
17 DC = 16 cm, AD = 12 cm, OC = OB = OD = 10 cm,
AC = BD = 20 cm *18*

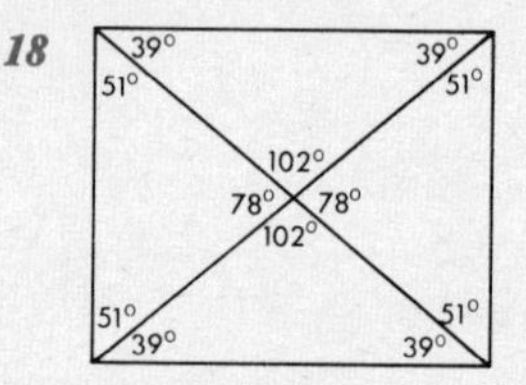

19 (0, 6) *20* (6, 3), (2, 5)
21 (5, 5), (6, 6), (1, 1), (2, 2), etc.; 12·5 square units *22* 3 cm
24 A′(4, 8), B′(7, 4). They are equal and parallel; the same.

Page 151 Cumulative Revision Exercise B

1 a sphere *b* cone *c* cylinder *2* 52 cm
3 16 cm by 16 cm, or 8 cm by 8 cm
4 a 36 *b* 24 *c* 15 *d* 4 *e* 3
5 a 90°, 270° *b* $22\frac{1}{2}$°, $337\frac{1}{2}$° *c* 160°, 200°
6 a 000°, 180°, 225° *b* 122°
7 a 24° *b* 140° *c* 90°
d x, x, y, y make a straight angle, so x, y make a right angle.

8 84 m

9b (1, 2), (4, 8), (5, 10), (8, 16)

c (6, 3), (7, 2)

10a a rectangle

b (11, 3), (11, 6)

11b x is greater than 10

c R lies to the left of the line

12b s is less than 5

c P lies above the line

13 Because MA = MB = MC = MD (diagonals bisect each other and are equal)

14

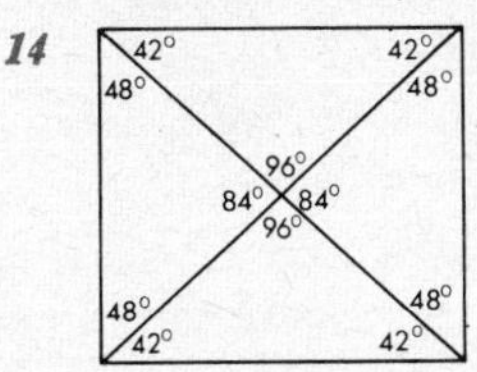

2, 4, 4

15 Four. (2, 1), (4, 1), (6, 1), (6, 3), (6, 5), (4, 5), (2, 5), (2, 3)

16 a rectangle; 12 square units; $(5, 4\frac{1}{2})$

17 24 square units; 96 square units

18a 45°, 45° *b* 60°, 30° *c* 18°, 72° *d* $67\frac{1}{2}°$, $22\frac{1}{2}°$

19 80°, 80°, 20°

20a no *b* yes *c* no

22 5 cm; 30 square units

23 Midpoint of AD or BC; no; AC and BD are equal and parallel

24 48 cm²; approximately 34 cm

Arithmetic—Answers to Chapter 1

Page 157 Exercise 1A

1 83 *2* 135 *3* 427 *4* 202

5 2346 *6* 43701 *7* 60+8 *8* 200+30+4

9 1000+300+20+4

10 20000+4000+100+20+4

11b 1000000, 100000, 10000

12a 400000, 40000, 4000

b 50000, 5000, 500

c 6000000, 600000, 60000

d 7000, 700, 70

13b $\frac{1}{10}, \frac{1}{100}, \frac{1}{1000}$

14a $\frac{4}{10}, \frac{4}{100}, \frac{4}{1000}$

b $\frac{5}{100}, \frac{5}{1000}, \frac{5}{10000}$

c $6, \frac{6}{10}, \frac{6}{100}$

d $\frac{7}{1000}, \frac{7}{10000}, \frac{7}{100000}$

15 $10^3, 10^2, 10, 1, \frac{1}{10}, \frac{1}{10^2}, \frac{1}{10^3}$

Page 158 Exercise 1B

1 323 *2* 707 *3* 4301 *4* 20506

5 500+40+2 *6* 1000+500+6 *7* 2000+20

8 90000+8000+700+60+5

9 a 800000, 80000, 8000; $\frac{8}{10}, \frac{8}{100}, \frac{8}{1000}$

b 2000000, 200000, 20000; $2, \frac{2}{10}, \frac{2}{100}$

c 9000, 900, 90; $\frac{9}{1000}, \frac{9}{10000}, \frac{9}{100000}$

10a	$(3\times10^2)+(2\times10)+3$	*b*	5×10^3	*c*	$(6\times10^3)+(7\times10)+9$
d	$(2\times10^4)+(5\times10^3)+(8\times10^2)+(4\times10)+6$				

Page 159 Exercise 2A

9	5·9, $5\frac{9}{10}$	*10*	8·5, $8\frac{1}{2}$	*11*	12·3, $12\frac{3}{10}$	*12*	23·6, $23\frac{3}{5}$
13	5·87, $5\frac{87}{100}$	*14*	104·7, $104\frac{7}{10}$	*15*	23·12, $23\frac{3}{25}$	*16*	0·1
17	0·3	*18*	0·5	*19*	0·7	*20*	7·9
21	3·09	*22*	1·009	*23*	0·2	*24*	$\frac{1}{2}$
25	$\frac{3}{5}$	*26*	$1\frac{2}{5}$	*27*	$2\frac{7}{10}$	*28*	$8\frac{4}{5}$
29	$\frac{3}{20}$	*30*	$\frac{7}{20}$	*31*	$\frac{7}{25}$	*32*	$\frac{3}{4}$
33	$\frac{1}{20}$	*34*	$2\frac{17}{50}$	*35*	$1\frac{2}{25}$		

Page 159 Exercise 2B

9	83·2, $83\frac{1}{5}$	*10*	249·1, $249\frac{1}{10}$	*11*	30·5, $30\frac{1}{2}$
12	102·9, $102\frac{9}{10}$	*13*	28·09, $28\frac{9}{100}$	*14*	543·21, $543\frac{21}{100}$

15	0·9	*16*	0·03	*17*	0·005	*18*	1·3
19	2·07	*20*	30·15	*21*	17·023	*22*	10·02
23	$\frac{7}{10}$	*24*	$1\frac{9}{10}$	*25*	$3\frac{3}{5}$	*26*	$\frac{23}{100}$
27	$\frac{79}{100}$	*28*	$\frac{3}{50}$	*29*	$1\frac{1}{50}$	*30*	$\frac{1}{8}$
31	$\frac{357}{10000}$	*32*	$12\frac{3}{8}$	*33*	$11\frac{403}{500}$	*34*	$20\frac{1}{200}$

Page 161 Exercise 3

1 a m *b* cm *c* km *d* mm *e* km *f* m

4 a	3·0 cm	*b*	5·7 cm	*c*	5·1 cm	*d*	2·4 cm
5 a	4·0, 3·2, 2·4 cm			*b*	3·5, 2·2, 4·6, 2·3, 1·8 cm		

Page 163 Exercise 4

1	16, 8, 6, 89 cm	*2*	18, 65, 14, 19 kg
3	11·2, 23·8, 56·4, 9·9 s	*4*	46, 1048, 234, 8 km
5	126, 80, 19, 7 ml	*6*	11·5, 14·7, 0·9, 11·0 cm
7	51, 100, 8, 6047 cm²	*8*	10·15, 54·61, 28·02 s

Page 164 Exercise 5

1	4·7, 9·7, 27·8, 0·1, 8·9, 8·4, 6·8			*2*	5·33, 0·05, 1·05, 0·01, 3·15, 8·02		
3	6·543, 12·026, 0·079, 0·536, 1·008						
4 a	0·1428	*b*	0·143	*c*	0·14	*5*	0·33
6	0·67, 0·667			*7*	0·64, 0·636, 0·6364		

Page 164 Exercise 6

1	2	*2*	2	*3*	3	*4*	3	*5*	1	*6*	1
7	3	*8*	3	*9*	2	*10*	3	*11*	4	*12*	3
13	1	*14*	3	*15*	5						

Page 165 Exercise 7

1	2	*2*	2	*3*	3	*4*	4	*5*	1	*6*	2
7	3	*8*	4	*9*	3	*10*	4	*11*	1	*12*	4
13	3	*14*	2	*15*	4	*16*	4	*17*	3	*18*	3
19	4	*20*	4								

Page 165 Exercise 8

1 9·6, 1·9, 13, 26, 5·0, 3·0 *2* 48·2, 2·46, 0·491, 0·499

3 1·286, 0·1296, 0·05378, 124·7, 0·005816

4 a	3·14	*b*	3·142	*c*	3·1416	*5a*	2·72	*b*	2·718	*c*	2·7183
6 a	6·1	*b*	5·01	*c*	19000	*d*	18900	*e*	0·0052	*f*	5
g	10·00	*h*	3·14								

Page 166 Exercise 9A

1	6·5	*2*	8·41	*3*	3·55	*4*	11·25
5	1·755	*6*	9·572	*7*	4·7	*8*	4·86
9	0·129	*10a*	12·61	*b*	11·04	*c*	0·56
11	14·1 cm	*12*	1·9 litres	*13*	22·7 cm	*14*	£3·69½
15	6·4 cm						

Page 167 Exercise 9B

1	14·34	*2*	23·76	*3*	160·77	*4*	16·89
5	180·38	*6*	501·661	*7*	3·811	*8*	10·587
9	0·064	*10a*	6·52	*b*	19·82	*c*	1·117
11	10·15 litres	*12*	157·23, 154·08, 3·15			*13*	194·4 km
14	3·81 dollars	*15*	C by 0·9 cm				

Page 169 Exercise 10A

1 34, 176, 8·4, 2931, 0·007 *2* 194, 270, 6, 3166, 97·2

3 247, 2470, 24700; 3·4, 34, 340; 19·876, 198·76, 1987·6; 0·065, 0·65, 6·5

4	11, 110, 1100	*5*	2·72, 27·2, 272	*6*	18, 180
7	11·5, 115, 1150	*8*	4·83	*9*	14·04
10	6·25	*11*	0·117	*12*	0·0952
13	2·4723	*14*	0·02461	*15*	912·5
16	12·366, 12366	*17*	3·055, 30·55		

Page 169 Exercise 10B

1 134, 1340, 13400; 68·5, 685, 6850; 128·5, 1285, 12850; 0·4, 4, 40; 6·4, 64, 640; 0·082, 0·82, 8·2

2 8·4, 84, 8400; 11·25, 112·5, 11 250; 37·02, 370·2, 37 020; 0·261, 2·61, 261; 0·0156, 0·156, 15·6

3 9, 90, 9000; 20·76, 207·6, 20 760; 54·36, 543·6, 54 360; 1·254, 12·54, 1254; 0·0222, 0·222, 22·2

4 5·5, 550; 36·5, 3650; 62·5, 6250; 3·9, 390

5 31·36 *6* 222·358 *7* 0·2144 *8* 0·347 76

9 14·093, 1409·3, 1·4093 *10* 1·9482, 19 482, 0·194 82

Page 170 Exercise 11

1 4, 1, 3, 0·9, 40, 200 *2* 20·16 cm² *3* 22·1 cm²

4 7·2 cm, 3·24 cm² *5* 19·25 hours *6* 1·92 cm³

7 2·744 cm³ *8* 13·4 cm, 9·9 cm² *9a* 22·5 cm *b* 506·25 cm²

10a 48 cm *b* 87·12 cm² *c* 45·36 cm³ *11* 79·6 cm

12a 30 cm, 20·25 cm² *b* 29·8 cm, 12·6 cm² *c* 27·6 cm, 10·4 cm²

Page 173 Exercise 12A

1 5·73, 0·621, 36·5, 0·04, 0·0009 *2* 24·87, 0·6381, 0·019, 2·431, 0·0006

3 286·4, 28·64, 2·864; 35·76, 3·576, 0·3576; 0·285, 0·0285, 0·00285; 0·004, 0·0004, 0·00004 *4* 230, 23, 2·3

5 1·9, 0·19, 0·019 *6* 16·2, 1·62, 0·162 *7* 45·4

8 2·2 *9* 2·5 *10* 0·34 *11* 14

12 38·4 *13* 5·97 *14* 0·158 *15* 16·667

Page 173 Exercise 12B

1 168·4, 16·84, 1·684; 3·465, 0·3465, 0·034 65; 0·293, 0·0293, 0·002 93; 0·074, 0·0074, 0·000 74; 0·0002, 0·00002, 0·000 002

2 87, 8·7, 0·087; 4·2, 0·42, 0·0042; 1·14, 0·114, 0·001 14; 0·0024, 0·000 24, 0·000 002 4

3 85, 8·5, 0·085; 2·44, 0·244, 0·002 44; 0·306, 0·0306, 0·000 306; 0·008, 0·0008, 0·000 008 *4* 71·9 *5* 2·5 *6* 2·25

7 620, 6 200 000 *8* 211, 2·11 *9* 1·38

10 0·62 *11* 0·061 *12* 0·000 30

Page 174 Exercise 13

1 a 16·8 *b* 290 *2* 15 *3* 40

4 0·7 cm *5* 0·6 cm *6* 1·4 cm *7* 0·045 cm

8 1546 *9a* 6·8 hours *b* 4·5 hours

10a 0·7 m *b* 0·49 m² *11* 2·4 cm *12* 1·6

Page 175 Exercise 14

1 183·3 *2* 30·85 *3* 2435·5 *4* 25·913 *5* 9418

6 3·2 *7* 0·0155 *8* 0·249 *9* 101·96 *10* 48·8

11 22260 *12* 22·26 *13* 853·83 *14* 582·845 *15* 2007
16 99·98 *17* 9·8 *18* 2·6 *19* 1·9 *20* 0·2

Page 176 Exercise 15

1 0·5 *2* 0·25 *3* 0·75 *4* 0·3 *5* 0·9
6 0·875 *7* 0·2 *8* 0·6 *9* 0·31 *10* 0·8
11 0·375 *12* 0·07 *13* 0·125 *14* 0·1875 *15* 0·333
16 0·143 *17* 0·182 *18* 0·0769
(Answers to questions ***15–18*** are rounded off to 3 significant figures.)

Page 177 Exercise 16A

1 2, 3, 7, 3, 2, 100, 1, 4, 2, 5
2 a 25, second *b* 27, third, 3 *c* 10000, 10000, fourth, 10 *3* 3^2
4 a 81 *b* 125 *c* 64 *d* 10 000 000
5 a 1000 *b* 100000 *c* 1000000 *d* 10000 *e* 1000 000 000
6 a 200 *b* 12 000 *c* 250
d 12900 *e* 37 000 *f* 580 000
g 91000000 *h* 86 000 *i* 285
j 61 200 *k* 671000000 *l* 12150000000
m 40 *n* 46 *o* 46·8 *p* 46·85

Page 177 Exercise 16B

1 a 64, sixth *b* 625, 625, fourth, 5
c 1000000000, 1000000000, ninth, 10
2 a 81 *b* 216 *c* 1024 *d* 100000000 *3* 3^6, 5^4, 2^9
4 a 100 *b* 1000000 *c* 10000000000
d 10000 *e* 10000000
5 a 5000 *b* 5700 *c* 830000 *d* 836 000
e 490 *f* 110 000 *g* 96000000 *h* 37000
i 375 *j* 8510 000 *k* 314200 *l* 314·2
m 90 *n* 79 *o* 57·9 *p* 15·79

Page 179 Exercise 17A

1 a 300 *b* 2400 *c* 425000 *d* 5000000
2 a 5×10^1 *b* 5×10^2 *c* 5×10^3 *d* 5×10^4
e 2×10^2 *f* $5{\cdot}6 \times 10^1$ *g* $3{\cdot}579 \times 10^3$ *h* $2{\cdot}34 \times 10^1$
i $5{\cdot}07 \times 10^4$ *j* 6×10^3 *k* $7{\cdot}5 \times 10^6$ *l* 1×10^8
3 a $3{\cdot}8 \times 10^5$ *b* $9{\cdot}4 \times 10^6$ *c* $2{\cdot}16 \times 10^8$
d $3{\cdot}24 \times 10^8$ *e* 4×10^{13}
4 a $6{\cdot}34 \times 10^3$ *b* $5{\cdot}00 \times 10^2$ *c* $1{\cdot}0 \times 10^2$
d $1{\cdot}2 \times 10^1$ *e* $8{\cdot}0 \times 10^1$
5 a $2{\cdot}47 \times 10^3$ *b* $3{\cdot}03 \times 10^5$ *c* $2{\cdot}34 \times 10^2$ *d* $1{\cdot}36 \times 10^4$
e $1{\cdot}01 \times 10^4$ *f* $1{\cdot}73 \times 10^3$ *g* $3{\cdot}39 \times 10^1$ *h* $3{\cdot}40 \times 10^1$
6 108000

Page 180 Exercise 17B

1 a 8000000 *b* 997 *c* 70000 *d* 70·0

2 a 1×10^3 *b* $2{\cdot}3 \times 10^1$ *c* $4{\cdot}87 \times 10^4$ *d* $9{\cdot}999 \times 10^3$
e $4{\cdot}86 \times 10^1$ *f* $4{\cdot}863 \times 10^1$ *g* $8{\cdot}1 \times 10^5$ *h* 6×10^{12}

3 a $2{\cdot}3 \times 10^6$ *b* $1{\cdot}07 \times 10^8$ *c* $2{\cdot}5 \times 10^3$ *d* $1{\cdot}33 \times 10^{15}$

4 a $3{\cdot}84 \times 10^4$ *b* $6{\cdot}500 \times 10^3$ *c* $2{\cdot}0 \times 10^2$ *d* $1{\cdot}81 \times 10^2$

5 a $3{\cdot}8 \times 10^4$ *b* $2{\cdot}9 \times 10^1$ *c* $3{\cdot}0 \times 10^1$ *d* $1{\cdot}7 \times 10^8$

6 a $2{\cdot}0 \times 10^2$ *b* $1{\cdot}1 \times 10^4$ *c* $1{\cdot}3 \times 10^7$ *d* $2{\cdot}0 \times 10^4$ *e* $1{\cdot}3 \times 10^2$

Page 182 Exercise 18A

1 a 152 cm *b* 147 cm *c* 509 cm *d* 6·8 cm *e* 25·2 cm
f 19·9 cm

2 a 6·48 m *b* 40·5 m *c* 17·08 m

3 a 5·283 km *b* 10·035 km *c* 1·001 km *4 a* 5·25 fr *b* 81·09 fr

5 a 10·10 *b* 57·05 dollars *6a* 2·486 *b* 5·048 kg

7 a 3·673, *b* 2·002 litres *8* 2486 g

9 185 cents *10* 2125 ml

11 138·5 cm *12* 585 centimes, 5·85 fr

13a 15·5 *b* 372 tonnes

Page 183 Exercise 18B

1 a 4·7 cm *b* 227·4 cm *c* 502·9 cm

2 a 3·47 m *b* 27·07 m *c* 31·02 m

3 a 5·297 km *b* 11·023 km *c* 7·007 km

4 a 3·25 marks *b* 854·05 marks *5a* 35·64 francs *b* 128·03 francs

6 a 24·135 kg *b* 82·007 kg *7a* 5·278 litres *b* 0·876 litres

8 335 cents *9* 1034·3 m *10* 7·257 kg

11 23153 ml *12* 16·2 kg *13* 8

Page 184 Exercise 19A

1 £1·18 *2* £19·70 *3* £17·20 *4* £22·44

5 £2·98 *6* £49·25 *7* £15·18 *8* £255·45

Page 184 Exercise 19B

1 £1·46 *2* £38·22 *3* 12·8 m *4* 24·3 km

5 2·86 g *6* £0·48 *7* £19·80 *8* £13·48

Arithmetic—Answers to Chapter 2

Page 189 Exercise 1A

1 a 2 tens and 7 units *b* 2 hundreds, 3 tens and 4 units
c 7 hundreds, 0 tens and 2 units
d 2 thousands, 1 hundred, 3 tens and 5 units

2 a 1 four and 1 unit; 5 *b* 1 four and 1 two; 6
c 1 eight and 1 four; 12 *d* 1 eight, 1 four, 1 two, 1 unit; 15

3 a 1 *b* 3 *c* 4 *d* 7 *e* 9 *f* 11 *g* 13 *h* 17

4 a 1 *b* 2 *c* 4 *d* 8 *e* 16
They form a sequence; they are powers of two.

5 1, 10, 11, 100, 101, 110, 111, 1000, 1001, 1010, 1011, 1100

6 a 1101 *b* 1011 *c* 1010 *d* 10100

7 a 1000 *b* 1010 *c* 1001 *d* 1101

Page 189 Exercise 1B

1 a 7 tens and 5 units *b* 2 hundreds, 3 tens and 1 unit
c 1 thousand, 2 hundreds, 3 tens and 4 units
d 5 ten-thousands, 0 thousands, 7 hundreds, 2 tens, 1 unit

2 a 1 eight; 8 *b* 1 eight, 1 four and 1 unit; 13
c 1 sixteen, 1 eight and 1 two; 26
d 1 sixteen, 1 four, 1 two and 1 unit; 23

3 a 15 *b* 14 *c* 34 *d* 63

4 a 1 *b* 3 *c* 7 *d* 15 *e* 31
The numbers differ by powers of 2; each number is 1 less than a power of 2.

5 111111, 1111111, 11111111; 63, 127, 255

6 1100, 1101, 1110, 1111, 10000, 10001, 10010, 10011, 10100

7 a 11111, 100001 *b* 101001, 101011
c 100100, 100110 *d* 111110, 1000000

Page 190 Exercise 2A

1	11	*2*	101	*3*	110	*4*	1001
5	1011	*6*	1110	*7*	10010	*8*	11000
9	11110	*10*	100101	*11*	1000000	*12*	1000111

Page 190 Exercise 2B

1	10011	*2*	10000	*3*	10100	*4*	11101
5	100000	*6*	110010	*7*	111111	*8*	1000001
9	1001110	*10*	1011110	*11*	10000001	*12*	100000100

Page 193 Exercise 3A

1	111	*2*	1110	*3*	1111	*4*	11011
5	1100	*6*	1101	*7*	11011	*8*	11001
9	10110	*10*	100110	*11*	10010	*12*	101010
13	100	*14*	1000	*15*	1100	*16*	11001
17	1100	*18*	11100	*19*	100001	*20*	10000

21 1011 (7+4 = 11) *22* 11110 (21+9 = 30)

23 100100 (27+9 = 36) *24* 100010 (29+5 = 34)

Page 194 Exercise 3B

1	111	*2*	1111	*3*	11111	*4*	110
5	10110	*6*	11100	*7*	10100	*8*	110101
9	101101	*10*	1000101	*11*	110110	*12*	1011110
13	100000	*14*	11100	*15*	100100	*16*	101110
17	100000 (19+13 = 32)			*18*	1010010 (63+19 = 82)		
19	1100101 (59+42 = 101)			*20*	1101100 (47+61 = 108)		

Page 195 Exercise 4A

1	100	*2*	101	*3*	1000	*4*	10001
5	101	*6*	11	*7*	10	*8*	1101
9	10110	*10*	1011	*11*	1011	*12*	110
13	1	*14*	11	*15*	10	*16*	1001

Page 196 Exercise 4B

1	1	*2*	10000	*3*	110	*4*	100010
5	1010	*6*	1010	*7*	10100	*8*	1001
9	10001	*10*	1011	*11*	10101	*12*	10001
13	11110	*14*	101001	*15*	10001	*16*	10110

Page 197 Exercise 5A

1	11011	*2*	10010	*3*	11110	*4*	11010
5	110110	*6*	1100011	*7*	111111	*8*	101101
9	10101	*10*	10000100	*11*	1101001	*12*	1001011

Page 197 Exercise 5B

1	1110	*2*	100111	*3*	110111	*4*	1001110
5	1110101	*6*	1110011	*7*	101111010	*8*	10111010
9	1000110						

Page 198 Exercise 6A

1	11	*2*	10	*3*	111	*4*	100
5	100	*6*	11	*7*	11	*8*	10
9	110	*10*	11, rem. 1	*11*	100, rem. 1	*12*	101, rem. 1
13	1001	*14*	111	*15*	111, rem. 11		

Page 198 Exercise 6B

1	101	*2*	1001	*3*	101
4	110, rem. 1	*5*	1001, rem. 1010	*6*	101, rem. 1100
7	111	*8*	1100	*9*	1001

Page 202 Exercise 8

1 16 (including 0)

2 2+3−1 = 4 (end of message)

6 Each column of squares can show only 16 different symbols.

Page 203 Exercise 9A

1 a 10000, 1011 *b* 11000, 10001 *c* 11010, 10101

2 a 13, yes *b* 26, no *c* 31, yes *d* 21, no

3

1	0	1		1
1		1		1
1	0	1	1	1
0		1		
	1	0	1	1

4 1111 cm^2, 10000 cm

5 $A = \{0, 11, 110, 1001, 1100, 1111\}$
$B = \{100, 101, 110, 111\}$. $A \cap B = \{110\}$

6 10, 11, 101, 111, 1011, 1101

7 a F *b* T *c* F *d* T *e* T

Page 204 Exercise 9B

1 a 11001, 11000, 10110, 10101 *b* 11111, 11101, 11011, 10111

2 a T *b* F *c* F *d* T

3

1	0	1		1	0
1		1		1	
	1	1	1	0	1
1	1		1		0
0			1		1
1	0	0	0	0	0

4a Its last digit is 0.
b Its last two digits are 00.

5 11001 cm^2, 10100 cm

6 a 8 *b* 29 *c* 70; 111 *7* 9

Book 2 Arithmetic—Answers to Chapter 3

Page 209 Exercise 1A

1 a 6, 6×10^6; 4, 4×10^6; 15, $1{\cdot}5 \times 10^7$; 15, $1{\cdot}5 \times 10^7$; 10, $1{\cdot}0 \times 10^7$
b Spain

3 a Jan, Mar, Jun *b* Apr, May, Aug, Sep, Dec
c 9, 4, 4, 5, 6, 6, 5, 5, 7, 4, 5, 2 *d* 4, 6, 9, 2, 1, 7, 4, 2, 0, 7, 4, 5
e 35% *f* 27%

Page 211 Exercise 1B

2 400, 850, 1000, 1400

Page 212 Exercise 2A

5 a 7; 8 *b* science and technical *c* native language and PE *6* VEE

Page 214 Exercise 2B

4 *a* (*1*) 6 (*2*) 7·25 *b* April; December
c Totals: 248, 187·5, 186, 201·5, 203, 217, 292·5
6 *a* 250000, 260000 *b* Ford
c 20000

Page 216 Exercise 3A

1 30%, 35%, 15%, 20%

2 *a* drama, education, sport, entertainment, films, news, children's, documentaries, religion

3 *a* income tax *b* $\frac{1}{6}$
c income tax, other taxes, other income, tobacco, oil, spirits, purchase tax
d about £5000 million

7 11200, 3600, 3800, 1400 million units

Page 218 Exercise 3B

3 22630000, 13870000, 11680000, 16790000, 5110000, 2920000 tonnes

Page 220 Exercise 4A

1 *a* £20–£40

2 *a* 141 cm *b* between 15 and 15½ *c* his 17th *d* about 18

3 *a* no *b* 245 thousand, 280 thousand; no
c depression in the 1920s; possibly emigration *d* no

4 *a* Inverness; Mar, Jul/Aug; Eskdalemuir; May, Jan
b Feb: 48, 117 mm; Sep: 69, 134 mm
c 60 mm *d* May–Jun

5 *a* Dec *b* Apr–May–Jun

Page 223 Exercise 4B

2 *a* 127
b 143 tonnes; increased mechanization and closing of unproductive pits
c 9 or 10 *d* 24 or 25 tonnes

4 *a* £4½ million, £6 million *b* 1966; about £1¼ million *c* no

Arithmetic—Answers to Revision Exercises

Page 227 Revision Exercise 1A

1 4·3, $4\frac{3}{10}$; 6·72, $6\frac{18}{25}$; 10·4, $10\frac{2}{5}$; 234·56, $234\frac{14}{25}$

2 3·14, 8·25, 12·01, 0·12, 0·34, 15·10, 8·56, 10·11, 0·05, 0·01

3 28·4, 4·96, 0·982, 0·703, 3700, 0·00569, 0·0410, 69·9, 0·100

4 (rough estimates) 80 or 100, 200, 48, 2 or 3, 0·8, 0·06, 10, 3, 0·2, 0·04, 2

5 *a* 21 cm, 20 cm² *b* 34 cm, 61 cm² 6*a* 38·5 *b* 110

7 0·9, 0·5, 0·75, 0·8, 0·125, 0·875

8 0·67, 0·71, 0·44, 0·15 *9* 500, 15000, 280, 3140

10 $5{\cdot}37 \times 10$, $3{\cdot}76 \times 10^2$, $5{\cdot}042 \times 10^3$, 3×10^6, $5{\cdot}621 \times 10^2$, $2{\cdot}76 \times 10^3$, $2{\cdot}46 \times 10^5$, $1{\cdot}765 \times 10$

11 4800 *12* 61·2 kg *13* 14 km

14a 3·15 m by 1·25 m *b* 3·94 m²

15a £4·65 *b* £1·27 *c* 31p *d* £2·63 *16* 330·75 marks

Page 228 Revision Exercise 1B

1 (rough estimates) 40, 600, 0·4, 0·02, 4, 0·4, 0·04, 0·3, 0·01, 10, 0·005

2 64·3, 8·73, 374, 207, 0·0605, 0·0882; 64·3, 8·7, 373·5, 206·5, 0·1, 0·1

3 a 14 cm, 7·5 cm² *b* 23 cm, 14 cm² *4* 10

5 $2{\cdot}85 \times 10^2$ g, $2{\cdot}08 \times 10$ litres, $3{\cdot}4 \times 10^3$ m, $2{\cdot}8 \times 10^5$ km, $1{\cdot}567 \times 10^5$ cm, $1{\cdot}86 \times 10^8$

6 0·3, 0·21, 0·25, 0·6, 0·375, 0·3125

7 0·86, 0·33, 0·11, 0·91 *8* $3{\cdot}1 \times 10^3$

9 $q = 2{\cdot}05$, $n = 7$ *10a* 0·04 *b* Hearts by 0·35 *c* 64·7, 17·6, 17·6

11 £9990 million, £10 480 million *12* 465·05 francs

Page 230 Revision Exercise 2A

1 a 10001 *b* 10101 *c* 100011 *d* 1000011

2 a 11 *b* 23 *c* 45

3 a 100100 *b* 101110 *c* 100110 *d* 110 *e* 1100 *f* 110

4 a 100111 *b* 1000001 *c* 101101 *d* 1110, rem. 1
e 1001, rem. 10 *f* 1010

5 13, 17, 25, 29; 25 *6* 10010 cm², 10010 cm

7 11 cm *8a* 4 *b* 100, 11 m

Page 231 Revision Exercise 2B

1 a 100111 *b* 111011 *c* 1011111 *d* 11111111 *e* 100000100

2 a 27 *b* 30 *c* 59 *d* 125

3 a 111001 *b* 1111011 *c* 101011 *d* 111001
e 11110011 *f* 100100110 *g* 111 *h* 101, rem. 11

4 By 10, by 100 *a* 1010, 1100, 110110, 1100, 101100
b 1100, 10100, 10000, 110100 cm *c* 1010, 10000, 101100 cm

5 1001, 11001, 10000, 10101001 cm²

6 10010 m², 10010 m; 1101110 m², 101010 cm

7 a 11 cm *b* 10101 cm *c* 110 m

8 22, 201, 1000

+	0	1	2
0	0	1	2
1	1	2	10
2	2	10	11

×	0	1	2
0	0	0	0
1	0	1	2
2	0	2	11

42 pence

Page 232 Answers—Revision Exercise 3A

1 53, 57; 21, 17; 13, 12; 6, 4; 1, 3; 6, 9

5 a 5·3 m, 6·4 m *b* 11.06 pm, 8·2 m; 5.06 pm, 1·8 m
c 2.45 pm and 7.35 pm

6 a Others, Hausa, Ibo, Yoruba, Fulani *b* about 0·4

Page 235 Answers—Revision Exercise 3B

1 a The scale does not start at zero *b* 50p, 39p; 5 years; war costs
c 5p, 1947

4 a 85°C, 80°C, 80°C, 44°C, 30°C *b* $13\frac{1}{2}$, 15, 28, 32 minutes

Page 240 Cumulative Revision Exercise A

1 F *2* T *3* T *4* F *5* T *6* F *7* F

8 F *9* A *10* C *11* D *12* B *13* D

14 60 *15a* 25, 36, 49 *b* 64, 125, 216

16a 23, 27, 31 *b* 13, 11, 9 *c* 32, 64, 128 *d* $\frac{1}{6}, \frac{1}{7}, \frac{1}{8}$

17 60 *18a* 0, 0 *b* 1, 1 *c* 6, 6; commutative, associative

19 $2^3 \times 13$, $2^4 \times 3^2$ *20a* 60 *b* 63 *21* 75×45

22a $58\frac{1}{2}$ m² *b* 9·8 cm² *c* 24 000 m² *23* 27 cm³, 4000

24a 30, 45 *b* 0·5, 0·52 *c* 4, 3·6 *d* 5, 4·6
e 0·2, 0·22 *f* 15, 16 *25* 3·88 km, 23·52 km

26 17·45 kg *27a* 31p *b* £1·82 *c* 19p

28a 11000 *b* 110 *c* 1000110 *d* 11 *e* 1011

29 £2·19½, £2·80½ *30* £21·20

31a 1760 *b* 9300000 *c* 48000 *d* 80000000

32a 3×10^3 *b* $2{\cdot}86 \times 10^5$ *c* $7{\cdot}5 \times 10^7$ *d* 9×10^{12}

Page 243 Cumulative Revision Exercise B

1 T *2* F *3* T *4* T *5* F *6* F *7* T

8 F *9* C *10* A *11* D *12* A *13* D

14 131 *15a* 71, 64, 57 *b* 125, 216, 343 *c* 37, 50, 65
d 19, 23, 29 *16a* 72 *b* 61 *17a* commutative *b* true *c* 6

18 $2^2 \times 7^2$, $2^7 \times 3^2$ *19* 115 cm², 23 : 16, $143\frac{3}{4}$

20 250, 15000 cm³ *21a* 20, 29·6 *b* 0·02, 0·0181 *c* 80, 85·4
d 0·1, 0·130 *e* 1, 1·03 *22* 25, 37·5 m *23* 308

24 £31·22 *25* £559 *26* 111000 cm, 1110011 cm²

27 £13·41 *28* 373·89 marks, £42

29a 130000 *b* 840000000 *c* 70000000000

30a $2{\cdot}24 \times 10^3$ *b* $8{\cdot}7 \times 10^{10}$ *c* 2×10^9 *d* 1×10^9